王者辉　主编

Pharmaceutical Polymers

药用高分子

化学工业出版社

·北京·

《Pharmaceutical Polymers（药用高分子）》基于现代药剂学的发展，依据国内外最新资料，结合作者多年来的教学科研实践编写而成，系统讲述了高分子在药物制剂中的应用。全书共分为十章，具体内容包括：药用高分子化学基础理论（第一章）；药用高分子的原理与结构（第二章）；天然来源与合成药用高分子（第三章）；天然来源高分子在药学领域的应用（第四章）；疏水性高分子聚合物的药用价值（第五章）；高分子聚合物在药物传递中的应用（第六章）；医药纳米技术（第七章）；药用高分子的热分析（第八章）；原子力显微镜在药用高分子中的应用（第九章）；药物高分子的流变学研究（第十章）。

《Pharmaceutical Polymers（药用高分子）》可供普通高等医药院校药学、药物制剂、制药工程、高分子材料与工程等相关专业学生及从事药物制剂生产、研究的人员参考使用。

图书在版编目（CIP）数据

药用高分子＝Pharmaceutical Polymers：英文/王者辉主编. —北京：化学工业出版社，2018.8
ISBN 978-7-122-32483-2

Ⅰ.①药… Ⅱ.①王… Ⅲ.①高分子材料-药剂-辅助材料-英文 Ⅳ.①TQ460.4

中国版本图书馆 CIP 数据核字（2018）第 136498 号

责任编辑：褚红喜　宋林青　　　　装帧设计：关　飞
责任校对：宋　夏

出版发行：化学工业出版社(北京市东城区青年湖南街 13 号　邮政编码 100011)
印　　刷：大厂聚鑫印刷有限责任公司
装　　订：三河市宇新装订厂
787mm×1092mm　1/16　印张 19¼　字数 600 千字　2018 年 10 月北京第 1 版第 1 次印刷

购书咨询：010-64518888(传真：010-64519686)　售后服务：010-64518899
网　　址：http://www.cip.com.cn
凡购买本书，如有缺损质量问题，本社销售中心负责调换。

定　　价：88.00 元

前　言

药用高分子（Pharmaceutical polymers）是具有生物相容性、经过安全评价且应用于药物制剂的一类高分子量药用辅料。药用高分子作为药物缓释载体，包覆在药物表面保护小分子药物使其在短时间内不会被生命体吸收，同时随血液流动到特定区域，在特定区域药物表面的高分子材料溶解到血液中，释放出药物，而本身最终随体液排出。药用高分子的主要应用有：①载体，用以控制药物缓慢释放，一般由有机硅橡胶、聚甲基丙烯酸酯类等制成封闭细管、微囊或薄膜。②主药，有长效的特点，其结构有三类：（a）在高分子侧基上连接小分子药物，这类结构最受重视，即高分子亲和药物，在高分子侧链上除含有药物结构外，在较远端侧链上还含有亲水结构、配基或酶，以提高药用高分子在体内分布的专一性；（b）在高分子主链中含小分子药物；（c）高分子中不含小分子药物，其本身可作为药物或功能试剂。

本书结合现代药剂学的发展以及作者多年来的教学科研实践，根据国内外最新资料，简述了有关高分子化学和物理的基本内容，概述了与药剂学相关的高分子材料基本知识，重点介绍国际上经法定程序验证及实际生产中已被采用的药用高分子材料的原理和应用，充实了国外近年来备受瞩目的给药系统用天然来源药用高分子及可生物降解的合成药用高分子和复合物的有关内容，此外，还对热分析、原子力显微镜和流变学在药用高分子领域中的应用进行了阐述。本书可供普通高等医药院校药学、药物制剂、制药工程、高分子材料与工程等相关专业学生及从事药物制剂生产、研究的人员参考使用。

全书共分为十章，参与本书编写的人员及具体编写分工如下：第一章由孙红、王者辉编写；第二章由张宏志、孙立平编写；第三章由巩学勇、刘凤霞编写；第四章由李庆杰、常海涛编写；第五章由郑超、张倩编写；第六章由刘光耀、江栋编写；第七章由卫振华、刘爱军编写；第八章由王者辉、李宝庆编写；第九章由杜江山、刘振亮编写；第十章由姜洪丽、孟宪锋编写；全书由王者辉统一整理并任主编。

由于水平所限，书中不足之处在所难免，诚恳希望广大读者加以批评指正。

编　者

2018 年 5 月

Preface

Pharmaceutical polymers (both synthetic and natural-based polymers) have found their way into the pharmaceutical industries and their applications are growing at a fast pace. The desirable polymer properties in pharmaceutical applications are film forming (coating), thickening (rheological modifier), gelling (controlled release), adhesion (binding), pH-dependent solubility (controlled release), solubility in organic solvents (taste masking), and barrier properties (protection and packaging). Major applications of polymers in current pharmaceutical field are for controlled drug release. From the solubility point of view, pharmaceutical polymers can be classified as water-soluble and water-insoluble (oil soluble or organic soluble). Many water-soluble pharmaceutical polymers including polyethylene glycol, polyvinyl alcohol, polyethylene oxide, polyvinyl pyrrolidone, polyacrylate or polymethacrylate esters, cellulose based polymers, hydrocolloids, and many plastics and rubbers will be tabulated and their anionic and cationic functionalities are summarized and discussed. PEGylation is defined as the covalent attachment of poly ethylene glycol (PEG) chains to bioactive substances.

Pharmaceutical polymer, a kind of macromolecular compound with bioco-mpatibility, safety evaluation, is used as a kind of high molecular mass pharma-ceutical excipients for drug preparation. As a sustained release drug carrier, the pharmaceutical polymer is coated on the surface of the drug to protect small molecular weight drugs not to be absorbed by the living body in a short time. At the same time, with the flow of blood in a specific area, the polymer materials on the specific area of the drug dissolve into the blood and release the drug, and the drug itself is eventually discharged with the body fluid. The main applications of pharmaceutical polymers include: (1) the carrier, which is used to control the slow release of drugs, usually made of organosilicon rubber, polymethacrylate and others in the form of closed tubes, microcapsules or films. (2) the main drug with characteristic long time efficacy has three kinds of structures: (i) polymeric affinity drug, with the hydrophilic groups, the ligands or enzymes on the far end of the chain, have small molecular drugs connecting on the side of the polymer chains, which is the most important structure with pharmaceutical polymer uniform distribution inside the body; (ii) polymeric drug contains small molecule drugs in the main chain; (iii) polymers do not contain small molecular drugs, it act as medicine or functional reagent.

This book, combined with the development of modern pharmaceutics and many years practice of teaching and scientific research, briefly describes the fundamentals of polymer chemistry and physics, and summarizes the basic knowledge of polymers related to pharmaceutics. According to the latest information at home and abroad, this book mainly introduces the principles and applications of pharmaceutical polymers that have been used in international legal procedure verification and actual production, and enriches the related contents of natural sources of pharmaceutical polymers and biodegradable synthetic pharmaceutical polymers and complexes in recent years. The applications of thermal analysis, atomic force microscopy and rheology in the field of pharmaceutical polymers are included. This book can be used as a reference for readers of pharmaceutical engineering, polymer materials and engineering and related majors and people engaged in the production and research of pharmaceutical preparations.

The personnel involved in this book include: the first chapter by Hong Sun and Zhehui Wang; the second chapter by Hongzhi Zhang and Liping Sun; the third chapter by Xueyong Gong and Fengxia Liu; the fourth chapter by Qingjie Li and Haitao Chang; the fifth chapter by Chao Zheng and Qian Zhang; the sixth chapter by Guangyao Liu and Dong Jiang; the seventh chapter by Zhenhua Wei and Aijun Liu; the eighth chapter by Zhehui Wang and Baoqing Li; the ninth chapter by Jiangshan Du and Zhenliang Liu; the tenth chapter by Hongli Jiang and Xianfeng Meng; this whole book is edited by Zhehui Wang.

As to the limitation of the authors' level, all kinds of criticisms are welcome for various mistakes in the book, and we sincerely hope that readers will be able to inform us for correction.

Zhehui Wang

2018.5

Content

Chapter 1

Fundamentals of Pharmaceutical Polymer Chemistry

Polymers are high molecular mass compounds formed by polymerization of monomers. The simple reactive molecule from which the repeating structural units of a polymer are derived is called a monomer. A polymer is chemically described by its degree of polymerization, molar mass distribution, tacticity, copolymer distribution, the degree of branching, by its end-groups, crosslinks, crystallinity and thermal properties such as its glass transition temperature and melting temperature. Polymers in solution have special characteristics with respect to solubility, viscosity and gelation. Schematically polymers are subdivided into biopolymers and synthetic polymers according to their origin. Each one of these classes of compounds can be subdivided into more specific categories in relationship to their use, properties and physicochemical characteristics. The biochemistry and industrial chemistry are disciplines that are interested in the study of polymer chemistry.

1.1 The Concept of a Polymer

1.1.1 Historical Introduction

The differences between the properties of crystalline organic materials of low molecular mass and the more indefinable class of materials referred to by Graham in 1861 as 'colloids' has long engaged the attention of chemists. This class includes natural substances such as gum acacia, which in solution are unable to pass through a semi-permeable membrane. Rubber is also included among this class of material.

The idea that the distinguishing feature of colloids was that they had a much higher molecular mass than crystalline substances came fairly slowly. Until the work of Raoult, who developed the cryoscopic method of estimating molecular mass, and Van't Hoff, who enunciated the solution laws, it was difficult to estimate even approximately the polymeric state of materials. It also seems that in the nineteenth century there was little idea that a colloid could consist, not of a product of fixed molecular mass, but of molecules of a broad band of molecular masses with essentially the same repeat units in each. Vague ideas of partial valence unfortunately derived from inorganic chemistry and a preoccupation with the idea of ring formation persisted until after 1920. In addition scientists did not realize that a process such as ozonisation virtually destroyed a polymer as such, and the molecular mass of the ozonide, for example of rubber, had no idea on the original molecular mass.

The theory that polymers are built up of long molecular chains was vigorously advocated by Staudinger from 1920 onwards. He extended this in 1929 to the idea of a three-dimensional network copolymer to account for the insolubility and infusibility of many synthetic polymers, for by that time technology had by far outstripped theory. Continuing the historical outline, mention must be made of Carothers, who from 1929 began a classical series of experiments which indicated that polymers of definite structure could be obtained by the use of classical organic chemical reactions, the properties of the polymer being controlled by the starting compounds. While this was based on research in condensation compounds, the principles hold good for addition polymers.

The last four decades have seen major advances in the characterization of polymers. Apart from increased sophistication in methods of measuring molecular mass, such as the cryoscopic and vapor pressure methods, almost the whole range of the spectrum has been called into service to elucidate polymer structure. Ultraviolet and visible spectroscopy, infrared spectroscopy, Raman and emission spectroscopy, photon correlation spectroscopy, nuclear magnetic resonance and electron spin resonance all play a part in the understanding of the structure of polymers; X-ray diffraction and small-angle X-ray scattering have been used with solid polymers. Thermal behavior in its various aspects, including differential thermal analysis and high-temperature pyrolysis followed by gas-liquid chromatography, has also been of considerable value. Other separation methods include size exclusion and hydrodynamic chromatography. Electron microscopy is of special interest with particles formed in emulsion polymerization. Thermal and gravimetric analysis give useful information in many cases. There are a number of standard works that can be consulted.

1.1.2 Definitions

A polymer in its simplest form can be regarded as comprising molecules of closely related composition of molecular mass at least 2000, although in many cases typical properties do not become obvious until the mean molecular mass is about 5000. There is virtually no upper end to the molecular mass range of polymers since giant three-dimensional networks may product crosslinked polymers of a molecular mass of many millions.

Polymers (macromolecules) are built up from basic units, sometimes referred to as 'mers'. These units can be extremely simple, as in addition polymerization, where a simple molecule adds on to itself or other simple molecules, by methods that will be indicated subsequently. Thus ethylene $CH_2{=}CH_2$ can be converted into polyethylene, of which the repeating unit is $—CH_2CH_2—$, often written as $—(CH_2CH_2)_n—$, where n is the number of repeating units, the nature of the end groups being discussed later. The major alternative type of polymer is formed by condensation polymerization in which a simple molecule is eliminated when two other molecules condense. In most cases the simple molecule is water, but alternatives include ammonia, an alcohol and a variety of simple substances. The formation of condensation polymer can best be illustrated by the condensation of hexamethylenediamine with adipic acid to form the polyamide which is best known as nylon:

$$\begin{aligned} &H_2N(CH_2)_6NH_2 + HOOC(CH_2)_4COOH + H_2N(CH_2)_6NH_2 \\ &\quad = H_2N(CH_2)_6NHOC(CH_2)_4CONH(CH_2)_6NH_2 + 2H_2O \end{aligned} \tag{1-1}$$

This formula (1-1) has been written in order to show the elimination of water. The product of condensation can continue to react through its end groups of hexamethylenediamine and adipic acid, and thus a high molecular mass polymer is prepared.

Monomers such as adipic acid and hexamethylenediamine are described as bifunctional, because they have two reactive groups. As such they can only form linear polymers. Similarly, the simple vinyl monomers such as ethylene ($CH_2{=}CH_2$) and vinyl acetate ($CH_2{=}CHOOCCH_3$) are considered to be bifunctional. If the functionality of a monomer is greater than two, a branched structure may be formed. Thus the condensation of glycerol [$HOCH_2CH(OH)CH_2OH$] with adipic acid [$HOOC(CH_2)_4COOH$] will give branched structure. It is represented diagrammatically below:

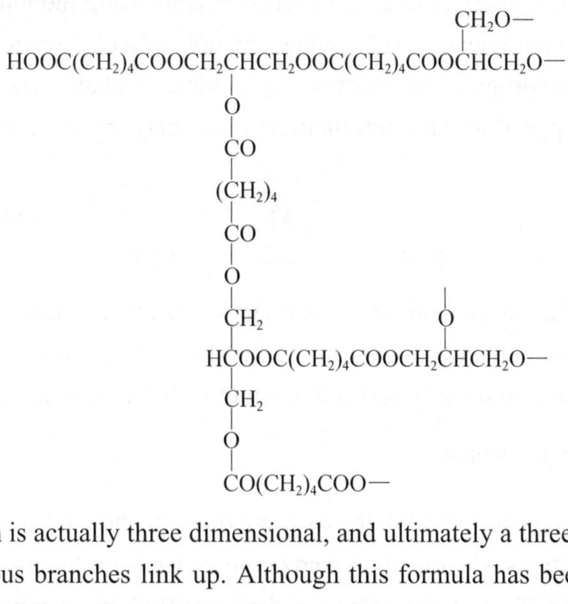

```
                                                 CH2O—
                                                 |
HOOC(CH2)4COOCH2CHCH2OOC(CH2)4COOCHCH2O—
                |
                O
                |
                CO
                |
                (CH2)4
                |
                CO
                |
                O
                |                       |
                CH2                     O
                |                       |
               HCOOC(CH2)4COOCH2CHCH2O—
                |
                CH2
                |
                O
                |
                CO(CH2)4COO—
```

The condensation is actually three dimensional, and ultimately a three-dimensional structure is formed as the various branches link up. Although this formula has been idealized, there is a statistical probability of the various hydroxyl and carboxyl groups combining. This results in a network being built up, and while it has to be illustrated on the plane of the paper, it will not necessarily be planar. As functionality increases, the probability of such networks becoming interlinked increases, as does the probability with increase in molecular mass. Thus a gigantic macromolecule will be formed which is insoluble and infusible before decomposition. It is only comparatively recently that structural details of these crosslinked or 'reticulated' polymers have been elucidated with some certainty.

Addition polymers are normally formed from unsaturated carbon-to-carbon linkages. This is not necessarily the case since other unsaturated linkages including only one carbon bond may be polymerized. Addition polymerization of a different type takes place through the opening of a ring, especially the epoxide ring in ethylene oxide $\left(\overset{CH_2-CH_2}{\diagdown_{O}\diagup}\right)$. This opens as $—CH_2CH_2O—$; ethylene oxide thus acts as a bifunctional monomer forming a polymer as $H(CH_2CH_2O)_nCH_2CH_2OH$, in this case a terminal water molecule being added. A feature of this type of addition is that it is much easier to control the degree of addition, especially at relatively low levels, than in the vinyl polymerization described above. Addition polymerization from which polymer emulsions may be available occurs with the silicones and diisocyanates. These controlled addition polymerizations are sometimes referred to as giving 'stepwise' addition polymers. This term may also refer to condensation resins.

1.2 Addition Polymerization

Addition polymerization, the main type with which this volume is concerned, is essentially a chain reaction, and may be defined as one in which only a small initial amount of initial energy is required to start an extensive chain reaction converting monomers, which may be of different formulae, into polymers. A well-known example of a chain reaction is the initiation of the reaction between hydrogen and chlorine molecules. A chain reaction consists of three stages: initiation, propagation and termination, and may be represented simply by the progression:

Activation	+M		+M		+nM	
M----- M*	------	M_2*	----------	M_3*	---------	M_{n+3} etc.

The termination reaction depends on several factors, which will be discussed later.

The mechanism of polymerization can be divided broadly into two main classes: free radical polymerization and ionic polymerization, although there are some others.

1.2.1 Ionic Polymerization

Ionic polymerization was probably the earliest type to be noted, and is divided into cationic and anionic polymerizations. Cationic polymerization depends on the use of catalysts which are good electron acceptors. Typical examples are the Friedel-Crafts catalysts such as aluminium chloride ($AlCl_3$) and boron trifluoride (BF_3). Monomers that polymerize in the presence of these catalysts have substituents of the electron releasing type. They include styrene ($C_6H_5CH{=}CH_2$) and the vinyl ethers ($CH_2{=}CHOC_nH_{2n+1}$).

Anionic initiators include reagents capable of providing negative ions, and are effective with monomers containing electronegative substituents such as acrylonitrile ($CH_2{=}CHCN$) and methyl methacrylate [$CH_2{=}C(CH_3)COOCH_3$]. Styrene may also be polymerized by an anionic method. Typical catalyst include sodium in liquid ammonia, alkali metal alkyls, Grignard reagents and triphenylmethyl sodium [$(C_6H_5)_3C\text{-}Na$]. Among other modern methods of polymerization are the Ziegler-Natta catalysts and group transfer polymerization catalysts. Ionic polymerization is not of interest in normal aqueous polymerization since in general the carbonium ions by which cationic species are propagated and the corresponding carbanions in anionic polymerizations are only stable in media of low dielectric constant, and are immediately hydrolyzed by water.

1.2.2 Free Radical Polymerization

1.2.2.1 Free Radical

A free radical may be defined as an intermediate compound containing an odd number of electrons, but which do not carry an electric charge and are not free ions. The first stable free radical, triphenylmethyl, $(C_6H_4)_3C\bullet$, was isolated by Gomberg in 1900, and in gaseous reactions the existence of radicals such a methyl ($CH_3\bullet$) was postulated at an early date.

The decomposition of oxidizing agents of the peroxide type, as well as compounds such as azodiisobutyronitrile, N N N N, which decomposes into two radicals, CN (CH$_3$)$_2$C•

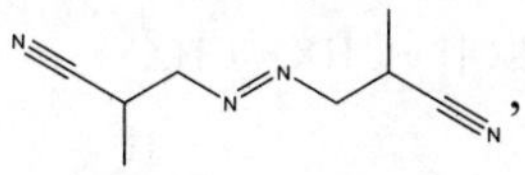

and nitrogen (N_2), is well known. Thus a free radical mechanism is the basis of addition polymerization where these types of initiator are employed. For a transient free radical, the convention will be used of including a single dot after or over the active element with the odd electron.

A polymerization reaction may be simply expressed as follows. Let R• be radical from any source. $CH_2{=}CHX$ represents a simple vinyl monomer where X is a substituent, which may be H as in ethylene ($CH_2{=}CH_2$), Cl as in vinyl chloride ($CH_2{=}CHCl$), $OOCCH_3$ as in vinyl acetate ($CH_2{=}CHOOCCH_3$) or many other groups, which will be indicated in lists of monomers.

The first stage of the chain reaction, the initiation process, consists of the attack of the free radical on one of the doubly bonded carbon atoms of the monomer. One electron of the double bond pairs with the odd electron of the free radical to form a bond between the latter and one carbon atom. The remaining electron of the double bond shifts to the other carbon atom which now becomes a free radical. This can be expressed simply in equation form:

$$R\cdot + CH_2{=}CHX \longrightarrow RCH_2\underset{X}{\overset{H}{C}}\cdot \tag{1-2}$$

The new free radical can, however, in its turn add on extra monomer units, and a chain reaction occurs, representing the propagation stage:

$$RCH_2\underset{X}{\overset{H}{C}}\cdot + nCH_2{=}CHX \longrightarrow R(CH_2CHX)_n\,CH_2\underset{X}{\overset{H}{C}}\cdot \tag{1-3}$$

The final stage is termination, which may take place by one of several processes. One of these is combination of two growing chains reacting together:

$$R_1(CH_2CHX)_nCH_2\underset{X}{\overset{H}{C}}\cdot + \cdot\underset{X}{\overset{H}{C}}CH_2(CHXCH_2)_mR_2 \longrightarrow R_1(CH_2CHX)_nCH_2\underset{X}{\overset{H}{C}}{-}\underset{X}{\overset{H}{C}}CH_2(CHXCH_2)_mR_2 \tag{1-4a}$$

An alternative is disproportionation through transfer of a hydrogen atom:

$$R_1(CH_2CHX)_nCH_2\underset{X}{\overset{H}{C}}\cdot + \cdot\underset{X}{\overset{H}{C}}CH_2(CHXCH_2)_mR_2 \longrightarrow R_1(CH_2CHX)_nCH_2\underset{X}{\overset{H}{C}}H + \underset{X}{\overset{H}{C}}H(CHXCH_2)_mR_2 \tag{1-4b}$$

A further possibility is chain transfer. This is not a complete termination reaction, but it ends the propagation of a growing chain and enables a new one to commence. Chain transfer may take place via a monomer, and may be regarded as a transfer of a proton or of a hydrogen atom:

$$Z\text{-}CH_2\underset{H}{\overset{X}{C}}\cdot + CH_2CHX \longrightarrow Z\text{-}CH{-}CHX + CH_3\underset{H}{\overset{X}{C}}\cdot \tag{1-5}$$

where Z is a polymeric chain.

Chain transfer takes place very often via a fortuitous impurity or via a chain transfer agent which is deliberately added. Alkyl mercaptans with alkyl chains C_8 or above are frequently added for this purpose in polymerization formulations. A typical reagent is *t*-dodecyl mercaptan, which reacts as in the following equation:

$$R(CH_2CHX)_nCH_2\underset{X}{\overset{H}{C}}\cdot + t\text{-}C_{12}H_{25}SH \longrightarrow R(CH_2CHX)_nCH_2CH_2X + t\text{-}C_{12}H_{25}S\cdot \tag{1-6a}$$

Chlorinated hydrocarbons are also commonly used as chain transfer agents, and with carbon tetrachloride it is a chlorine atom rather than a hydrogen atom that takes part in the transfer:

$$R(CH_2CHX)_nCH_2\underset{X}{\overset{H}{C}}\cdot + CCl_4 \longrightarrow R(CH_2CHX)_nCH_2CHXCl + Cl_3C\cdot \tag{1-6b}$$

Most common solvents are sufficiently active to take part in chain transfer termination, the aliphatic straight-chain hydrocarbons and benzene being among the least active. The effect of solvents is apparent in the following equation, where SolH denotes a solvent:

$$R(CH_2CHX)_nCH_2\overset{H}{\underset{X}{C}}\cdot + SolH \longrightarrow R(CH_2CHX)_nCH_2CH_2X + Sol\cdot \qquad (1\text{-}6c)$$

In all the cases mentioned, the radicals on the right-hand side of the equations must be sufficiently active to start a new chain; otherwise they act as a retarder or inhibitor. Derivatives of allyl alcohol ($CH_2{=}CHCH_2OH$), although polymerizable by virtue of the ethylenic bond, have marked chain transfer properties and produce polymers of low molecular mass relatively slowly. Stable intermediate products do not form during a polymerization by a free radical chain reaction, and the time of formation of each polymer molecule *i* of the order of 10^{-3} s.

Kinetic equations have been deduced for the various processes of polymerization. These have been explained simply in a number of treatises. A useful idea which may be introduced at this stage is that of the order of addition of monomers to a growing chain during a polymerization. It has been assumed in the elementary discussion that if a growing radical M—$CH_2C\cdot$ is considered, the next unit of monomer will add on to produce:

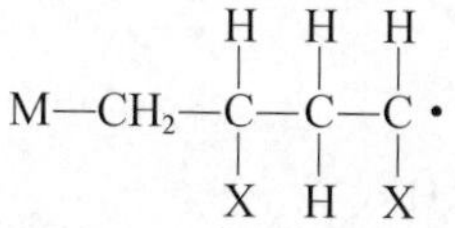

$$M{-}CH_2{-}\overset{H}{\underset{X}{C}}{-}\overset{H}{\underset{H}{C}}{-}\overset{H}{\underset{X}{C}}\cdot$$

It is theoretically possible, however, for the next unit of monomer to add on, producing

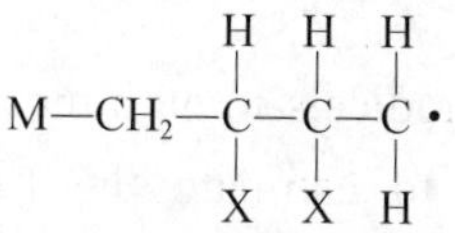

$$M{-}CH_2{-}\overset{H}{\underset{X}{C}}{-}\overset{H}{\underset{X}{C}}{-}\overset{H}{\underset{H}{C}}\cdot$$

The latter type of addition in which similar groups add in adjacent fashion is known as 'head-to-head' addition in contrast to the first type above, known as 'head-to-tail' addition. The head-to-tail addition is much more usual in polymerizations, although in all cases head-to-head polymerization occurs at least to some extent.

There are various ways of estimating head-to-head polymerization, both physical and chemical. Nuclear magnetic resonance data should be mentioned among the former. The elucidation of polyvinyl acetate $(CH_2CHOOCCH_3)_n$ may be taken as representative of a chemical investigation. A head-to-tail polymer when hydrolyzed to polyvinyl alcohol would typically produce units of ($CH_2CHOHCH_2CHOH$). A head-to-head unit is ($CH_2CHOHCHOHCH_2$). In the latter case there are two hydroxyl groups on adjacent carbon atoms, and the polymer is therefore broken down by periodic acid HIO_4, which attacks this type of unit. It is possible to estimate the amount of head-to-head addition from molecular mass reduction or by estimation of the products of oxidation.

1.2.2.2 Retardation and Inhibition

If the addition of a chain transfer agent to a polymerizing system works efficiently, it will both slow the polymerization rate and reduce the molecular mass. This is because the free radical formed in the equivalent of equation (1-6a) may be much less active than the original radical in starting new chains, and when these are formed, they are terminated after a relatively short growth.

In some cases, however, polymerization is completely inhibited since the inhibitor reacts with radicals as soon as they are formed. The most well known is *p*-benzoquinone.

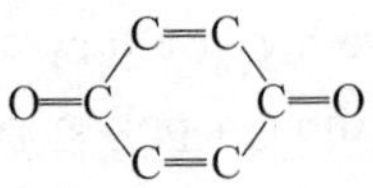

$$\mathrm{O{=}C}\langle\begin{smallmatrix}\mathrm{C{=}C}\\ \mathrm{C{=}C}\end{smallmatrix}\rangle\mathrm{C{=}O}$$

This produces radicals that are resonance stabilized and are removed from a system by mutual combination or disproportionation. Only a small amount of inhibitor is required to stop polymerization of a system. A calculation shows that for a concentration of azodiisobutyronitrile of 1×10^{-3} mol·L^{-1} in benzene at 60℃, a concentration of 8.6×10^{-5} mol·L^{-1} of inhibitor is required. *p*-Hydroquinone [$C_6H_4(OH)_2$], probably the most widely used inhibitor, only functions effectively in the presence of oxygen which converts it to a quinone-hydroquinone complex giving stable radicals. One of the most effective inhibitors is the stable free radical 2,2-diphenyl-1-picryl hydrazyl:

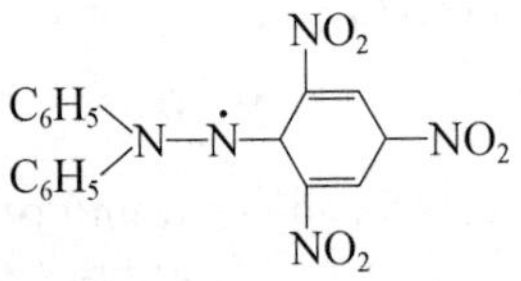

$$(C_6H_5)_2N{-}\dot{N}{-}C_6H_2(NO_2)_3$$

This compound reacts with free radicals in an almost quantitative manner to give inactive products, and is used occasionally to estimate the formation of free radicals. Aromatic compounds such as nitrobenzene ($C_6H_5NO_2$) and the dinitrobenzenes (*o*-, *m*-, *p*-)$C_6H_4(NO_2)_2$ are retarders for most monomers, e.g. styrene, but tend to inhibit vinyl acetate polymerization, since the monomer produces very active radicals which are not resonance stabilized. Derivatives of allyl alcohol such as allyl acetate are a special case. While radicals are formed from this monomer, the propagation reaction competes with that shown in the following equation:

$$M_x + CH_2{=}CHCH_2OOCCH_3 \longrightarrow M_xH + H_2\dot{C}CH{=}CHOOCCH_3 \qquad (1\text{-}7)$$

In this case the allylic radical is formed by removal of an alpha hydrogen from the monomer, producing an extremely stable radical which disappears through bimolecular combination. Reaction (1-7) is referred to as a degradative chain transfer.

1.2.2.3 Free Radical Initiation

Initiators of the type required for vinyl polymerizations are formed from compounds with relatively weak valency links which are relatively easily broken thermally. Irradiation of various wavelengths is sometimes employed to generate the radicals from an initiator, although more usually irradiation will generate radicals from a monomer as in the following equation:

$$CH_2CHX \xrightarrow{h\nu} CH_2\dot{C}HX \qquad (1\text{-}8)$$

The activated molecule then functions as a starting radical. Since, however irradiation is not normally a method of initiation in emulsion polymerization, it will only be given a brief mention. The decomposition of azodiisobutyronitrile has already been mentioned, and it may be noted that the formation of radicals from this initiator is accelerated by irradiation.

Another well-known initiator is dibenzoyl peroxide, which decomposes in two stages:

$$C_6H_5COOOCC_6H_5 \longrightarrow 2\,C_6H_5COO\bullet \qquad (1\text{-}9a)$$

$$C_6H_5COO\bullet \longrightarrow C_6H_5\bullet + CO_2 \qquad (1\text{-}9b)$$

Studies have shown that under normal conditions the decomposition proceeds through to the second stage, and it is the phenyl radical ($C_6H_5\bullet$) that adds on to the monomer. Dibenzoyl peroxide decomposes at a rate suitable for most direct polymerizations in bulk, solution and aqueous media, whether in emulsion or bead form, since most of these reactions are performed at 60~100℃. Dibenzoyl peroxide has a half-life of 5 h at 77℃。

A number of other diacyl peroxides have been examined. These include *o*-, *m*- and *p*-bromobenzoyl peroxides, in which the bromine atoms are useful as markers to show the fate of the radicals. Dilauroyl peroxide ($C_{11}H_{23}COOOOCC_{11}H_{23}$) has been used technically.

Hydroperoxides as represented by *t*-butyl hydroperoxide [$(CH_3)_3COOH$] and cumene hydroperoxide [$C_6H_5C(CH_3)_2OOH$] represent an allied class with technical interest. The primary dissociation

$$RCXOOH \longrightarrow RCXO\bullet + \bullet OH$$

is by secondary decompositions, which may include various secondary reactions of the peroxide induced by the radical in a second-order reaction and by considerable chain transfer. These hydroperoxides are of interest in redox initiators. Dialkyl peroxides of the type di-*t*-butyl peroxide [$(CH_3)_3COOC(CH_3)_3$] are also of considerable interest, and tend to be subject to less side reactions except for their own further decomposition, as shown in the equation below:

$$(CH_3)_3COOC(CH_3)_3 \longrightarrow 2(CH_3)_3CO\bullet \qquad (1\text{-}10a)$$

$$(CH_3)_3CO\bullet \longrightarrow (CH_3)_2CO + CH_3\bullet \qquad (1\text{-}10b)$$

These peroxides are useful for polymerizations that take place at 100~120℃, while di-*t*-butyl peroxide, which is volatile, has been used to produce radicals for gas phase polymerizations.

A number of peresters are in commercial production, e.g. *t*-butyl perbenzoate [$(CH_3)_3COOOCC_6H_5$], which acts as a source of *t*-butoxy radicals at a lower temperature than di-*t*-butyl hydroperoxide, and also as a source of benzoyloxy radicals at high temperatures. The final decomposition, apart from some secondary reactions, is probably mainly

$$(CH_3)_3COOOCC_6H_5 \longrightarrow (CH_3)_3CO\bullet + CO_2 + C_6H_5\bullet \qquad (1\text{-}11)$$

For a more detailed description of the decomposition of peroxides, a monograph should be consulted. While some hydroperoxides have limited aqueous solubility, the water-soluble initiators are a major type utilized for polymerizations in aqueous media. In addition, some peroxides of relatively high boiling point such as *t*-butyl hydroperoxide are sometimes added towards the end of emulsion polymerizations to ensure a more complete polymerization. These peroxides are also sometimes included in redox polymerization, especially to ensure rapid polymerization of the remaining unpolymerized monomers.

Hydrogen peroxide (H_2O_2) is the simplest compound in this class and is available technically as a 30%~40% solution. (This should not be confused with the 20%~30% volume solution available in pharmacies.) Initiation is not caused by the simple decomposition $H_2O_2 \longrightarrow 2OH\bullet$, but the presence of a trace of ferrous ion, of the order of a few parts per million of water present, seems to be essential, and radicals are generated according to the Haber-Weiss mechanism:

$$H_2O_2 + Fe^{2+} \longrightarrow HO\bar{\bullet} + Fe^{3+} + HO\bullet \tag{1-12}$$

The hydroxyl radical formed commences a polymerization chain in the usual manner and is in competition with a second reaction that consumes the radical:

$$Fe^{2+} + HO\bullet \longrightarrow Fe^{3+} + OH\bar{\bullet} \tag{1-13}$$

When polymerizations are performed, it seems of no consequence whether the soluble iron compound is in the ferrous or ferric form. There is little doubt that an equilibrium exists between the two states of oxidation, probably due to a complex being formed with the monomer present.

The other major class of water-soluble initiators consists of the persulfate salts, which for simplicity may be regarded as salts of persulfuric acid ($H_2S_2O_8$).

Potassium persulfate ($K_2S_2O_8$) is the least soluble salt of the series, between 2% and 5% according to temperature, but the restricted solubility facilitates its manufacture at a lower cost than sodium persulfate ($Na_2S_2O_8$) or ammonium persulfate [$(NH_4)_2S_2O_8$]. The decomposition of persulfate may be regarded as thermal dissociation of sulfate ion radicals:

$$S_2O_8^{2-} \longrightarrow 2SO_4^{-}\bullet \tag{1-14}$$

A secondary reaction may, however, produce hydroxyl radicals by reaction with water, and these hydroxyls may be the true initiators:

$$SO_4^{-}\bullet + H_2O \longrightarrow HSO_4^{-} + HO\bullet \tag{1-15}$$

Research using ^{35}S-modified persulfate has shown that the use of a persulfate initiator may give additional or even sole stabilization to a polymer prepare in emulsion. This may be explained by the polymer having ionized end group from a persulfate initiator, e.g. $ZOSO_3Na$, where Z indicates a polymer residue. A general account of initiation methods for vinyl acetate is applicable to most monomers.

1.2.3 Redox Polymerization

The formation of free radicals, which has already been described, proceeds essentially by a unimolecular reaction, except in the case where ferrous ions are included. However, radicals can be formed readily by a bimolecular reaction, with the added advantage that they can be formed in situ at ambient or even subambient temperatures. These systems normally depend on the simultaneous reaction of an oxidizing and a reducing agent, and often require in addition transition element that can exist in several valency states. The Haber-Weiss mechanism for initiation is the simplest case of a redox system.

Redox systems have assumed considerable importance in water-based systems, since most components in systems normally employed are water soluble. This type of polymerization was developed simultaneously during the Second World War in Great Britain, the United States and Germany, with special reference to the manufacture of synthetic rubbers. For vinyl polymerizations, as distinct from those where dienes are the sole or a major component, hydrogen peroxide or a persulfate is the oxidizing moiety, with a sulfur salt as the reductant. These include sodium metabisulfite ($Na_2S_2O_5$), sodium hydrosulfate [also known as hyposulfite or dithionite ($Na_2S_2O_4$)], sodium thiosulfate ($Na_2S_2O_3$) and sodium formaldehyde sulfoxylat [$Na(HSO_3 \cdot CH_2O)$]. The last named is one of the most effective and has been reported to initiate polymerizations, in conjunction with a persulfate, at temperatures as low as 0℃. In almost all of these redox polymerizations, a complete absence of oxygen seems essential, possibly because of the destruction by oxygen of the intermediate radicals that form.

However, in redox polymerizations operated under reflux conditions, or in otherwise unsealed reactors, it is often unnecessary in large-scale operations to continue the nitrogen blanket after polymerization has begun, probably because monomer vapour acts as a sealant against further oxygen inhibition. There have been relatively few detailed studies of the mechanisms of redox initiation of polymerization. A recent survey of redox systems gives a number of redox initiators, especially suitable for vinyl acetate, most of which are also suitable for other monomers.

Since almost all such reactions take place in water, a reaction involving ions may be used as an illustration. Hydrogen peroxide is used as the oxidizing moiety, together with a bisulfite ion:

$$H_2O_2 + S_2O_5^{2-} \longrightarrow HO\bullet + HS_2O_6^{2-}\bullet \tag{1-16}$$

The $HS_2O_6^{2-}\bullet$ represented here is not the dithionate ion, but an ion radical whose formula might be

```
   O⁻    O⁻
   |     |
 —S—O—S • OH
   ‖     ‖
   O     O
```

Alternatively, a hydroxyl radical may be formed together with an acid dithionate ion. Some evidence exists for a fragment of the reducing agent rather than the oxidizing agent acting as the starting radical for the polymerization chain. This seems to be true of many phosphorus-containing reducing agents; e.g. hypophosphorous acid with a diazonium salt activated by a copper salt when used as an initiating system for acrylonitrile shows evidence of a direct phosphorus bond with the polymer chain and also shows that the phosphorus is present as one atom per chain of polymer. Many of the formulations for polymerization quoted in the various application chapters are based on redox initiation.

1.2.4 Copolymerization

1.2.4.1 The Mechanism of Copolymerization

There is no reason why the process should be confined to one species of monomer. In general, a growing polymer chain may add on most other monomers according to a general set of rules, with some exceptions will be enunciated later.

If we have two monomers denoted by M_1 and M_2 and M_1• and M_2• denote chain radicals having M_1 and M_2 as terminal groups, irrespective of chain length four reactions are required to describe the growth of polymer:

$$M_1\bullet + M_1 \xrightarrow{K_{11}} M_1\bullet$$

$$M_1\bullet + M_2 \xrightarrow{K_{12}} M_1\bullet$$

$$M_2\bullet + M_1 \xrightarrow{K_{21}} M_2\bullet$$

$$M_2\bullet + M_2 \xrightarrow{K_{22}} M_2\bullet$$

where K has the usual meaning of a reaction rate constant. These reactions reach a 'steady state' of copolymerization in which the concentration of radicals is constant; i.e. the copolymerization is constant and the rates of formation of radicals and destruction of radicals by chain termination are constant. Under these conditions the rates of formation of each of the two radicals remain constant and without considering any elaborate mathematical derivations we may define the monomer reactivity ratios r_1 and r_2 by the expressions:

$$r_1 = \frac{K_{11}}{K_{12}} \quad \text{and} \quad r_2 = \frac{K_{22}}{K_{21}}$$

These ratios represent the tendency of a radical formed from one monomer to combine with itself rather than with another monomer. It can be made intelligible by a practical example. Thus, for styrene ($C_6H_5CH{=}CH_2$)(r_1) and butadiene [$CH_2{=}CHCH{=}CH_2$](r_2), $r_1 = 0.78$ and $r_2 = 1.39$. These figures tend to indicate that if we start with an equimolar mixture, styrene radicals tend to copolymerize with butadiene rather than themselves, but butadiene has a slight preference for its radicals to polymerize with each other. This shows that if we copolymerize an equimolar mixture of styrene and butadiene, a point occurs at which only styrene would remain in the unpoly-merized state. However, for styrene and methyl methacrylate, $r_1 = 0.52$ and $r_2 = 0.46$ respectively. These two monomers therefore copolymerize together in almost any ratio. As the properties imparted to a copolymer by equal weight ratios of these two monomers are broadly similar, it is often possible to replace one by the other on cost alone, although the inclusion of styrene may cause yellowing of copolymer films exposed to sunlight.

Nevertheless, if an attempt is made to copolymerize vinyl acetate with styrene, only the latter will polymerize, and in practice styrene is an inhibitor for vinyl acetate. The reactivity ratios, r_1 and r_2 for styrene and vinyl acetate respectively have been given as 55 and 0.01. However, vinyl benzoate ($CH_2{=}CHOOCC_6H_5$) has a slight tendency to copolymerize with styrene, probably because of a resonance effect. If we consider the case of vinyl acetate and *trans*-dichlorethylene (TDE) *trans*-$CH_2Cl{=}CH_2Cl$, r_1(vinyl acetate) = 0.85 and r_2 = 0. The latter implies that TDE does not polymerize by itself, but only in the presence of vinyl acetate. Vinyl acetate, on the other hand, has a greater tendency to copolymerize with TDE than with itself, and therefore if the ratios are adjusted correctly all of the TDE can be copolymerized.

Let us consider the copolymerization of vinyl acetate and maleic anhydride:

```
        O
       //
CH—C
||     >O    r1=0.055. r2=0
CH—C
       \\
        O
```

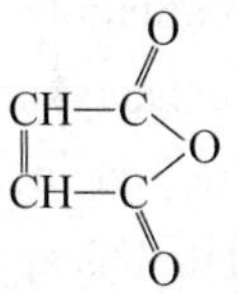

r_1=0.055. r_2=0

Sometimes a very low r_2 is quoted for maleic anhydride, e.g. 0.003. Vinyl acetate thus has a strong preference to add on to maleic anhydride in a growing radical rather than on to another vinyl acetate molecule, while maleic anhydride, which has practically no tendency to add on to itself, readily adds to a vinyl acetate unit of a growing chain. (Note that homopolymers of maleic anhydride have been made by drastic methods.) This is a mathematical explanation of the fact that vinyl acetate and maleic anhydride tend to alternate in a copolymer whatever the starting ratios. Excess maleic anhydride, if present, does not homopolymerize. Surplus vinyl acetate, if present, forms homopolymer, a term used to distinguish the polymer formed from a single monomer in contradistinction to a copolymer. Styrene also forms an alternating copolymer with maleic anhydride.

Only in one or two exceptional cases has both r_1 and r_2 been reported to be above 1. Otherwise it is a general principle that at least one of the two ratios is less than 1. It will be readily seen that in a mixture of two monomers the composition of the copolymer gradually changes unless an 'azeotropic' mixture is used, i.e. one balanced in accord with r_1 and r_2, provided that r_1 and r_2 are each less than 1.

Polymers of fixed composition are sometimes made by starting with a small quantity of monomers, e.g. 2%~5% in the desired ratios, and adding a feedstock which will maintain the original ratio of reactants. This is especially noted, as will be shown later, in emulsion polymerization. If it is desired to include the more sluggishly polymerizing monomer, and an excess is used, this must be removed at the end, by distillation or extraction.

However, as a general principle it should not be assumed that, because two or more monomers copolymerize completely, the resultant copolymer is reasonably homogeneous. Often, because of compatibility variations among the constantly varying species of polymers formed, the properties of the final copolymer are liable to vary very markedly from those of a truly homogeneous copolymer.

The term 'copolymer' is sometimes confined to a polymer formed from two monomers only. In a more general sense, it can be used to cover polymers formed from a larger number of monomers, for which the principles enunciated in this section apply. The term 'terpolymer' is sometimes used when three monomers have been copolymerized.

When copolymerization takes place in a heterogeneous medium, as in emulsion polymerization, while the conditions for copolymerization still hold, the reaction is complicated by the environment of each species present. Taking into account factors such as whether the initiator is water or monomer soluble (most peroxidic organic initiators are soluble in both), the high aqueous solubility of monomers such as acrylic acid ($CH_2{=}CHCOOH$) and, if partition between water-soluble and water-insoluble monomers is significant, the apparent reactivities may differ markedly from those in a homogeneous medium. Thus, in an attempted emulsion polymerization, butyl methacrylate [$CH_2{=}C(CH_3)COOC_4H_9$] and sodium methacrylate [$NaOOCC(CH_3){=}CH_2$] polymerize substantially independently. On the other hand, methyl methacrylate and sodium methacrylate will copolymerize together since methyl methacrylate has appreciable water solubility.

More unusually vinyl acetate and vinyl stearate ($CH_2{=}CHOOCC_{17}H_{35}$) will only copolymerize in emulsion if a very large surface is present due to very small emulsion particles (of order <0.1 μm) or a class of emulsifier known as a 'solubilizer' is present, which has the effect of solubilizing vinyl stearate to a limited extent in water, increasing the compatibility with vinyl acetate which is about 2.3% water soluble.

1.2.4.2 The *Q*, *e* Scheme

Several efforts have been made to place the relative reactivities of monomers on a chemical-mathematical basis. The chief of these has been due to Alfrey and Price. Comparison of a series of monomers with a standard monomer is most readily made by using the reciprocal of r with respect to that monomer; i.e. the higher the value of $1/r$ the poorer the copolymerization characteristics. Thus, taking styrene as an arbitrary 1.0, methyl methacrylate 2.2 and acrylonitrile 20, vinyl acetate is very high on this scale. However, the relative scale of reactivities is not interchangeable using different radicals as references.

It has been observed that the product r_1r_2 tends to be smallest when one of the two monomers concerned has strongly electropositive (electron-releasing) substituents and there are electronegative (electron-attracting) substituents of the other. Thus alternation tends to occur when the polarities of the monomers are opposite.

Alfrey and Price therefore proposed the following equation:

$$K_{12} = P_1Q_2 \exp(-e_1e_2) \quad (1\text{-}17)$$

where P_1 and Q_2 are constants relating to the general activity of the monomers M_1 and M_2 respectively, and e_1 and e_2 are proportional to the residual electrostatic charges in the respective reaction groups. It is assumed that each monomer and its corresponding radical has the same reactivity. Hence, from the reactions in Section 1.2.4.1,

$$r_1r_2 = \exp[-(e_1-e_2)^2] \quad (1\text{-}18)$$

The product of the reactivity ratios is thus independent of Q. The following equation is also useful:

$$Q_2 = \frac{Q_1}{r_1}\exp[-e_1(e_1-e_2)] \quad (1\text{-}19)$$

A series of Q and e values has been assigned to a series of monomers by Price. Typical e values are −0.8 for butadiene, −0.8 for styrene, −0.3 for vinyl acetate, +0.2 for vinyl chloride, +0.4 for methyl methacrylate and +1.1 for acrylonitrile.

while the Q, e scheme is semi-empirical, it has proved highly useful in coordinating otherwise disjointed data.

1.3 Chain Branching; Graft and Block Copolymers

1.3.1 Chain Branching

Occasionally chain transfer (Section 1.2.2.1) results in a hydrogen atom being removed from a growing polymer chain. Thus in a chain that might be represented as $(CH_2CHX)_n$, the addition of further units of $CH_2{=}CHX$ might produce an intermediate as $(CH_2CHX)_nCH_2CHXCH_2CHX$. A short side chain is thus formed by hydrogen transfer. For simplicity, this has been shown on the penultimate unit, but this need not be so; nor is there any reason why there should only be one hydrogen abstraction per growing chain. From the radicals formed branched chains may grow.

Chain branching occurs most readily from a tertiary carbon atom, i.e. a carbon atom to which only one hydrogen atom is attached, the other groupings depending on a carbon to carbon attachment, e.g. an alkyl or an aryl group. The mechanism is based on abstraction of a hydrogen atom, although of course abstraction can also occur with a halogen atom. With polyvinyl acetate, investigations have shown that limited chain transfer can occur through the methyl grouping of the acetoxy group ($-OOCCH_3$). The result of this type of branching is a drastic reduction of molecular mass of the polymer during hydrolysis, since the entire branch is hydrolyzed at the acetoxy group at which branching has occurred, producing an extra fragment for each branch of the original molecule. It has also been shown that in a unit of a polyvinyl acetate polymer the ratio of the positions marked (1), (2) and (3) is 1 : 3 : 1.

$$-CH_2- \quad \dot{C}HOOCCH_3$$
$$(1) \quad (2) \quad (3)$$

It is now known that there is significant chain transfer on the vinyl H atoms of vinyl acetate.

Another method of forming branched chains involves the retention of a vinyl group on the terminal unit of a polymer molecule, either by disproportionation or by chain transfer to monomer. The polymer molecule with residual unsaturation could then become the unit of a further growing chain. Thus a polymer molecule of formula $CH_2CHX-(M_p)$, where M_p represents a polymer chain, may become incorporated into another chain to give a structure $(M_q)CH_2CHX-(M_p)(M_r)$, where M_q and M_r represent polymer chains of various lengths, that may be of the same configuration or based on different monomers, depending on conditions. Ethylene, $CH_2{=}CH_2$, which is normally a gas (b.p. 760 mmHg/104℃, critical temperature

9.5℃) is prone to chain branching when polymerized by the free radical polymerization process at high temperatures and pressures, most branches having short chains. In this case intra-molecular formation of short chains occurs by chain transfer, and is usually known as 'back biting'.

$$\begin{array}{c} H_2 \\ H_2C\diagup^{C}\diagdown CH_2 \\ | \qquad\quad | \\ -CH_2 \quad \cdot CH_2 \end{array} \longrightarrow \begin{array}{l} H_2C{-}CH_3 \\ \;| \\ H_2C \\ \;| \\ H_2C \\ \;| \\ -\cdot CH \end{array}$$

$$\begin{array}{l} H_2C{-}CH_3 \\ \;| \\ H_2C \\ \;| \\ H_2C \\ \;| \\ -\cdot CH \end{array} + nC_2H_4 \longrightarrow \begin{array}{l} H_2C{-}CH_3 \\ \;| \\ H_2C \\ \;| \\ H_2C \\ \;| \\ -\cdot CH{-}(CH_2CH_2)_nE \end{array}$$

where E represents an end group. The carbon with the asterisk is the same throughout to illustrate the reaction. Excessive chain branching can lead to crosslinking and insolubility. It is possible for chain branching to occur from completed or 'dead' molecules by hydrogen abstraction, and although this impinges on grafting, it is treated as a chain branching phenomenon if it occurs during a polymerization.

1.3.2 Graft and Block Copolymers

The idea of a graft copolymer is a natural extension of the concept of chain branching and involves the introduction of active centers in a previously prepared chain from which a new chain can grow. In most cases this is an added monomer, although two-polymer molecules can combine directly to form a graft. The graft base need not be an ethylene addition polymer. Various natural products, including proteins and water-soluble gums, have been used as a basis for graft copolymers by formation of active centers.

A block copolymer differs from a graft in only that the active centre is always at the end of the molecule. In the simplest case, an unsaturated chain end arising from a chain transfer can act as the basis for the addition of a block of units of a second monomer, while successive monomers or the original may make an additional block. Another possibility is the simultaneous polymerization of a monomer which is soluble only in water with one which is water insoluble, provided that the latter is in the form of a fine particle size emulsion. Whether the initiator is water soluble or monomer soluble, an extensive transfer through the surface is likely, with the continuation of the chain in the alternate medium.

There are a number of ways of achieving active centers, many of which depend on an anionic or cationic mechanism, especially the former. However, since in water-based graft polymerization only free radical polymerizations and possibly a few direct chemical reactions involving an elimination are of interest, the discussion here will be confined to these topics.

Graft centers are formed in much the same manner as points of branching, with the difference that the graft base is preformed. It may be possible to peroxidize a polymer directly with oxygen, to provide hydroperoxide (—OOH) groups directly attached to carbon. This is facilitated, particularly, where numerous tertiary carbon occur as, for example, in polypropylene $(\text{—}CH_2\underset{\displaystyle CH_3}{\underset{|}{C}}H\text{—})$. In other cases the direct use of a peroxide type of initiator encourages the formation of free radicals on existing polymer chains. Particularly useful in this respect is *t*-butyl hydroperoxide, *t*-C_4H_9OOH, because of the strong tendency of the radical formed from it to abstract hydrogen atoms. Dibenzoyl peroxide $[(C_6H_5COO)_2]$ is also frequently used as a graft initiator. In aqueous systems initiators such as *tert*-butyl hydroperoxide may be used in conjunction with a salt of a sulfur-reducing acid to lower the temperature at which radicals are generated.

Graft methods make it possible to add to polymers such as butadiene-styrene chains of a monomer that is not normally polymerizable, such as vinyl acetate. The polymerization medium in which a graft can take place is in general not restricted; the process may take place fairly readily in emulsion. There is a vast amount of literature available on the formation and properties of graft copolymers.

There are very often special considerations in respect of graft copolymerizations that take place in emulsion form, with particular reference to water-soluble stabilizers of the polyvinyl alcohol type. In some cases halogen atoms may be removed by a radical. This occurs particularly with polymers and copolymers based on vinyl chloride ($CH_2{=}CHCl$), vinylidene chloride ($CH_2{=}CCl_2$) and chloroprene ($CH_2{=}CClCH{=}CH_2$). Ultraviolet light and other forms of irradiation are particularly useful in this respect.

Properties of graft copolymers are sometimes unique, and not necessarily an intermediate or balance between those of polymers derived from the respective monomers. This is particularly noticeable with solubility properties and transition points. A brief reference may be made here to the more direct chemical types of graft formation that do not involve free radicals. These depend on the direct reaction of an active group on the polymer. The simplest group is hydroxyl (—OH), which under suitable conditions may react with carboxyl (—COOH), carboxyanhydride (—COOOC—) or carbochloride (—COCl) to form ester or polyesters depending on the nature of the side chain. Equally, hydroxyl groups may react with oxirane ($\underset{O}{CH_2CHX}$) groups. This applies especially with ethylene oxide ($\underset{O}{\bigtriangledown}$) to form oxyethylene side chains, giving graft copolymers of the type:

$$\text{—}CH_2OH\text{—}CH_2\text{—}\underset{\displaystyle OC_2H_4(OC_2H_4)_nOH}{\underset{|}{C}H}\text{—}CH_2\text{—}$$

This will be of special interest in dealing with emulsions.

1.4 Polymer Structure and Properties

1.4.1 Polymer Structure

The physical properties of a polymer are determined by the configuration of the constituent atoms, and to some extent by the molecular mass. The configuration is partly dependent on the main chain, and partly on the various side groups. Most of the polymers which we are considering are based on long chains of carbon atoms. In representing formula we are limited by the plane of the paper, but a three-dimensional structure must be considered. The C—C internuclear distance is 1.54 Å, and where free rotation occurs the C—C—C bond is fixed at 109° (the tetrahedral angle).

By tradition, we represent polyethylene chain in the full extended fashion: $\diagdown CH_2 \diagup CH_2 \diagdown CH_2 \diagup CH_2 \diagdown CH_2 \diagup$.

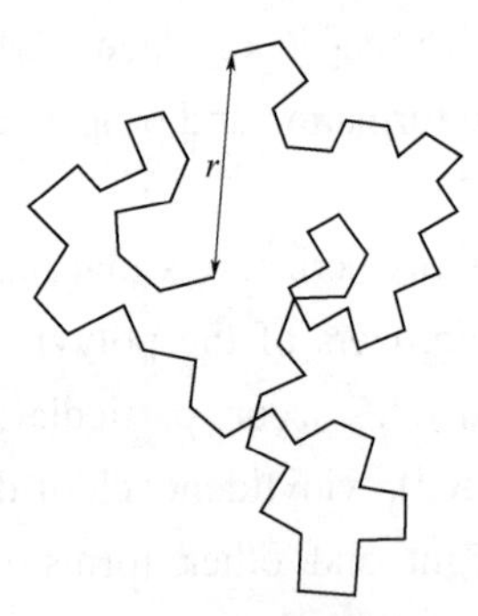

Fig.1.1 Diagrammatic molecular coil.

In practice the polymer is an irregular coil, as shown in Fig.1.1. The dimension most frequently used to describe an 'average' configuration is the 'root mean square', symbolized as r, which can be symbolized mathematically as

$$\frac{\left[\sum n_1 r_1^2\right]^{0.5}}{\left[\sum n_1\right]}$$

where there are n individual polymer molecules, and the distance apart of the chain ends is r_1, r_2, etc. This concept of root mean square is necessary in dealing with certain solution properties, and also certain properties of elasticity.

No real polymer molecule can have completely free and unrestricted rotation, although an unbranched polythene $(C_2H_4)_n$ approaches closest to this ideal. [The theoretical polymethylene $(CH_2)_m$ has been prepared by the polymerization of diazomethane (CH_2N_2), with elimination of nitrogen.] The properties of polyethylene over a wide range of molecular masses are, at ambient temperatures, those of a flexible, relatively inelastic molecule, which softens fairly readily. Chain branching hinders free rotation and raises the softening point of the polymer. Even a small number of crosslinks may, however, cause a major hindrance to the free rotation of the internal carbon bonds of the chain, resulting in a sharp increase in stiffness of the resulting product.

Many side chain groups cause steric hindrance and restrictions in the free rotation about the double bonds. A typical example is polystyrene, where the planar zigzag formulation is probably modified by rotations of 180° a round alternate double bonds to produce a structure of minimum energy, such as

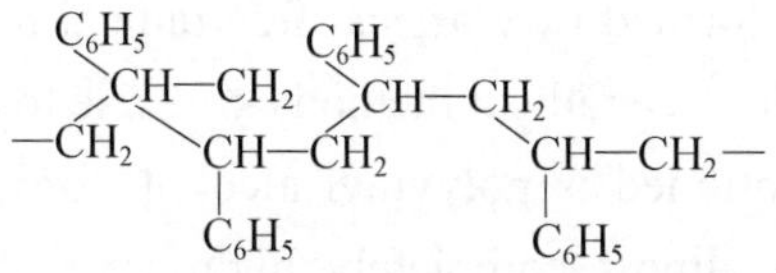

$$
\begin{array}{l}
\quad\quad C_6H_5 \quad\quad C_6H_5 \\
\quad\quad\;\; CH-CH_2 \;\; CH-CH_2 \\
-CH_2 \quad CH-CH_2 \quad\quad CH-CH_2- \\
\quad\quad\quad | \quad\quad\quad\quad\quad\quad | \\
\quad\quad\; C_6H_5 \quad\quad\quad\; C_6H_5
\end{array}
$$

Because of the steric hindrance, polystyrene is a much harder polymer than polyethylene.

Other molecular forces that effect the physical state of the polymers are the various dipole forces and the London or dispersion forces. If different parts of a group carry opposite charges, e.g. the carbonyl $\overset{O}{\overset{\|}{-C-}}$ and hydroxyl —OH, strong interchain attraction occurs between groups on different chains by attraction of opposite charges. This attraction is strongly temperature dependent. A special, case of dipole forces is that of hydrogen bonding, by which hydrogen atoms attached to electronegative atoms such as oxygen or nitrogen exert a strong attraction towards electronegative atoms on other chains. The principal groups of polymers in which hydrogen bonding occurs are the hydroxyl and the amino —NHX or amide $-CONH_2$ groups and are illustrated by the following:

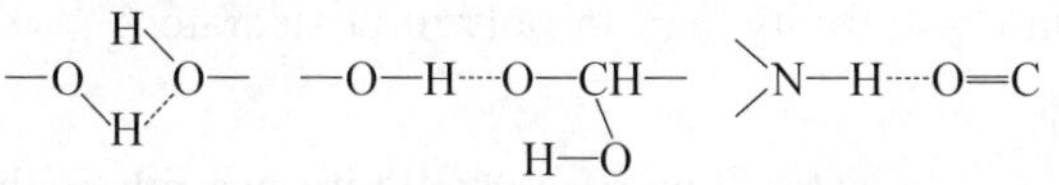

$$
-O\overset{H}{\underset{H}{\diagdown\!\!\diagup}}O- \qquad -O-H\cdots O-\underset{H-O}{\underset{\diagdown}{CH}}- \qquad \rangle N-H\cdots O{=}C
$$

The net effect of dipole forces, especially hydrogen bonding, is to stiffen and strengthen the polymer molecules, and in extreme cases to cause crystalline polymers to be formed. Examples of polymers with strong hydrogen bonding are polyvinyl alcohol $[-(CH_2CHOH)_n-]$, polyacrylamide $[\underset{CONH_2}{\underset{|}{-(CH_2CH)_n-}}]$ and all polymers including carboxylic acid groups, e.g. copo- lymers including units of acrylic acid ($CH_2{=}CHCOOH$) and crotonic acid ($CH_2{=}CHCH_2COOH$). The London forces between molecules come from time-varying dipole moments arising out of the continuously varying configurations of nuclei and electrons which must, of course, average out to zero. These forces, which are independent of temperature, vary inversely as the seventh power of the distance between the chains, as do dipole forces, and only operate at distances below 5 Å.

Forces between chains lead to a cohesive energy, approximately equal to the energies of vaporization or sublimation. A high cohesive energy is associated with a high melting point and may be associated with crystallinity. A low cohesive energy results in a polymer having a low softening point and easy deformation by stresses applied externally.

While inorganic materials often crystallize and solid organic polymers generally possess crystallinity, X-ray diffraction patterns have shown that in some polymers there are non-amorphous and crystalline regions, or crystallites. While crystallinity is a characteristic of natural products such as proteins and synthetic condensation products such as the polyamide fibers, crystallinity sometimes occurs in addition polymers. Even if we discount types prepared by special methods, such as use of the Ziegler-Natta catalysts, which will not be discussed further here since they are not formed by classical free radical reactions, a number of polymers prepared directly or indirectly by free radical methods give rise to crystallinity.

One of these already mentioned is polyvinyl alcohol, formed by hydrolysis of polyvinyl acetate. It must, however, be almost completely hydrolyzed, of the order of 99.5%, to be effectively crystalline, under which conditions it can be oriented and drawn into fibers. If hydrolysis is partial, the resulting disorder prevents crystallinity. This is the case with the so-called 'polyvinyl alcohol' of saponification value about 120, which is used for emulsion polymerization. This polymer consists, by molar proportions, of about 88% of vinyl alcohol and 12% of vinyl acetate units.

Polymers of vinylidene chloride (CH_2CCl_2) are strongly crystalline. Polymers of vinyl chloride (CH_2CHCl) and acrylonitrile (CH_2CHCN) are partially crystalline, but crystallinity can be induced by stretching the polymer to a fibre structure to induce orientation. Polyethylene, when substantially free from branching, is crystalline and wax-like because of the simple molecular structure. It does not, of course, have the other properties associated with crystallinity caused by hydrogen bonding, such as high cohesive strength. Another type of crystallinity found in polymers is side chain crystallinity, e.g. in polyvinyl stearate ($\underset{OOCC_{17}H_{35}}{\overset{—(CH_2CH)_n—}{|}}$) or polyoctadecyl acrylate $\underset{COOC_{16}H_{37}}{\overset{—(CH_2CH)_n—}{|}}$. This type of crystallinity has relatively little application, since the products tend to simulate the crystalline properties of a wax. However, this property may be useful in connection with synthetic resin-based polishes.

In considering the effect of side chains on polymer properties, it is convenient to take a series of esters based on acrylic acid and compare the derived polymers. These are most readily compared by the second-order transition points (T_g). Technical publications show some variation in these figures, probably because of variations in molecular mass. However, polymers prepared under approximately the same conditions have much the same degree of polymerization (DP), and emulsion polymers are preferred as standards in this connection.

Fig.1.2 shows the variations in T_g of a series of homologous polymers based on acrylic acid ($CH_2{=}CHCOOH$) and methacrylic acid [$CH_2{=}C(CH_3)COOH$]. The striking difference in T_g of the polymers based on the methyl esters should be noted, being almost 100℃. This is due to the steric effect of the angular methyl $\cdot CH_3$ group on the carbon atom to which the carboxyl group is attached. Polymethyl methacrylate is an extremely hard solid, used inter alia for 'unbreakable' glass.

The effect of the angular methyl group slowly diminishes as the alcohol side chains become longer; these latter keep the chains apart and reduce the polar forces. In consequence the T_g diminishes in the case of alkyl ester polymers of acrylic acid until the alkyl chain reaches about 10 carbon atoms. It then increases again with side chain crystallinity. The methacrylate ester polymers, however, continue to drop in T_g, usually until a C_{13} alkyl group is reached, since the steric effect of the angular methyl group on the main chain also prevents side chain crystallinity at first.

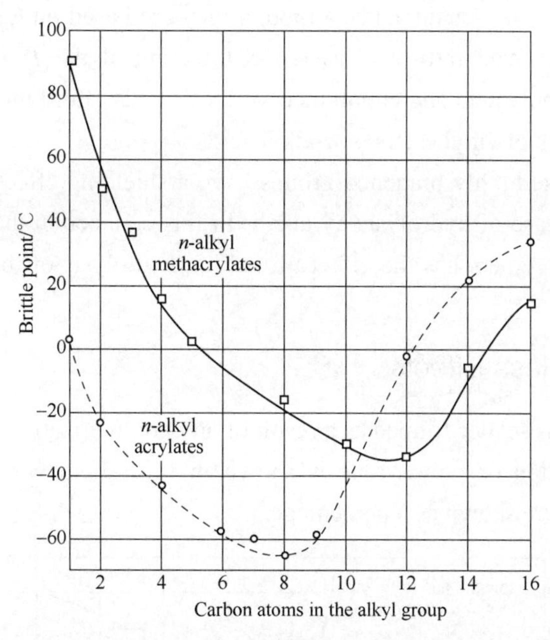

Fig.1.2 Brittle points of polymeric *n*-alkyl acrylates and methacrylates.

Similar conditions prevail in the homologous series of vinyl esters of straight chain fatty acids based on the hypothetical vinyl alcohol CH_2CHOH. From polyvinyl formate $(-CH_2\underset{\displaystyle OOCH}{CH}-)_n$ through polyvinyl acetate $(-CH_2\underset{\displaystyle OOCCH_3}{CH}-)_n$ to vinyl laurate $(-CH_2\underset{\displaystyle OOCC_{11}H_{23}}{CH}-)_n$, there is a steady fall in T_g, the polymers varying from fairly brittle films derived from a latex at ambient temperature to viscous sticky oils as the length of the alcohol chain increases. However, note that the polymerization and even copolymerization of monomers with long side chains, above about C_{12}, becomes increasingly sluggish.

The above examples, in both the acrylic and the vinyl ester series, have considered the effect of straight chains inserted as side chains in polymer molecules. The effect of branched chains, however, is different. As chain branching increases, the effect of the overall size of the side chain diminishes. Polyisobutyl methacrylate has a higher T_g than polybutyl methacrylate. Polymers based on *tert*-butyl acrylate or *tert*-butyl methacrylate have a higher softening point than the corresponding *n*-butyl esters. Another interesting example of the effect of branched chains is that of the various synthetic branched chain acids in which the carbon atom in the α position to the carbon of the carboxyl is quaternary, corresponding to a general formula $HOOCCR_1R_2R_3$, where R_1 is CH_3, R_2 is CH_3 or C_2H_5 and R_3 is a longer chain alkyl group, which may be represented as $C_{4\sim6}H_{9\sim13}$. These form vinyl esters which correspond in total side chain length to vinyl caprate $CH_2{=}CHOOCC_9H_{19}$ but do not impart the same flexibility in copolymers.

It is often more practical to measure the effect of monomers of this type by copolymerizing them with a harder monomer such as vinyl acetate and measuring the relative effects. Thus the vinyl esters of these branches chain acids, although they are based on C_{10} acids on average, are similar to a C_4 straight chain fatty acid as far as lowering of the T_g is concerned. It is also interesting to note that polymers and copolymers of these acids afford much greater resistance to hydrolysis than polymers of vinyl esters of *n*-alkyl acids.

In copolymers these highly branched groups have a shielding effect on neighbouring ester groups, reducing their ease of hydrolysis by alkali. In this connection the angular methyl group in methacrylate ester polymers has the effect of making hydrolysis of these products extremely difficult.

1.4.2 Molecular Mass Effects

The molecular mass scatter formed as a result of any polymerization is typical of a Gaussian type. Thus a fractionation of polystyrene is shown in Fig.1.3, in which the distribution and cumulative mass totals are shown as a percentage.

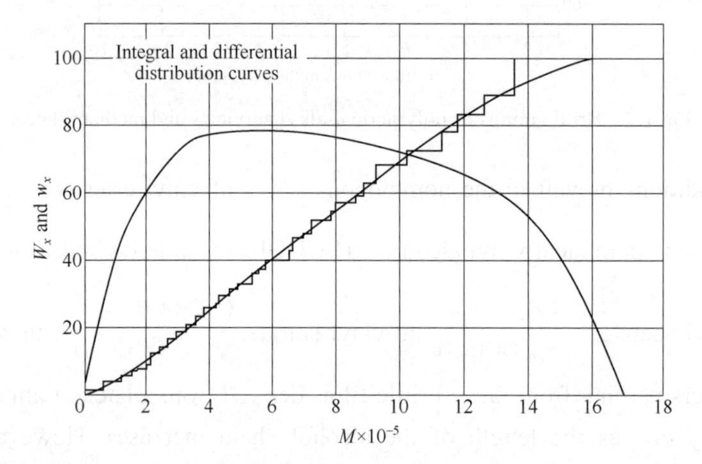

Fig.1.3 Molecular mass distribution for thermally polymerized polystyrene as established by fractionation.

Before discussing the general effect of molecular mass on polymer characteristics, some further definitions are desirable. The number average molecular mass M_n is the simple arithmetical average of each molecule as a summation, divided by the number of molecules, the 'popular' idea of an average. Another measurement of average is the 'weight' average, and is an expression of the fact that the higher molecular mass fractions of a polymer play a greater role in determining the properties than do the fractions of lower molecular mass.

Its definition is based on multiplying the number of identical molecules of molecular mass M_n by the overall weight of molecules of that weight and dividing by the sum total of the weights. Mathematically, this is given by

$$M_w = \frac{\sum w_1 M_1}{w_1} \tag{1-20}$$

where w_1 represents the overall weight of molecules of molecular mass M_1; the weight average molecular mass M_w is invariably greater than the number average as its real effect is to square the weight figure. For certain purposes, the *z* average is used in which M_1 in the equation above is squared, giving even higher prominence to the higher molecular mass fractions.

In practice all the viscosity characteristics of a polymer solution depend on M_w rather than M_n. Thus nine unit fragments of a monomer of molecular mass 100 individually pulled off a polymer of molecular mass 1 000 000 reduces its M_n to 100 000. The M_w is just over 999 000. This corresponds to a negligible viscosity change.

A number of methods of measuring molecular mass are used and are summarized here:

(a) Osmometry. This is a vapour pressure method, useful for polymers of molecular mass up to about 25 000; membrane osmometry is used for molecular masses from 20 000 to 1 000 000. These are number average methods.

(b) Viscometry. This is a relative method, but the simplest, and its application is widespread in industry. Viscometry is approximately a weight average method.

(c) Light scattering. This is a weight average method.

(d) Gel permeation chromatography. This is a direct fractionation method using molecular mass. It is relatively rapid and has proved to be one of the most valuable modern methods.

(e) Chemical methods. These usually depend on measuring distinctive end groups. They are number average methods.

In some cases, selective precipitation can be used to fractionate a polymer according to molecular mass. This is essentially a relative method based on known standards. This method also differentiates between varying species in a copolymer.

The properties of polymers are governed to some extent by molecular mass as well as molecular structure. Properties also depend partly on the distribution of molecular masses, and in copolymers on the distribution of molecular species. The differences in solubility in solvents in exploited in fractionation where blended solvents are used, only one being a good solvent for the polymer. The added poor solvent will tend to precipitate the higher molecular mass fractions first. Thus polyvinyl acetate may be fractionated by the gradual addition of hexane C_6H_{14} to dilute solutions of the polymer in benzene.

In some cases, molecular mass variations have an extreme effect of polymer properties. This is particularly significant in the polyvinyl ethers ($\begin{array}{c}-CH_2CH \\ | \\ OC_nH_{2n+1}\end{array}$), in which a polymer can vary from an oil at a molecular mass of about 5000 to a rubbery material if the molecular mass is above 100 000. The polyvinyl ethers, however, are not prepared as homopolymers by a free radical mechanism. The differences are usually illustrated by the change in the second-order transition point. The softening points, which correspond approximately to melting ranges, and which are estimated by standard methods, are also affected by molecular mass. The overall effect of solvents on polymers is too complex to be considered here.

1.4.3 Transition Points

Although when dealing with a crystalline substance, there is a sharp melting point, sometimes denoted T_{m}; when dealing with a polymer containing molecules with a range of molecular masses it is not possible to describe the changes in state on heating in a similar manner.

Amorphous materials, unless crosslinked or decomposing at a relatively low temperature, will soften gradually, and although a softening point or range may be quoted, this depend on an arbitrarily chosen test, usually on the time taken for a steel ball to penetrate a known thickness of the polymer.

However, an amorphous polymer has a number of physical changes of condition, the most important being the second-order transition point, usually referred to as T_g, already mentioned previously. Physically this transition point is connected with the mobility of the polymer chains. Below T_g, the chains may be regarded as substantially immobile, except for movements around an equilibrium position. Above this temperature, appreciable movement of segments occurs in the polymer chains. Below the T_g, the polymer is a hard, brittle solid; above this temperature, increased flexibility and possibly rubberlike characteristics are observed.

The second-order transition point may be measured in various ways; e.g. the rate of change of polymer density varies with temperature, as does the rate of change of other properties such as specific heat. Most useful is differential thermal analysis (DTA), which indicates the differential in the heating capacity of a substance. Modern DTA instruments are extremely sensitive. It may be noted that an alternative temperature, known as the minimum film formation temperature (MFT) is frequently a substitute for T_g. This is the lowest temperature at which a drying emulsion containing polymer particles will form a continuous film. Because of the conditions of film formation, this temperature is usually 3~5℃ higher than the T_g. Differential calorimetrical scanning (DSC) results have shown that many polymers have transition points other than T_g. These are associated with the thermal motion of the molecules.

In many cases where a polymer has practical utility, it may be desirable that T_g should be reduced to achieve reasonable flexibility for the polymer. This is accomplished by plasticization which reduces T_g to a level below ambient, or below the MFT in the case of a latex. As an example of plasticization, about 40 parts of di-2-ethylhexyl phthalate are required to transform 60 parts of polyvinyl chloride from a hard, horny material to a flexible sheet. 'Internal plasticisation' is a term used for the formation of a copolymer, the auxiliary monomer of which gives increased flexibility to the polymer formed from the principal monomer.

1.5 Technology of Polymerization

Monomers may be polymerized by free radical initiation by one of five methods: polymerization in bulk, in solution, dispersed as large particles in water or occasionally in another non-solvent (suspension polymerization), or dispersed as fine particles, less than 1.5 μm, usually less than 1 μm in diameter. The last-named process is usually known as emulsion polymerization. A variant of suspension polymerization may be described as solution precipitation. It is often applied to copolymers, e.g. a copolymer of methyl methacrylate and methacrylic acid. In concentrated solution, the acid solubilizes the methyl methacrylate. On polymerization, a fine water-insoluble powder is produced, which, depending on the monomer ratios, is usually alkali soluble.

In the past two decades, a variation of emulsion polymerization has been introduced, the polymers being known as 'dispersymers'. To form dyspersymers, a liquid monomer forms an emulsion-like product in an organic liquid, usually a liquid hydrocarbon, in which the polymer is insoluble. The final emulsion closely resembles an aqueous polymer emulsion in physical appearance.

1.5.1 Bulk Polymerization

While in most cases laboratory experiments may be performed on undiluted monomers, or on controlled dilutions with solvents do not affect the polymerization seriously, this process produces difficulties in large-scale production, which may be of the order of 5 tonnes in a single batch. Problems are caused by an increase in viscosity of the mass during polymerization, and in particular the removal of the heat of polymerization, which for most monomers is of the order of 20 kg•cal•gram^{-1}•mole^{-1}. Special equipment with a high surface-volume ratio is desirable, as with the polymerization of methyl methacrylate which is polymerized in thin sheets with a very low initiator ratio. Bulk polymerization of vinyl acetate was described in reports of German factory production after the Second World War. In this case, the hot polymer is sufficiently fluid to be discharged directly from a cylindrical reactor. In an alternative continuous process, the monomer is passed down a polymerization tower. Processes have been developed for the bulk polymerization of vinyl chloride which is insoluble in its own monomer.

1.5.2 Solution Polymerization

Polymerization with a solvent diluent can be readily accomplished as the major problems of bulk polymerization are overcome with increasing dilution. However, some practical problems persist. Commercial solvents are seldom pure and the impurities may have an inhibiting or retarding effect on polymerization; this is especially so with monomers such as vinyl acetate which are not resonance stabilized. In addition, many solvents have a chain transfer effect.

Towards the end of a polymerization, the degree of dilution of the monomer is extremely high; the efficiency of initiator therefore falls, and it is lost by chain transfer with the solvent or by mutual destruction of the radicals. Thus several repeat initiations are necessary towards the end of a practical polymerization in solvent to achieve the 99% polymerization generally desired.

Practical experience has shown that molecular masses in solution polymerization are also susceptible to a number of other factors, such as the type and nature of stirring, and the type and nature of the reactor, including its shape and the surface-volume ratio. There is likely to be a 'wall effect', which may terminate growing radicals. In addition, stirring conditions affect the rate of attainment of equilibrium, while the amount of reflux, where present, also affects the nature of the final polymer. Since it is not usually desirable and may be difficult to distill unpolymerized monomer, even if a satisfactory azeotrope with the solvent exists, direct solvent polymerization has limited practical application and is of principal interest where the solutions are used directly, either as solvent-based coatings or as adhesives. Solvent polymerization can normally be used only to prepare polymers of relatively low molecular mass.

1.5.3 Suspension Polymerization

Suspension polymerization may be described as a water-cooled bulk polymerization, although initiators that are water soluble may create some variations. The fundamental theory is simple and depends on the addition to the water of a dispersing agent. This may be a natural water-soluble colloid such as gum acacia, gum tragacanth, a semi-synthetic such as many cellulose derivatives or a fully synthetic polymer. These include polyvinyl alcohol, or alternatively a water-soluble salt derived from a styrene/maleic anhydride 1 : 1 copolymer

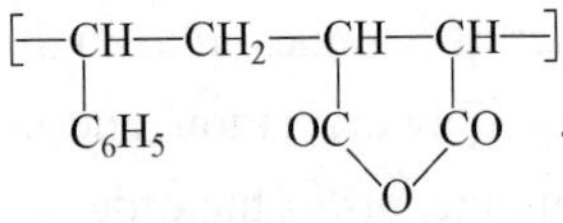

$$\begin{array}{llll} [-CH- & CH_2- & CH- & CH-] \\ \quad | & & \;| & \;| \\ C_6H_5 & & OC & CO \\ & & \quad \backslash & O \;/ \end{array}$$.

These dispersing agents may be mixed; occasionally a small quantity of surfactant of the order of 0.01% is added. The normal concentration of dispersing agents is about 0.1%, based on the water present. Monomer or monomers are added so that overall concentration is 25%~40%, although occasionally specific formulations claim 50%.

The function of the dispersing agent is that of forming an ‘envelope’ around the beads as formed by stirring and preventing their coagulation and fusion during polymerization. An intermediate or ‘sticky’ state occurs in almost all polymerizations in which a solution of polymer in monomer of high viscosity is formed, and the beads would fuse together very readily, except for the energy supplied by the stirring in keeping them apart and the stabilizing action of the dispersing agent. Certain monomers, e.g. vinyl chloride, which are not solvents for their own monomers are an exception to this rule, but the same principle applies to the dispersant acting as a particle stabilizer. This type of polymerization is sometimes referred to as ‘bead’ polymerization.

There is a very clear distinction between suspension polymerization and emulsion polymerization. while emulsion polymerization produces particles usually 1 μm in diameter, occasionally up to 2.5 μm, suspension particles are at least ten times larger in diameter, often of the order of 1 mm, although they are not necessarily spherical in shape. The kinetics of polymerization of the two types are often quite different. To ensure that beads or ‘pearls’ (another term used) are formed, the second-order transition point T_g must be below the ambient temperature; otherwise the beads will flow together as soon as stirring is stopped. This tendency can be reduced somewhat, e.g. by coating a bead dispersion of low molecular mass polyvinyl acetate with cetyl alcohol, which is present during the polymerization. Bead polymerization is only practicable as a general rule, where the T_g is above about 25℃ and preferably above 35℃. A summary of some suspension polymerization processes for vinyl chloride is available.

It is possible to perform suspension polymerization using solid dispersants. Thus styrene may be polymerized in suspension with organic peroxide initiators with a tricalcium ortho-phosphate dispersant and sodium dodecylbenzene sulfonate.

1.6 The Principal Monomers and Their Polymers

1.6.1 Hydrocarbons

The simplest hydrocarbon capable of free radical addition polymerization is ethylene (C_2H_4), which as a gas is treated under pressure. Higher aliphatic hydrocarbons such as propylene ($CH_3CH{=}CH_2$), 1-butene ($CH_3CH_2CH{=}CH_2$) and a number of longer chain aliphatic ethenes cannot in general be polymerized by themselves by free radical, as distinct from ionic methods, because of their allylic character. However, they are capable of copolymerization, and some specifications have claimed their copolymerization with vinyl acetate in emulsion. Only hydrocarbons with their unsaturation in the 1-position can be copolymerized satisfactorily in this manner.

Styrene ($C_6H_5CH{=}CH_2$) is the simplest aromatic hydrocarbon monomer. Others are vinyl-toluene or *o*-, *m*- and *p*-methylstyrene ($CH_3C_6H_4CH{=}CH_2$). *α*-Methylstyrene [$C_6H_4C(CH_3){=}CH_2$] is also a technical product, but its polymerization has problems because it has a low ceiling temperature; i.e. the propagation and depropagation rates during formation tend to become equal and hence no polymer is formed unless a low-temperature initiator system is used.

The divinyl benzene, written as $CH_2{=}CH{-}C_6H_4{-}CH{=}CH_2$, a notation used when it is desired to leave the positions of the substituents undecided, are a by-product of styrene manufacture and are used for crosslinking. There have been a few other specialized monomers based on condensed rings, but as they are generally solids their use in emulsion systems in very limited, if at all.

1.6.2 Vinyl Esters

Vinyl esters are derived from the hypothetical vinyl alcohol, $CH_2{=}CHOH$, an isomer of acetaldehyde (CH_3CHO), which is normally formed when an attempt is made to prepare the monomer. The esters, however, whether derived from acetylene or ethylene, are of major importance in many latex applications. The principal ester of commerce is vinyl acetate ($CH_2{=}CHOOCCH_3$), a liquid which is fairly readily hydrolyzed, of b.p. 73℃. Vinyl acetate has the advantage of being one of the cheapest monomers to manufacture. Vinyl propionate ($CH_3CH_2COOCH{=}CH_2$) is fairly well established as a monomer, probably by direct acetylene preparation. Other esters are encountered less frequently, and in most cases are probably prepared by vinylolysis rather than directly, using vinyl acetate as an intermediate.

Vinylolysis is not the same as *trans*-esterification and involves a mercury salt such as *p*-toluene sulfonate as an intermediate. Thus vinyl caprate is prepared by reacting vinyl acetate with capric acid in the presence of a mercuric salt, using an excess of vinyl acetate; the reaction is reversible:

$$CH_3COOCH{=}CH_2 + C_9H_{19}COOH \rightleftharpoons C_9H_{19}COOCH{=}CH_2 + CH_3COOH$$

The capric acid can be conveniently removed with sodium carbonate after removal of excess acetic acid with sodium bicarbonate, which does not react with the higher fatty acids. The vinyl esters of mixed C_8, C_{10} and C_{12} fatty acids have been used technically in forming copolymers with vinyl acetate.

Vinyl butyrate, $CH_3CH_2CH_2OOCCH{=}CH_2$, is referred to in the literature, but is probably not in commercial production. Vinyl laurate, $CH_3(CH_2)_8COOCH{=}CH_2$, has been in technical production in Germany. Of other esters of fatty acids, only vinyl stearate $C_{17}H_{35}COOCH{=}CH_2$, a solid, has been manufactured on a technical scale. The most interesting vinyl ester have been derivatives of pivalic acid $(CH_3)_3CCOOH$, the simplest branched chain fatty acid in which the carbon atom adjacent to the carboxyl group is quaternary. In these vinyl esters, one methyl group may be replaced by ethyl, and a second by a longer alkyl chain, and thus the general formula for the esters is $CH_3C_{7\sim8}H_{14\sim16}C(CH_3)_2C_{1\sim2}H_{2\sim4}OOCCH{=}CH_2$.

Vinyl chloroacetate ($ClCH_2COOCH{=}CH_2$) is occasionally quoted. Because of the relatively labile atom, copolymers including this monomer take part in a number of crosslinking reactions.

Vinyl benzoate, $C_6H_5COOCH{=}CH_2$, is the only aromatic vinyl ester that finds some application if relatively hard polymers with some alkali resistance are required .

1.6.3 Chlorinated Monomers

Vinyl chloride ($CH_2{=}CHCl$), a gas, is the cheapest monomer of the series, and has widespread commercial use. It may be polymerized in bulk with specialized apparatus, but it is also polymerized both in suspension and in emulsion. It is frequently copolymerized, especially with vinyl acetate.

Vinylidene chloride, $CH_2{=}CCl_2$, a low boiling liquid, is also a relatively low cost monomer. It forms a polymer with a marked tendency to crystallize because of its relatively symmetrical structure. In most cases it is copolymerized, especially with vinyl chloride or methyl acrylate. The corresponding symmetrical compound in the trans form, transdichloroethylene ($CHCl{=}CHCl$), has been used in limited quantities as a comonomer with vinyl acetate. Trichloroethylene ($CHCl{=}CCl_2$), although not normally considered as a monomer, may take part in some copolymerizations, especially with vinyl acetate.

Chloroprene, $CH_2{=}CClCH{=}CH_2$, a diene , is mainly used in the formulation of elastomers, but occasionally a polymer containing chloroprene is used as an alternative to a hydrocarbon diene.

Vinyl bromide, $CH_2{=}CHBr$, is technically available, and finds some application in specialist polymers with fire-resistant properties. The boiling point is 15.8℃. A number of highly chlorinated or brominated alcohol esters of acrylic and methacrylic acids have been described, and are probably in limited production, often for captive use for the production of fire-resistant polymers.

Vinyl fluoride (CH_2CHF), vinylidene fluoride ($CH_2{=}CF_2$), tetrafluoroethylene ($F_2C{=}CF_2$) and hexafluoropropylene ($CF_3CF{=}CF_2$) have found industrial applications in recent years. The monomers are gaseous.

1.6.4 Acrylics

1.6.4.1 Acrylic and Methacrylic Acids

The most numerous class of monomers are the acrylics, viz. esters of acrylic acid ($CH_2{=}CHCOOH$) and methacrylic acid [$CH_2{=}C(CH_3)COOH$]. Both are crystalline solids at low ambient temperatures, becoming liquid at slightly higher temperatures. These acids polymerize and copolymerize extremely readily, being frequently employed in copolymers to obtain alkalisoluble polymers. While both acids are water soluble, methacrylic acid, as might be expected because of its angular methyl group, is more soluble in ester monomers, and to some extent in styrene, and as such is more useful in copolymerization, especially if water based.

While esters of acrylic acid give soft and flexible polymers, except for those with long alkyl chains, methyl methacrylate polymerizes to an extremely hard polymers. The polymers in this series become softer with increasing alkyl chain lengths up to C_{12}. The highest alkyl chain acrylics in both series tend to give side chain crystallisation.

1.6.4.2 Individual Acrylic and Methacrylic Esters

A range of esters of acrylic acid are available commercially from methyl acrylate through ethyl acrylate to *n*-heptyl acrylate and 2-ethylhexyl acrylate [$CH_2{=}CHCOOCH_2CH(C_2H_5)(C_4H_9)$]. They vary from the fairly volatile, pungent liquids of the lowest member of the series to characteristic, but not necessarily unpleasant, odour of the higher members of the series. The highest members require distillation under reduced pressure to avoid simultaneous decomposition and polymerization.

The methacrylic ester series closely parallels the acrylics, but boiling points tend to be somewhat higher, especially with the short chain esters. Methyl methacrylate [$CH_2{=}C(CH_3)COOCH_3$] is by far the most freely available and least costly of the monomers of the series.

As an alternative to the simple alkyl esters, several alkoxyethyl acrylates are available commercially, e.g. ethoxyethyl methacrylat [$CH_2{=}C(CH_3)COOC_2H_4OC_2H_5$] and the corresponding acrylate. The ether oxygen which interrupts the chain tends to promote rather more flexibility than a simple carbon atom.

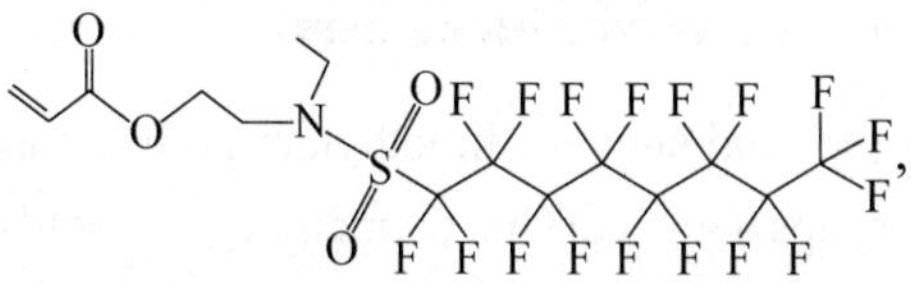

Some technical perfluorinated alkyl acrylates have been described. They include *N*-ethyl-perfluorooctanesulfonamido ethyl acrylate, (*n* approximately 7.5, fluorine content 51.7%), the corresponding methacrylate and the corresponding butyl derivatives. The ethyl derivatives are waxy solids, the ethyl acrylate and the corresponding methacrylate derivative having a melting range of 27~42℃. The butyl acrylic derivative is a liquid, freezing at 10℃.

Various glycol diacrylates and dimethacrylates are available. Ethylene glycol dimethacrylate [$CH_2{=}C(CH_3)COOC_2H_4OOC(CH_3)C{=}CH_2$] is extremely reactive, and is sometimes marketed as a solution in methyl methacrylate. It polymerizes extremely readily and acts as a powerful crosslinking agent. The dimethacrylates of triethylene glycol and higher glycols, some of which are also readily available, are less reactive, retain better flexibility and are more controllable in their polymerization characteristics.

Glycidyl methacrylate, 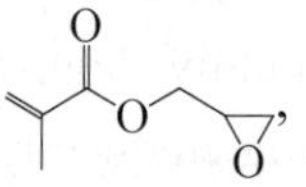 , has two reactive groups, the epoxide group being distinct in nature from the vinyl double bond. The epoxide group is only slowly reactive in water, and even in emulsion polymerization does not hydrolyze excessively. However, the presence of the group makes the methacrylate moiety much more prone to ready polymerization.

The half esters of both ethylene glycol and propylene glycol are now monomers of commerce, propylene glycol monoacrylate, $CH_3CHOHCH_2OOCCH{=}CH_3$, being typical (the primary alcohol unit is the active one in the formula). These monomers provide a source of the mildly reactive and hydrophilic groups on polymer chains. A problem with these monomers is that traces of a glycol dimethacrylate may be present as impurities at a low level.

1.6.4.3 Acrylics Based on the Amide Group

Acrylamide ($CH_2{=}CHCONH_2$) and methacrylamide [$CH_2{=}C(CH_3)CONH_2$] are articles of commerce, especially the former. Polymers of the former are water soluble, but the solubility of the latter depends on conditions of preparation, e.g. molecular mass of the polymers. Both are very frequently used in copolymerization. Polyacrylamide is often used as a flocculating agent. Certain derivatives, viz. methylolacrylamide ($CH_2{=}CHCONHCH_2OH$), methoxymethylacrylamide ($CH_2{=}CHCONHCH_2OCH_3$) and isobutoxyacrylamide ($CH_2{=}CHCONHCH_2OC_4H_9$-*iso*), are of interest in crosslinking. The last named has the advantage of being monomer soluble but water insoluble, making it more amenable to handling in emulsion polymerization.

Diacetoneacrylamide *N*-(1,1,-dimethyl-3-oxobutyl) acrylamide [also known as 1-dimethyl-3-oxobutyl acrylamide], $CH_2{=}CHCONHC(CH_3)_2COCH_3$ or 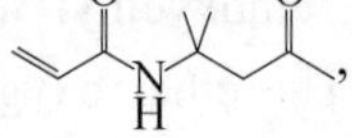 , has the advantage of both water and monomer solubility.

1.6.4.4 Cationic Acrylic Monomers

If a compound such as dimethylamino ethyl alcohol [$(CH_3)_2NC_2H_4OH$] is esterified via the hydroxyl groups with acrylic or methacrylic acids instead of neutralizing the amino group, a cationic monomer, e.g. $(CH_3)_2NC_2H_4OOCCH{=}CH_2$, is formed. At acid pH levels this monomer is cationic, with the amino group forming salts that polymerize and copolymerize in the normal way via the acrylic double bond. Another typical monomer is *t*-butylamino ethyl methacrylate [$t\text{-}C_4H_9NHC_2H_4OOCC(CH_3){=}CH_2$]. At neutral or higher pH levels, ionisation of this weak base is suppressed, and it acts as a nonionic monomer. However, hydrolysis tends to be rapid in aqueous media at high pH, forming the alkanolamine salts of the acids. Cationic monomers from the corresponding aminopropyl compound are also known.

1.6.4.5 Acrylonitrile

Acrylonitrile ($CH_2{=}CHCN$), and the less frequently used methacrylonitrile [$CH_2{=}C(CH_3)CN$], give extremely hard polymers and are employed as comonomers to give solvent resistance. Acrylonitrile monomer, like vinyl chloride, is not a solvent for its own polymer and is about 7% soluble in water, although its polymer is insoluble, making it of interest in theoretical studies. These monomers are unusually toxic because of the nitrile group.

1.6.5 Polymerizable Acids and Anhydrides

Besides acrylic and methacrylic acid, crotonic acid (strictly the *cis*-acid) ($CH_3CH{=}CHCOOH$), a white powder, often takes part in copolymerizations, especially with vinyl acetate, but it only self-polymerizes at low pH and with great difficulty. Itaconic acid (methylenesuccinic acid), $CH_2{=}C(CH_3COOH)COOH$, a water-soluble solid, also readily takes part in copolymerization, although it will only homopolymerize at about pH = 2.

Maleic acid (*cis*-$HOOCCH{=}CHCOOH$), the simplest dibasic acid, is rarely copolymerized on its own, but frequently as the anhydride maleic anhydride: [structure: O, O, O], which is much more reactive. However, it cannot be directly polymerized in water, although its rate of hydrolysis is slow. It readily forms copolymers, e.g. with styrene, ethylene or vinyl acetate, most readily as alternating (equimolar) copolymers, irrespective of the initial molar ratios. These copolymers are water soluble in their alkaline form after hydrolysis and frequently occur as stabilizers in emulsion polymerization.

Fumaric acid *trans*-$HOOCCH{=}CHCOOH$, the isomer of maleic acid and thermodynamically the most stable form, is occasionally used as a comonomer, although there is some doubt as to its reactivity, and it may do little more than provide end groups, thus acting as a chain transfer agent.

Aconitic acid, an unsaturated carboxylic acid of formula $HOOCCH_2C(COOH){=}CHCOOH$, obtained by removing the elements of water from citric acid, is occasionally quoted as a monomer in patents and theoretical studies. Citraconic acid (methylmaleic acid) $CH_3C(COOH){=}CHCOOH$, its isomer mesaconic acid (methylfumaric acid) and citraconic anhydride are also occasionally used for copolymerization. The acids in this paragraph are not articles of commerce as far as has been ascertained.

Various alkyl and alkoxy diesters of itaconic acid have been introduced, but as far as is known, production has not been sustained, although they are extremely good internal plasticizers for polyvinyl acetate. Their relatively high cost mitigated against their use.

1.6.6 Self-emulsifying Monomers

A number of monomers have the property of stabilizing emulsions without the assistance of emulsifiers or with a minimal quantity. Their polymers are generally water soluble and often so are their copolymers, depending on monomer ratios. These monomers contain strongly hydrophilic groups, the sulfonate $—SO_3Na$ being the most usual. They usually copolymerize readily with most of the standard monomers used in emulsion polymerization.

Among the earliest was sodium vinyl sulfonate $CH_2=CHSO_3Na$, which was in use in Germany in the 1940s. Other monomers of this class include sodium sulfoethyl methacrylate $CH_2=C(CH_3)COOCH_2CH_2SO_3Na$.

Of unusual interest is 2-acrylamido-2-methylpropanesulfonic acid (AMPS monomer®), $CH_2=CHCONHC(CH_3)_2CH_2SO_3H$, normally used as the sodium salt. This monomer also copolymerizes readily. A monograph describes these compounds in greater detail.

The salts of the polymerizable acids have appreciable self-emulsifying powers when used as comonomers, especially when they are about 10% or more by weight. The alkylolamine unsaturated esters , when used in the form of their alkali or amine salts, come into this category.

1.6.7 Esters for Copolymerization

Esters of maleic and fumaric acids are often used in copolymerization, both the diesters and more unusually the monoesters being reacted. The fumarate diesters, which are rather non-volatile liquids, have a feeble tendency to form homopolymers on prolonged heating with initiators, but little, if any, evidence exists to suggest that maleic esters can homopolymerize. Copolymerization characteristics of fumarate esters are more favourable than those of maleate esters, and they are mainly copolymerized with vinyl acetate to impart internal plasticisation. It has been suggested that maleate and fumarate esters isomerise to identical products during a polymerization reaction, but this has not been proved. Although in theory the units entering a polymer should become identical with the disappearance of the double bond, there are many steric factors associated with the polymer molecules as a whole. The principal esters are those of n-butyl alcohol, 2-ethylhexyl alcohol, a technical mixture of $C_{9\sim11}$ alcohols and 'nonyl alcohol', which is 1,3,3-trimethylhexanol.

The half esters of maleic acid and their salts are occasionally quoted in patents and other technical literature, and seem, probably because of their polar-non-polar balance to polymerize fairly readily. The methyl half ester *cis*-$CH_3OOCCH=CHCOOH$ is a solid; some of the higher alkali half esters are liquids. Half esters of other polymerizable acids such as fumaric and itaconic acids have been reported, but are more difficult to prepare. The half esters tend to disproportionate fairly readily, especially in the presence of water, to the free acid and the diester:

$$2CH_3OOCCH{=}CHCOOH \longrightarrow HOOCCH{=}CHCOOH + CH_3OOCCH{=}CHCOOCH_3$$

A number of successful copolymerizations in emulsion of half esters of long chain alcohol and sterically hindered alcohols have been disclosed.

1.6.8 Monomers with Several Double Bonds

Unsaturated hydrocarbons containing two double bonds constitute a special class of monomer. The principal representatives of this class are butadiene ($CH_2{=}CH{-}CH{=}CH_2$), isoprene [$CH_2{=}C(CH_3){-}CH{=}CH_2$] and chloropren ($CH_2{=}CClCH{=}CH_2$).

When a monomer contains more than one double bond which can polymerize approximately equally freely, crosslinking can occur readily and small quantities of this type of monomer are added to other polymerising systems to obtain controlled crosslinking. Examples are *p*-divinyl benzene, and ethylene glycol dimethacrylate [$CH_2{=}C(CH_3)COOCH_2CH_2OOCC(CH_3){=}CH_2$].

In these cases the radicals formed are resonance stabilized, so that two chains can form simultaneously, and when a biradical is added to a growing chain, two points occur from which the chain can continue, resulting in rapid branching and crosslinking.

The dienes are a special class in distinction to monomers such as the divinyl benzenes and the diesters such as a glycol acrylate. If a monomer such as butadiene is polymerized, the monoradical formed is highly stabilized by resonance.

The two resonance forms can be represented as $R{-}CH{=}CH{-}CH_3$ and $R{-}CH_2{-}CH_2{-}CH{=}CH_2$, where R represents a residual monovalent group.

As a result, two methods of addition are possible, one being known as 1, 2-addition, the other as 1, 4-addition, and may be represented by the following:

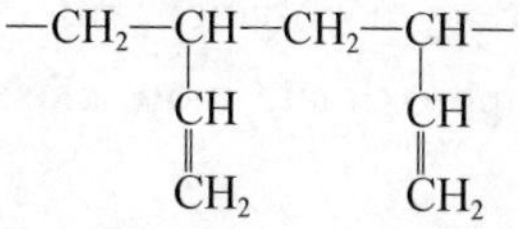

$$-CH_2-\underset{\underset{CH_2}{\overset{\|}{CH}}}{\overset{|}{CH}}-CH_2-\underset{\underset{CH_2}{\overset{\|}{CH}}}{\overset{|}{CH}}-$$

Bi-unit of a 1, 2-addition

$$-CH_2-CH{=}CH-CH_2-CH_2-CH{=}CH-CH_2-$$

Bi-unit of 1, 4-addition

During a free radical polymerization in emulsion, about 20% of a 1, 2-polymer addition and 80% of 1, 4-addition takes place. Copolymerization with other monomers such as styrene tends to increase 1, 2-units at the expense of 1, 4-units.

A further possibility of variation occurs because the 1,2-unit possesses an asymmetric carbon atom, while due to the double bond, 1, 4-addition may occur in the cis or trans positions, giving the following isomers:

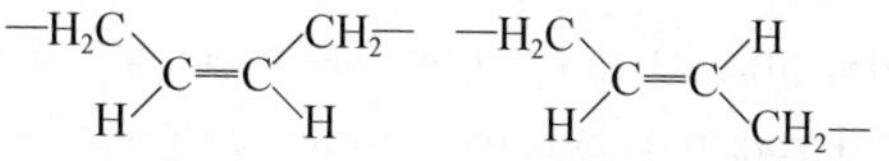

$$\begin{matrix} -H_2C & & CH_2- \\ & C{=}C & \\ H & & H \end{matrix} \qquad \begin{matrix} -H_2C & & H \\ & C{=}C & \\ H & & CH_2- \end{matrix}$$

cis *trans*

It has been found possible to deduce various structures by infrared absorption bonds, trans formation having been shown to decrease with temperature. During a polymerization including butadiene, there is a greater than usual tendency for side reactions to occur. These involve the residual double bonds in completed molecules or growing chains. This often causes gel formation, as measured by the insoluble fractions in acetone, or another standard solvent. Gel formation and other crosslinking reactions occur with increasing frequency as the degree of polymerization increases. In consequence, when solid products of controlled properties are required, polymerizations and copolymerizations involving butadiene are not taken to completion. The reaction is inhibited before polymerization is complete and surplus monomer is removed by distillation. Possibilities for isomerism in the polymerization of chloroprene and isoprene are even more complex than with butadiene.

The application of diene polymers and copolymers is largely associated with synthetic rubber, but these copolymers have other applications; e.g. copolymers with a styrene content of 40% and above have been used for coatings and for carpet backing. In these copolymers the residual double bonds render them prone to degradative oxidation.

The structure of butadiene copolymers is interesting and accounts for their physical properties. A polymer molecule may be considered to be a randomly coiled chain——an irregular spiral——in the unstretched state. Elastomers in the fully stretched state, particularly natural rubber, i.e. polyisoprene $(C_5H_8)_n$ tend to crystallise, this crystallisation being lost when the stress causing the extension is removed. Ideally a limited number of crosslinks is desirable for elastic recovery to occur. Because of their less regular structure, copolymers of butadiene do not tend to crystallize.

Modern work has shown that where polymerization takes place by methods that produce a highly stereoregular or stereospecific products, the tendency is for crystallisation to occur on stretching. In most copolymers that we will consider in these volumes, the high quantity of comonomer causes the normal plastic type of property to predominate, rather than the rubber-like extensibility. Thus the bulky phenyl $—C_6H_5$ groups in the styrene copolymers effectively prevent crystallisation, and the copolymers in film form tend to approximate more closely in properties to other vinyl-type polymers.

The double bonds in polymers involving dienes facilitate crosslinking, which in rubber technology is known as vulcanization. The utilization of the double bonds for crosslinking has increased in recent year.

1.6.9 Allyl Derivatives

Allyl alcohol ($CH_3{=}CHCH_2OH$) and its simple derivatives, such as allyl acetate ($CH_2{=}CH_2OOCCCCH_3$), have little practical application in vinyl polymerization, because of their powerful tendency to degradation chain transfer. Similar considerations apply to methallyl alcohol ($CH_2{=}CCH_3CH_2OH$) and its derivatives. A practical difficulty also arises with allyl alcohol and its more volatile derivatives because of their extreme lachrimatory character.

Certain other allyl derivatives, however, are of greater utility. Diallyl *o*-phthalate [*o*-$CH_2{=}C(CH_3)OOCC_6H_4COO(CH_3)CH{=}CH_2$] contains two vinyl groups, and as such the tendency to crosslink is in competition with that of chain transfer. While this diester is not normally used in emulsion polymerization, it is frequently included in the thermosetting polyesters, especially in conjunction with a monomer such as styrene, which will reduce the tendency to premature crosslinking. These derivatives find particular application in reinforced polyesters, viz. those reinforced with glass fibers.

Allyl derivatives containing epoxide groups seem to copolymerize somewhat more readily, probably because the nucleophilic epoxide group reduces the tendency to resonance. These derivatives are of interest as they are potentially crosslinking monomers. They include allyl glycidyl ether 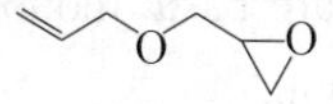 and allyl dimethyl glycidate 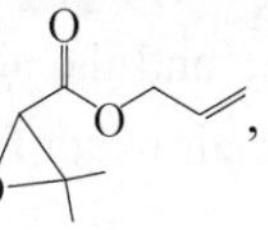 , which is formed by Darzen's reaction.

This little-known reaction would repay further study, at any rate as far as polymer production is concerned. It is fundamentally the reaction of a chlorinated ester, such as allyl chloroacetate with acetone in the presence of a stoichiometric quantity of alkali near 0℃, sodium hydride being particularly effective.

Monomers such as allyl methacrylate [$CH_2{=}CHCH_2OOC(CH_3)C{=}CH_2$] are occasionally quoted, having mild crosslinking properties.

1.6.10 Vinyl Ethers

While the vinyl ethers have long been known as monomers, they have been unimportant in aqueous polymerization. By themselves they only form copolymers, not homopolymers under free radical conditions, and ionic catalysts are used when a homopolymer is required. Although the vinyl ethers copolymerize readily with many vinyl monomers under free radical conditions, difficult arises during polymerization in the presence of water since they hydrolyse readily to acetaldehyde and alcohols below a pH of about 5.5. This make emulsion polymerization with a monomer such as vinyl acetate difficult, except under careful control of pH.

Except for the tendency to hydrolysis, the physical properties of the vinyl ether monomers closely resemble that of the corresponding saturated compounds. Available monomers, including vinyl methyl ether ($CH_2{=}CHOCH_3$), vinyl ethyl ether ($CH_2{=}CHOC_2H_5$), both *n*- and isobutyl vinyl ethers ($CH_2{=}CHOC_4H_9$) and a long-chain alkyl ether, vinyl cetyl ether ($CH_2{=}CHOC_{16}H_{33}$), have been available.

1.6.11 Miscellaneous Monomers Containing Nitrogen

N-Vinylpyrrolidone is a completely water-miscible cyclic monomer which can be regarded as a cyclic imide. It readily forms polymers and copolymers, the water soluble types being used as protective colloids. The monomer is

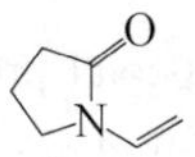

2-Vinylpyridine, 4-vinylpyridine and to a lesser extent 2-methyl-5-vinylpyridine have been prepared commercially, and polymerize to give products that are the basis of polymeric cationic electrolytes. They are most frequently in copolymers with butadiene and styrene in tyre cord

adhesives. The physical properties of the polymers tend to resemble those of styrene. The formulae of the monomers are shown below:

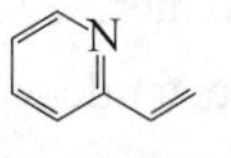

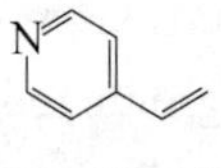

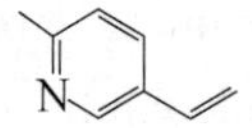

2-Vinylpyridine 4-Vinylpyridine 2-Methyl-5-vinylpyridine

1-Vinyl imidazole and the allied 1-vinyl-2-methylimidazole are basic monomers produced on a small scale, and their major function is improvement of adhesion:

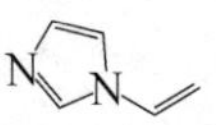

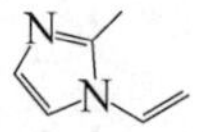

1-Vinyl imidazole 1-Vinyl-2-methylimidazole

Vinyl caprolactam is occasionally used as a reactive thinner:

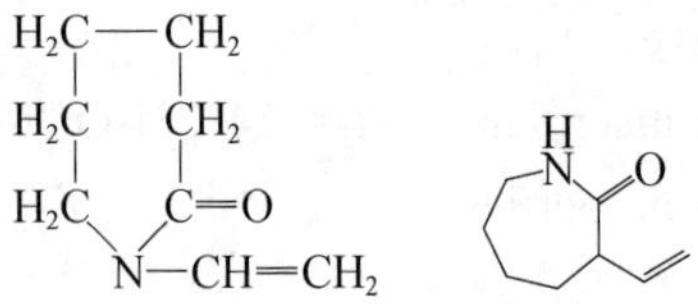

$H_2C—CH_2$

$H_2C\quad CH_2$

$H_2C\quad C=O$

$N—CH=CH_2$

It has a melting point of 34℃, and may be distilled under reduced pressure. It also has the property of improving adhesion.

Divinylethylene urea and divinylpropylene urea, with melting points of 66℃ and 65℃ respectively, find utilization as reactive thinners:

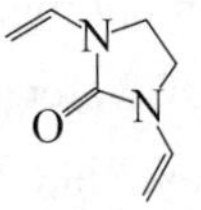

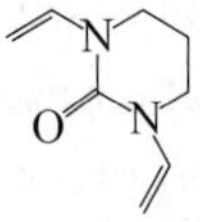

Divinylethylene urea Divinylpropylene urea

1.6.12 Toxicity and Handling

As a general rule, all the quoted monomers should be handled with at least the precautions associated with the corresponding saturated compounds. In some cases, e.g. the acrylic esters, the toxicity, in particular the vapour, is more toxic than the corresponding saturated esters. The lower acrylic esters, but not the methacrylic esters, have an extremely unpleasant odour, but the level of intolerance is well below the maximum safety level recommended. Some precautions are advised in handling acrylamide.

Acrylonitrile and methacrylonitrile have the characteristic toxicity of cyanides. On the laboratory scale, they should be handled in well-ventilated fume cupboards and prevented from coming into direct contact with the skin. Special precautions, including the wearing of oilskins and fresh air breathing apparatus, are required for large-scale manufacturing processes. Allyl alcohol and some of its derivatives are lachrimatory. The above comments are of a general character only. In all cases, manufacturers' literature and official literature should be consulted, safety information being obligatory in many countries.

The following are synthetic monomers based on the vinyl esters of mixed branched chain acids, known as Versatic® acids, the feature being that the carbon atom is in the alpha position of quaternary:

Veova 9 is the vinyl ester of acids averaging 9 carbon atoms; b.p.185~200℃, s.g. 0.89. The T_g of the homopolymer is 60℃. Veova 10 is the vinyl ester of acids averaging 10 carbon atoms and is less branched than Veova 9; b.p. 133~136℃, s.g. 0.875~0.885.

1.7 Physical Properties of Monomers

Table 1.1 is not intended to be exhaustive, but gives the b.p. and m.p. values where they are above about 5℃ and density (s.g.) of the principal monomers.

Table 1.1 Physical properties of principal monomers

Monomer	Formular	b.p./℃	m.p./℃	s.g. Pressure/mm ($d_{20/20}$)
Ethylene	$CH_2{=}CH_2$	−104		
Propylene	$CH_2{=}CHCH_3$	−31		0.5139 ($d_{20/4}$)
1-Butene	$CH_2{=}CHCH_2CH_3$	−6.3		0.5951 ($d_{20/4}$)
1-Hexene	$CH_2{=}CH(CH_2)_3CH_3$	63.5		0.6734 ($d_{20/20}$)
1-Octene	$CH_2{=}CH(CH_2)_5CH_3$	121.3		0.7194 ($d_{15.5/15.5}$)
Styrene	C_6H_5 $CH{=}CH_2$	145.2		0.905 ($d_{25/25}$)
Vinyl toluene	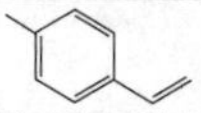	167.7		0.8930
o-Methyl styrene	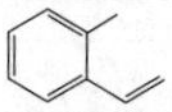	163.4		0.9062 ($d_{25/25}$)
Divinyl benzene (55% technical product)		195		
m-Diisopropenylbenzene	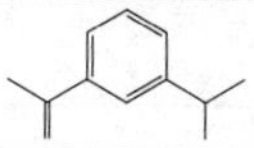	231		0.925
p-Diisopropenylbenzene (sublimes at 64.5℃)				
1,3-Butadiene	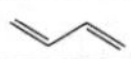	−4.7		0.6205
Isoprene	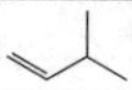	34.07		0.686 ($d_{15.6/15.6}$)
Vinyl chloride	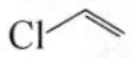 Cl	−14		0.912
Vinylidene chloride	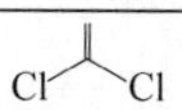 Cl Cl	31.7		1.1219 ($d_{20/4}$)
trans-Dichloroethylene	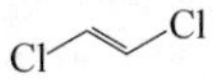 Cl Cl	49		1.265 ($d_{15/4}$)
Chloroprene	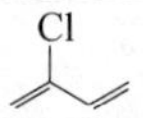 Cl	59.4		0.9583
Vinyl fluoride	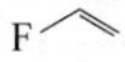 F	−57		
Vinylidene fluoride	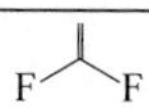 F F	−84		
Tetrafluoroethylene	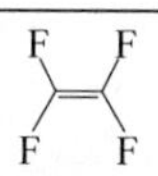 F F F F	−76		

续表

Monomer	Formular	b.p./℃	m.p./℃	s.g. Pressure/mm ($d_{20/20}$)
Vinyl formate	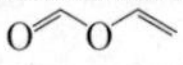	46.6		0.9651
Vinyl acetate	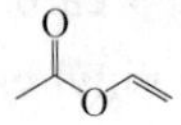	72.7		0.9338
Vinyl propionate	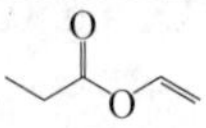	94.9		0.9173
Vinyl butyrate	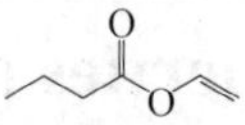	116.7		0.9022
Vinyl caprate	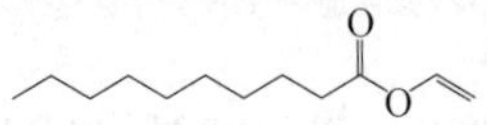	148/50		
Vinyl laurate	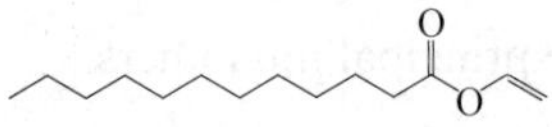	142/10		
Vinyl stearate	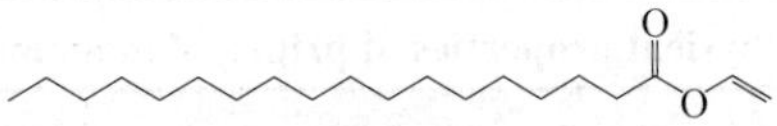	187~188	35~36	
Vinyl chloroacetate	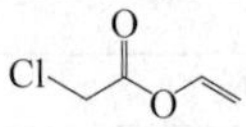		44~46	1.1888
Vinyl benzoate	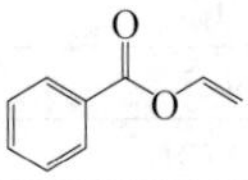	203		1.0703
Acrylic acid	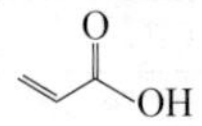	141.3	12.3	1.0472
Methacrylic acid	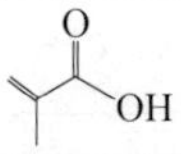	161	15	1.015
Methyl acrylate	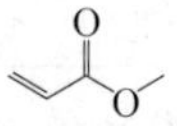	80		0.950
Ethyl acrylate	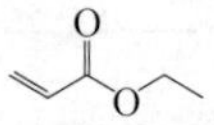	99.6		0.9230
n-Butyl acrylate	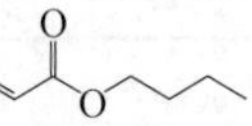	148.8		0.9015
n-Heptyl acrylate	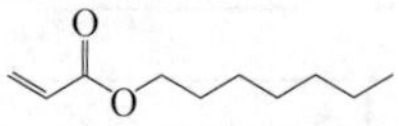	106/25		0.8794 ($d_{25/4}$)
2-Ethylhexyl acrylate	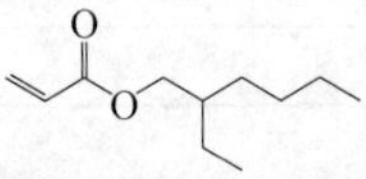	128/50		0.8869
Lauryl acrylate	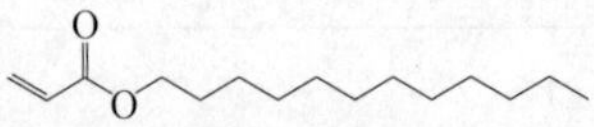			
Methyl methacrylate	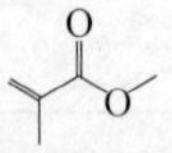	100.5		0.939
Ethyl methacrylate	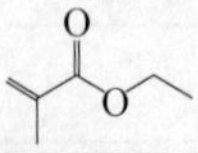	118.4		0.909
n-Butyl methacrylate	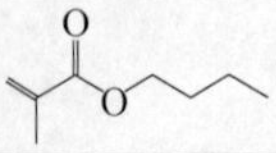	166		0.893

续表

Monomer	Formular	b.p./℃	m.p./℃	s.g. Pressure/mm ($d_{20/20}$)
Lauryl methacrylate	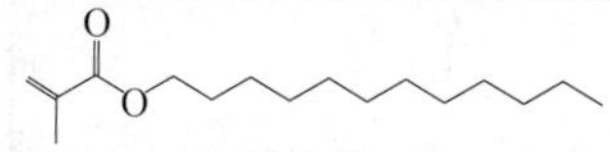		−8	0.868 ($d_{25/15.6}$)
Ethoxyethyl acrylate	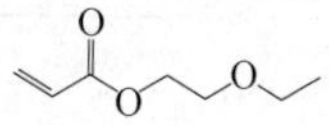	174.1		0.9834
Ethoxyethyl methacrylate	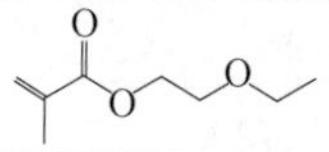	91~93/35		0.971 ($d_{15.5/15.5}$)
Ethylene glycol dimethacrylate (96% technical)	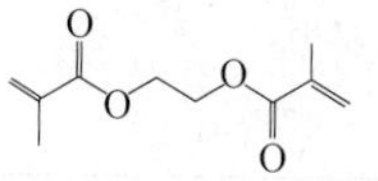	96~98/4		1.06 ($d_{15.5/15.5}$)
Glycidyl acrylate	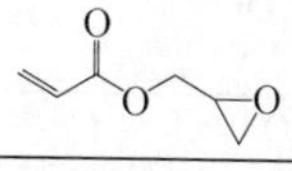	57/2		1.1074
Glycidyl methacrylate	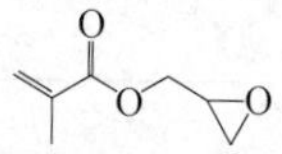	75/10		1.073 (25)
Ethylene glycol monoacrylate	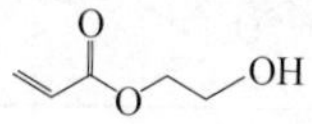	76/8		1.11
Ethylene glycol monomethacrylate	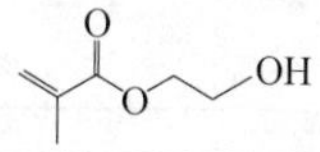	84/5		1.07
Propylene glycol monoacrylate	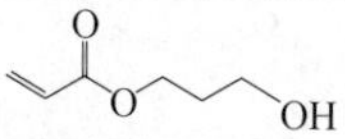	85/9		1.05
Propylene glycol monomethacrylate	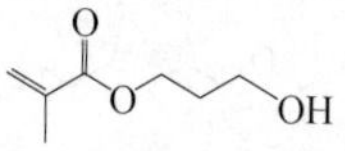	92/8		1.03
Acrylamide	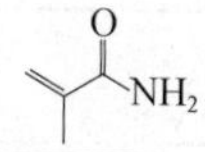	125/25	85	1.222
Methacrylamide	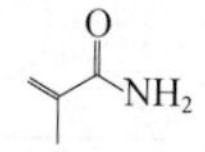		110	
Acrylonitrile		77.3		0.8060
Methacrylonitrile	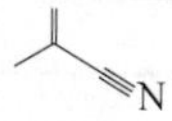	90.3		0.8001 ($d_{20/4}$)
Methylolacrylamide (available at 60% solution)	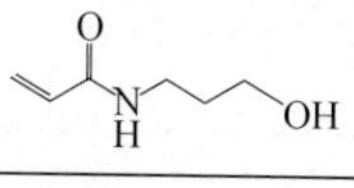			
Methylene diacrylamide	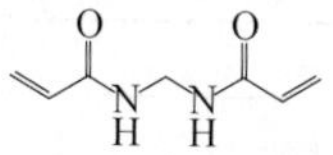	97.5/40		0.933 ($d_{25/4}$)
Diethylaminoethyl methacrylate	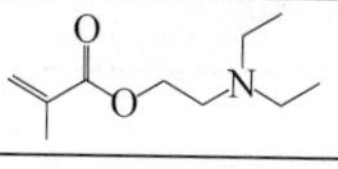	103/12		0.914 ($d_{20/4}$)
t-Butyl aminoethyl methacrylate	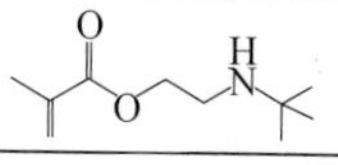	97.5/40		0.933
trans-Crontonic acid	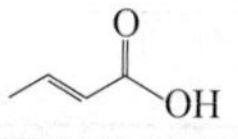	72		0.963 ($d_{80/4}$)

续表

Monomer	Formular	b.p./℃	m.p./℃	s.g. Pressure/mm ($d_{20/20}$)
Itaconic acid		167		1.6
Maleic acid		200	130	1.609 ($d_{20/4}$)
Fumaric acid		290	286	1.635 ($d_{20/4}$)
Aconitic acid			191	
Maleic anhydride		202	52.5	1.48
Di-*n*-butyl maleate		280.6		0.9964
Di-2-ethylhexyl maleate		209/10		0.9436 ($d_{15.5/15.5}$)
Di-nonyl maleate				0.9030 ($d_{15.5/15.5}$)
Di-*n*-butyl fumarate		138/8		0.9869 ($d_{20/4}$)
Di-2-ethylhexyl fumarate				
Methyl acid maleate				
Butyl acid maleate				
Dimethyl itaconate		91.5/10		1.27 ($d_{2/4}$)
Dibutyl itaconate		140/10		0.9833 ($d_{2/2}$)
Allyl alcohol		96		0.8540($d_{20/4}$)
Allyl chloride		45		0.9397 ($d_{20/4}$)
Allyl acetate		103.5		0.928
Diallyl phthalate		290		
Allyl glycidyl ether		50~52		0.967 ($d_{20/4}$)

续表

Monomer	Formular	b.p./℃	m.p./℃	s.g. Pressure/mm ($d_{20/20}$)
Allyl dimethyl glycidate		89/8		
Vinyl methyl ether		6		0.7500
Vinyl ethyl ether		35.5		0.7541
Vinyl *n*-butyl ether		94		0.7803
Vinyl isobutyl ether		83		0.7706
Vinyl cetyl ether				
N-Vinyl pyrrolidone		148/100	13.5	1.04
2-Vinyl pyridine		110/150		0.9746
4-Vinyl pyridine		121/150		0.988
2-Methyl-5-vinylpyridine		75/15		

References

[1] H. Staudinger et al., Ber., 53, 1073 (1929); Angew. Chem., 42, 37~40 (1929).

[2] W. H. Carrothers, in Collected Papers on High Polymeric Substances, H. Mark and G. Stafford Whitby (eds.), Interscience Publishers, 1940.

[3] C. Booth and C. Price (eds.). (1989) Comprehensive Polymer Science, Vol. 1, Pergamon Press.

[4] H. R. Allcock and F.W. Lampe. (1990) Contemporary Polymer Chemistry, Vol. 1, 2nd edn, PrenticeHall.

[5] M. P. Stevens. (1990) Polymer Chemistry, 2nd edn., Oxford University Press.

[6] G. Odian. (1991) Principles of Polymer Chemistry, 3rd edn., Wiley.

[7] P. E. M. Allen. (1963) The Chemistry of Cationic polymerization, Plesch (ed.), Pergamon Press.

[8] G. Natta and F. Danusso. (1967) Stereoregular Polymers and Stereospecific polymerization, Pergamon Press, Oxford, Y.V. Kissin in Handbook of Polymer Science and Technology, Chereminosoff (ed.), Vol. 8, Marcel Dekker, 1989, pp. 9~14.

[9] M. A. Doherty, P. Gores and A.H.E. Mueller, (1988) Polym. Prepr. (Am. Chem. Soc.), 29(2), 72~73.

[10] P.J. Flory. (1953) Principles of Polymer Chemistry, Cornell University Press.

[11] G. C. Eastmond, A. Ledwith, S. Russo and P. Sigwalt. (1989) Comprehensive Polymer Science, Sec. 1, Pergamon Press.

[12] F. W. Billmeyer. (1985) Textbook of Polymer Science, 3rd edn, Wiley.

[13] C. H. Bamford, (1988) Encyclopedia of Polymer Science and Engineering, 2nd edn, Vol. 13, pp.729~735 (refers specifically to retardation and inhibition).

[14] C. E. Schildknecht. Allyl Compounds and Their Polymers, Ch. 1, pp. 195 et seq.

[15] H. Warson. (1980) Per-Compounds and Per-Salts in Polymer Processes, Solihull Chemical Services.

[16] M. El-Aaser and J. W. Vanderhoff (eds.). (1981) Emulsion polymerization of Vinyl Acetate, Applied Science Publishers.

[17] G. S. Misra, and U. D.-N. Bajai. (1982) Progress in Polymer Science, Vol. 8, Pergamon Press.

[18] H. Warson. (1967) Makromol. Chemie, 105, 228~245.

[19] H. Warson and R.J. Parsons. (1959) J. Polym. Sci., 34(127), 251~269.

[20] H. Warson. (1967) Peintures, Pigments, Vernis, 43(7), 438~446.

Chapter 2

Pharmaceutical Polymers: Principles and Structures

Polymers have been widely used to develop innovative and high-through-put solutions for biomedical applications. As a function of specific bio-functionalities, they have been successfully employed to fabricate biocompatible devices (i.e., prostheses components, porous/nonporous scaffolds) with structural properties able to resist to external stimuli under physiological conditions (i.e., degradation from body fluids). This wide application field depends upon the peculiar polymer chemistry and physical properties (i.e., thermal, mechanical, etc.), which significantly concur to define the ultimate properties of the polymer for a specific biomedical use. This chapter aims at introducing the fundamental concepts regarding chemical and physical properties of polymers to provide the basic knowledge and better predict the structure occurring in manufacturing processes involving biomaterials. For this reason, we will discuss the main correlations occurring between physical and chemical structure to establish structure-property-function relationships. Meanwhile, we will describe the methods for the estimation and prediction of properties of polymers (i.e., thermal, mechanical) at different physical state, which are essential to set process conditions to properly design polymers in biomedical applications.

2.1 Introduction

Polymers are one of the major functional components in a number of biomedical devices for applications ranging from clinical diagnosis, extracorporeal procedures up to in vivo applications such as tissue engineering, biosensors and drug delivery. Their application in medical surgery fall into three broad categories:

(a) extracorporeal uses (catheters, tubing, and fluid lines; dialysis membranes/artificial kidney; ocular devices; wound dressings and artificial skin);

(b) permanently implanted devices (sensory devices; cardiovascular devices; orthopedic and dental devices);

(c) temporary implants (degradable sutures; implantable drug delivery systems; polymeric scaffolds for cell or tissue transplants; temporary vascular grafts and arterial stents; temporary small bone fixation devices, transdermal drug delivery) .

This incites chemists and materials scientists to increasingly investigate biopolymers with new functionalities to explore their properties under different biological conditions in order to satisfy the enormous demand in the area of medical science. Recently, a large number of biopolymers have been variously used in biomedical field, including natural polymers such as polysaccharides (starch, cellulose, chitin, alginate, hyaluronate, etc.) or proteins (collagens, gelatins, caseins, albumins) and/or synthetic and biodegradable polymers [Polyvinyl alcohol (PVA), Polyvinylpyrrolidone (PVP), Polyethyleneglycol (PEG), Polylactic acid (PLA), Polyhydroxy acid (PHA) etc.]. However, despite the apparent proliferation of biopolymers in clinical surgery, the modern science and technology of biopolymers is still in its early stages of development due to the large heterogeneity of chemical properties of polymers which makes difficult to completely understand dynamic structure-processing condition-property relationship required for the design of medical devices.

For instance, different approaches have been investigated to design novel devices which may be in turn able to directly or indirectly interact with the host tissues and to replace the functional tissue through the mimicry of morphological characteristics of natural systems. Starting from the basic principle of "learning from nature", polymers have been chemically designed to reproduce the native chemical/physical behavior of soft/hard tissues under the specific biomechanical and physiological conditions at the interface with the surrounding tissues. In particular, tailor-made polymers with different fluid transport and degradation properties have been synthesized to biomolecularly interact with cells: they may finely control basic cell functions, guiding the spatially and temporally complex multicellular processes of tissue formation and regeneration or facilitating the restoration of structure and function of damaged or dysfunctional tissues.

Still, several drawbacks of current synthesized polymers have been recognized in terms of durability of mechanical response over time so making their use in load bearing applications often unsatisfactory. This may be referred to a not efficient control of the material properties mediated by specific molecular interactions, i.e., side reactions coupled with the presence of unreacted pendant groups and physical bonds——which unclearly affect degradation mechanisms thus compromising the biomechanical performance after implantation, with potential effects on the required response of materials to the external stimuli. In this context, it is mandatory to identify and adequately investigate all the interactions which occur between macromolecules and macromolecular assemblies in order to properly address the properties to the specific application need.

As a general rule, it is possible to recognize two types of molecular interactions generated by intramolecular and intermolecular forces respectively. In detail, the macroscopic properties of polymer for biomedical use, are determined to a great extent by the physical/chemical characteristics of individual polymer chains, related to the presence of specific chemical moieties able to influence the elasticity or the mechanical stability of covalent bonds as well as of molecular conformations. Meanwhile, intermolecular forces between adjacent polymer chains in the bulk also contribute to the properties of the chains assembly, and influence the local interactions between specific groups along the backbone, thus negatively influence and sometimes compromise the materials biocompatibility for specific use. The intramolecular forces influence the properties of single polymer molecules, which result from the entropic and enthalpic elasticity of the polymers, as well as from specific structural changes along the stretched polymer chain, so, governing the elastic response of single polymer chains. Otherwise, intermolecular forces involve all processes in which the interaction is transmitted through the surrounding medium such as unbinding mechanisms of intermolecular aggregates based on non-covalent interactions (i.e., in ligand-receptor pairs, coordination complexes, hydrogen bonded systems, ion pairs or hydrophobically assembled structures). So, they govern the stability of intermolecular aggregates such as polymeric micelles or hydrophobic domains, folding of proteins or polymer crystals packing. In order to better understand the macroscopic properties of polymeric biomaterials and their potential use in biomedical applications, it is therefore necessary to properly frame the microscopic interactions occurring inter and intra molecular chains, so identifying the peculiar contribution of intramolecular and intermolecular forces which may optimize the ultimate properties of the biomedical device. In this chapter, it is proposed an accurate description of chemical and physical properties of biopolymers elucidating their origin from inter- or intra-molecular interactions (Fig.2.1).

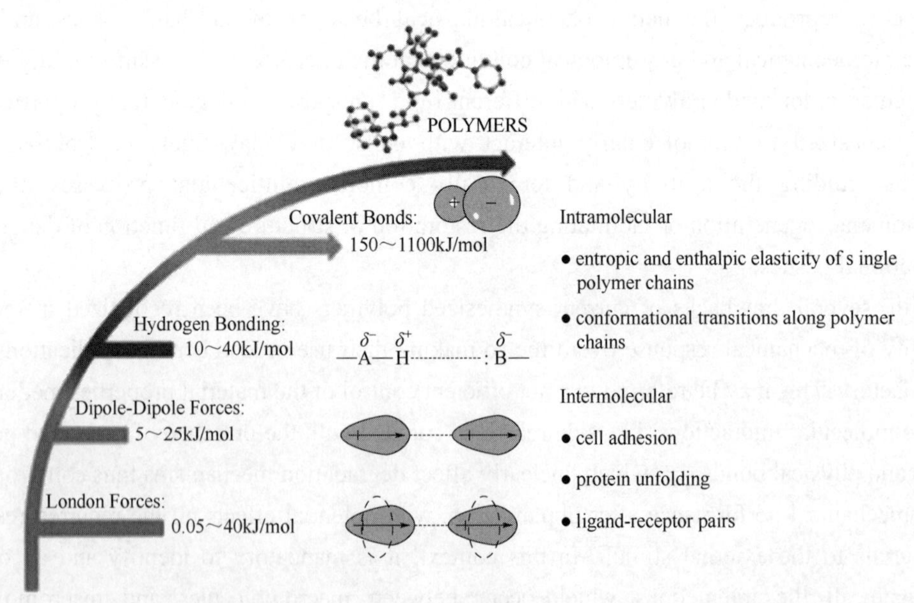

Fig.2.1 Intramolecular and Intermolecular forces in polymers for biomedical use.

2.2 Intra and Inter Polymer Molecular Interactions

A correct understanding of many physical and chemical properties of polymeric materials is linked to the knowledge of the intra and inter molecular interactions. All matter is held together by force and its property results from the properties of individual molecules and from how they act collectively. Preeminently, it is necessary to distinguish the types of forces acting within and between molecules: including intramolecular interactions that occur between macromolecular groups itself that are responsible for the complex chemical properties and the intermolecular interactions, responsible for the physical properties. At atomic level, considering the interaction between two isolated atoms as they are brought into close proximity from an infinite separation. At large distances interactions are negligible but as the atoms approach, each exerts forces on the other. These forces are of two types attractive force (F_a) and repulsive force (F_r) and the magnitude of each is a function of interatomic distance. The origin of attractive part depends on the particular type of bonding. The repulsion between atoms is related to the Pauli principle: when the electronic clouds surrounding the atoms, they start to overlap and the energy of the system increases suddenly. The sum of attractive and repulsive components is the net force F, and then when F_a and F_r become equal there is no net force and a state of equilibrium exists. Sometimes it is more convenient to work with potential energies between two atoms instead of forces considering that energy and force for atomic systems are mathematically related:

$$E = \int F dr \tag{2-1}$$

At equilibrium spacing r_0, the force F is zero and net energy corresponds to minimum energy E_0. The binding energy require to separate two atoms from their equilibrium spacing to an infinite distance apart is the minimum energy E_0. A large of material properties depend on E_0, curve shape and bonding type .

As already said, the potential curve is the result of two terms: the contribution of attractive force at long ranges (van der Waals force) and the repulsive term at short ranges. The curves indicate the strength of the bond based on the depth of the potential well: more deep is the well, more stable is the molecule.

By chemical bonding is meant, the effect that causes the energy of two atoms close to each other to be significantly lower than when they are apart. The effect arises from the sharing of the outer (valence) electrons of these atoms which, in the presence of two or more positively charged nuclei, rearrange themselves into spatial distributions (orbitals) where the electrons are, on the average, closer to the nuclei. The strength and length of a chemical bond are related to potential distance curve in which the minimum value of energy, the bond energy, is defined as the amount of energy required to increase the separation of the atoms forming the bond to infinity. However, what is lacking from the curves is the fact that chemical bonds, due to the redistributed electron clouds, have distinct directionality. The most widely used model for chemical bonding is the covalent bond or shared electron model. Different elements have different tendencies to draw electrons toward their nuclei. This ability is called electronegativity. Thus, in general the shared electron cloud is not evenly distributed in the space between the bonding nuclei. The difference in the electronegativities of the bonding atoms quantifies the extent how evenly or unevenly the bond electrons are shared.

A purely covalent bond (i.e., completely evenly shared electrons) can only be obtained when the bonding atoms are of the same element. The atom with a smaller electronegativity, can be viewed as having a partial positive charge $+\delta$, as the positive charge of its nuclei is not fully screened anymore by the valence electrons. The formation of one or more chemical bonds allows atoms to form energetically stable molecules which are characterized by their structure and composition. Since chemical bonds result in increased localized electron densities, it can be assumed that the spatial regions associated with different chemical bonds repel each other electrostatically. As became apparent from the discussion on chemical bonds above, many molecules contain polar chemical bonds that can be associated with electric dipoles, comprising of atomic partial charges.

The molecules, in turn, as collections of these partial charges, can be associated with electric dipoles, quadrupoles, and higher multipoles. Many chemical groups in molecules may also be charged, and then the electrostatic effects are important in a wide range of molecular systems. It turns out that relating the intricate intermolecular forces to simple electrostatic interactions between charges, dipoles, and higher multipoles works surprisingly well in understanding the structure and dynamics of condensed matter systems. The long chains are held together either by secondary bonding forces such as van der Waals and hydrogen bonds or primary covalent bonding forces through cross links between chains.

In general, the intermolecular interactions could be either a dipole-dipole interaction or dispersion (London) interaction. Dipole-dipole forces are referred to an electrostatic interaction acting between two permanent dipoles, averaged over different orientations due to the thermal motion of the molecules. These interactions tend to align the molecules in order to increase the attraction reducing potential energy. London dispersion interactions, instead, are of purely quantum mechanical origin. They are caused by correlated movements of the electrons in interacting molecules. Finally, hydrogen bond occurs between molecules that have a permanent net dipole resulting from hydrogen being covalently bonded to relatively electronegative atoms (fluorine, oxygen or nitrogen). Hydrogen bonds are a stronger intermolecular force than either dispersion forces or dipole-dipole interactions. Intermolecular and intramolecular interactions are generally present in several polymers largely used for biomedical applications, which have very long chain molecules, formed by covalent bonding along the backbone chain. In addition, each chain can have side groups, branches, and copolymeric chains or blocks which can also interfere with the long-range ordering of chains. In the case of branched polymers in which side chains are attached to the main backbone chain, chain packing is limited by the steric hindrance of side chains resulting in a not ordered structure.

2.3 Molecular Mass: Polydispersivity and Effects on Polymer Viscosity

Because polymers are long-chain molecules, their properties tend to be more complex than their short-chain counterparts. Thus, in order to choose a polymer type for a particular application, microscopic properties of polymers must be understood. Among them, polymer molecular mass is important because it determines many physical properties. Some examples include the transitions temperatures from liquids to rubbers, solids and mechanical properties such as stiffness, strength, viscoelasticity, toughness, and viscosity. If molecular mass is too low, the transition temperatures and the mechanical properties will generally not be adequate for the use in commercial applications in terms of transition temperatures or mechanical properties for load bearing application. Unlike small molecules, however, the molecular mass of a polymer is not one unique value. Rather, a given polymer will have a distribution of molecular masses that will depend on the way the polymer is produced. For polymers we should not speak of a molecular mass, but rather of the distribution of molecular mass or of the average molecular mass. There are many ways, however, to calculate an average molecular mass. The question therefore is how ones can define the average molecular mass for a given distribution of molecular masses.

The answer is that the type of property being studied will determine the desired type of average molecular mass. For example, strength properties may be influenced more by high molecular mass molecules than by low molecular mass molecules, and thus the average molecular mass for strength properties should be weighted to emphasize the presence of high molecular mass polymer. The molecular mass of polymer is described by two statistically useful definitions: the number average and weight average molecular masses. The number average molecular mass is correlated to the number of molecules present in the mixture. This average molecular mass follows the conventional definition for the mean value of any statistical quantity. In polymer science is called the number average molecular mass (M_N) that therefore is an average over the number of molecules.

$$M_N = \frac{\sum N_i M_i}{\sum N_i} \tag{2-2}$$

where N_i is the number of moles of species i, and M_i is the molecular mass of species i. By considering the polymer property which depends on the size or weight of each polymer molecule, it is possible to define the weight average molecular mass (M_w). It represents an average over the weight of each polymer chain.

$$M_w = \frac{\sum N^2{}_i M_i}{\sum N_i M_i} \tag{2-3}$$

where N_i is the number of moles of species i, and M_i, is the molecular mass of species i. In general, a series of average molecular masses can be defined by the equation:

$$M_k = \frac{\sum N_i M_i^{k+1}}{\sum N_i M_i^k} \tag{2-4}$$

where

$k = 0$ gives M_N

$k = 1$ gives M_w

$k = 2$ gives M_z

$k = 3$ gives M_{z+1}

One average molecular mass which does not fit into the general expression of M_k is the viscosity average molecular mass, M_v that is given by the Mark-Houwink and defined as:

$$M_v = \left(\frac{\sum N_i M_i^{a+1}}{\sum N_i M_i^a}\right)^{\frac{1}{a}} \tag{2-5}$$

Here a represents a constant that depends on the polymer/solvent pair used in the viscosity experiments. For all synthetic polydisperse polymers, the various average molecular masses always rank in the follow order:

$$M_N < M_v < M_w < M_z < M_{z+1} \tag{2-6}$$

For monodisperse polymers, all molecules have the same molecular mass, then all molecular mass averages are equal. To measure an average molecular mass, it is possible to use a colligative property which yields M_N or light scattering which yields M_w. A molecular mass distribution for a typical polymer is shown in Fig.2.2(a); it seems like a probability distribution curve.

Standard deviation of molecular mass is used in order to characterize the spread of the distribution function.

$$\sigma = M_N \sqrt{\frac{M_w}{M_N} - 1} \tag{2-7}$$

A key term in the standard deviation is the ratio of M_w to M_N. This term is known as the polydispersivity index and it is used as a measure of the breadness of the molecular mass distribution. The polydispersity index or dispersity D is commonly used to measure the distribution of molecular mass in a given polymer sample:

$$D = \frac{M_w}{M_N} \tag{2-8}$$

So, the dispersity calculated is the weight average molecular mass (M_w) divided by the number average molecular mass (M_N) indicating in this way the distribution of individual molecular masses in a batch of polymers.

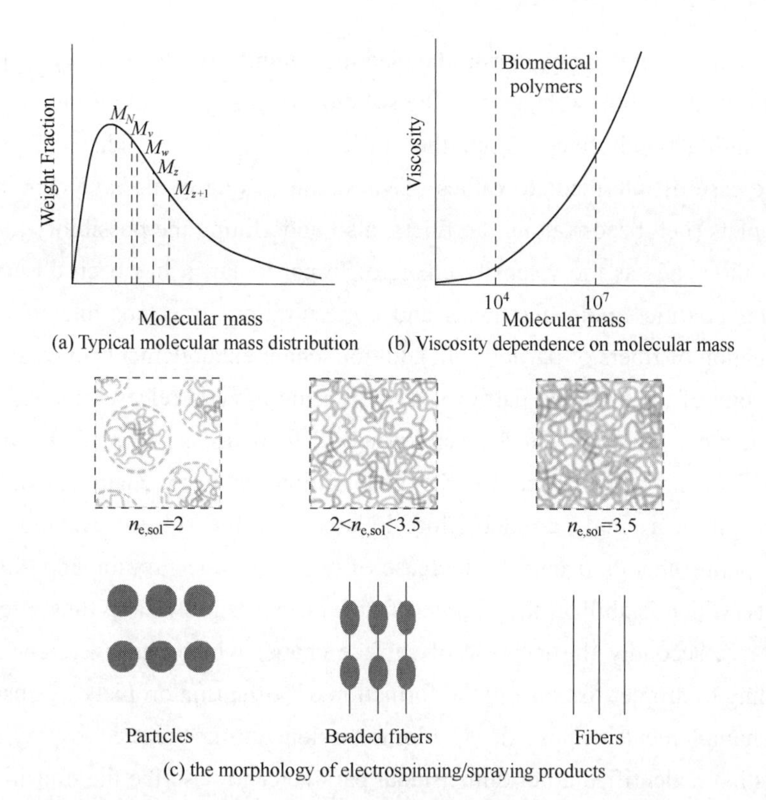

Fig.2.2 Typical molecular mass distribution (a) and viscosity dependence (b) for biomedical polymers. (c) Effect of entanglement molecular mass and viscosity on the morphology of electrospinning/spraying products.

Linear polymers used for biomedical applications generally have M_N in the range of 25,000~100,000 and M_N from 50,000 to 30,000. Higher or lower molecular masses may be necessary, depending on the ability of the polymer chains to exhibit secondary interactions such as hydrogen bonding which may concur to additional strength. In general, increase molecular mass corresponds to an increase of physical properties such as viscosity. Equally the intrinsic viscosity can be related to the molecular mass of the solute and gives an idea of molecular mass. For solutions of polymers with $M_N > 10,000$ Dalton, it is possible to consider the Mark-Houwink relation:

$$[\eta] = KM^{\alpha} \tag{2-9}$$

where K and α are constants characteristic of the particular system investigated. K is referred to the particular solvent-solute pair and 'α' is particularly related to the shape of the solute molecule. Values of these parameters have been determined for a large number of polymer-solvent pairs in the range of molecular masses normally of interest for characterization purposes. For instance molecular mass of the polymer also has an important effect on morphologies of electrospun scaffold. Here, the electric field forces permit to modify the size scale of fibers from micro to nanometer in order to obtain micro or nanostructured scaffolds. In these applications, molecular mass is correlated with the solution viscosity, then considering the same concentration of the same polymer increasing molecular mass is possible to switch from microspheres to fibers. Interestingly, during the electrospinning and electrospraying process, the morphology from beads to fibers changes as the solution viscosity: droplet shape in the case of lower viscosity until smooth fibers when the sufficient viscosity is achieved, as shown in Fig.2.2(b). In the case of intermediate values of viscosities, Zong et al. have demonstrated the formation of droplets (i.e., beads) along the fibers, also underlining the possibility to control the spacing between the beads as the viscosity changes. In particular, it has been observed that by properly choosing polymer molecular mass and concentration, it is possible to discriminate between the formation of fibers or particles. In addition, chain entanglements may strongly affect size and morphology of the obtained particles, thus offering tunable release kinetics suitable for different drug delivery systems. It is well-known that particle morphology obtained by electrospraying is mainly determined by the competition between chain entanglement and Coulomb fission within a single droplet. More in detail, as the solvent evaporates from the droplet, two competing factors occur: the increase of polymer concentration and, consequently, of entanglements which stabilize the droplet from further subdivision thus preserving its spherical shape and, secondly the increase of surface charge which can overcome the surface tension thus leading to droplet fission and the formation of "offspring droplets". Considering that polymer chain entanglements oppose the Coulomb fission during the electrospraying process, some researchers have identified a not dimensional parameter to describe the transition between fiber and bead formation and, consequently, between electrospraying and electrospinning process——as a function of the entanglements forming among chains in solution. This parameter $n_{e,sol}$ is defined as the ratio between polymer number average molecular mass (M_N) and M_e, which is the average molecular mass of the polymer segments between two entanglements in solution. So, it is possible to distinguish three different regimes in the electrospinning/spraying process of polymer solutions: $n_{e,sol} < 2$, in this case, particle formation occurs; $n_{e,sol} > 3.5$, regular electrospinning with fiber formation appears. In the intermediate case, $2 < n_{e,sol} < 3.5$, a mixed regime yielding beads and fibers takes place [Fig.2.2(c)].

Furthermore, molecular mass and viscosity are extremely important parameters for the processing of melt polymers at different temperatures. In scaffold design for tissue engineering, polymer blends have been largely investigated to obtain porous substrates with high control of pores interconnections and anisotropy. The mixing and extrusion of immiscible blends composed of polymers with different molecular mass, i.e., polycaprolactone (PCL) and poly (ethylene oxide) (PEO) is an interesting example where the sage balance of viscosity and relative composition play a crucial role on the final morphology of the scaffold. In this case, to reach a co-continuous phase organization, i.e., fully percolation of polymer phases, it is possible to study ab initio the viscosity ratio, depending upon the relative volume fraction of components in the binary polymer blend as follows:

$$\frac{\eta_1(\gamma)}{\eta_2(\gamma)} = \frac{\phi_1}{\phi_2} \tag{2-10}$$

In agreement with basic rules of co-continuous blends, a viscosity ratio equal to 1 should lead to a phase inversion at a 50:50 composition. However, when two components have different viscosities, the less viscous component (i.e., lower molecular mass) usually tends to encapsulate the more viscous one (i.e., higher molecular mass), thus pushing the phase inversion point toward a blend that *i* richer in the most viscous component. In the case the viscosity gap between the blend components is relatively modest, a phase inversion point close to 50/50 (% vol) may be expected. According to this idea, through the complex viscosity curve, it is possible to reach the isoviscosity condition that represents the essential requirement to obtain an optimal mixing of the polymer blend.

2.4 Thermal Properties: Effect of Crystalline and Amorphous Phases

Molecular shape arrangements are important factors in determining the physical properties of polymers. The molecular structure, conformation and orientation of the polymers have a relevant effect on the organization of molecules on the micro and macroscopic scale as they aggregate into more ordered structures. In order to explain this phenomena, crystalline and amorphous phase are generally recognized. In particular, polymer whose molecular structure lacks a definite repeating form, shape or structure is called amorphous polymer and has no definite shape while the polymer in which a unit structure repeats itself is called crystalline and has a definite shape, form and structure. There are some polymers that are completely amorphous, but most are a combination with the tangled and disordered regions surrounding the crystalline areas, called semicrystalline polymers. The presence of crystallites in the polymer usually leads to enhanced mechanical properties, unique thermal behavior, and increased fatigue strength. These properties make semicrystalline polymers desirable materials for biomedical applications.

As a function of the particular chain organization it is possible to identify three important physical transitions: glass transition, melting point and crystallization. These phenomena are important with respect to the design and processing of polymeric materials and maybe controlled by thermal conditions. Crystallization is the process by which, upon cooling, an ordered solid phase is produced from liquid melt having a highly random molecular structure. The melting transformation is the reverse process that occurs when a polymer is heated and the glass transition occurs with amorphous or non crystallizable polymers. Of course, during the physical steps of crystallization, melting and glass transition, it is possible to observe changes in physical and mechanical properties of polymers. In the case of semicrystalline polymers, the non-crystalline regions undergo the phenomenon of the glass transition, while the crystalline regions are affected by the melt phenomenon. The understanding of the mechanism and dynamics of crystallization in polymers is important since the degree of crystallinity affects the thermal and mechanical properties of these materials. Crystallization from a melt is the most fundamental of all phase transformations in materials that occurs through the nucleation and growth processes. Furthermore, the crystallization is a process associated with partial alignment of polymer molecular chains. These chains fold together and form ordered regions called lamellae, which compose larger spheroidal structures named spherulites. The crystallization proceeds with the formation of isolated spherulites, which then grow until their mutual impingement with further slow crystallization. In a sample of crystalline polymer there are billions of spherulities. There is an initial induction time required for the formation of spherulitic nuclei, followed by a period of accelerated crystallization in which spherulites grow in radius. When the spherulites begin to touch each other, crystallization rates slow down again. Complete crystallinity is almost never achieved, and the final degree of crystallinity is molecular-weight-dependent. Crystallization at low temperature nucleates a great number of spherulites which grow slowly while at high temperature crystallization results in rapid growth of relatively few spherulites, influencing in this manner the morphology of polymers.

Other two important parameters for polymers characterization are the glass transition temperature T_g and the melting point T_m. The term melting point, when applied to polymers suggests the transition from a crystalline phase or semicrystalline phase to a solid amorphous phase. The process of melting occurs in a specific range of temperature and not at a fixed temperature. Furthermore the melting behavior depends on the previous history of the polymers and, in particular, on the temperature at which crystallization occurs. Moreover, since the thickness of chain-folded lamellae depends on crystallization temperature; greater is the thickness of the lamellar structure much greater is also the melting temperature. Then, the polymeric materials react to thermal treatments with modifications of their structure and their properties. In addition, it is possible to get an increase in the lamellae thickness annealing the piece just below the melting temperature. The process of annealing, consecutively, increases the melting temperature of the polymer.

The phenomenon of the glass transition, instead, occurs in amorphous and semicrystalline polymers. It is due to a reduction in motion of large segments of molecular chains as the temperature decreases. Upon cooling, the glass transition corresponds to the gradual transformation from a liquid into a rubbery material, and then into a rigid solid; the last step corresponds to the glass transition. In particular, the temperature at which the polymer undergoes the transformation from a rubbery into a rigid state is called the glass transition temperature T_g. It is important to note that the glass transition does not occur suddenly, but usually takes place over a temperature range. The value of T_g depends on the mobility of the polymer chain, the more immobile the chain, the higher the value of T_g. Both melting temperature and glass transition temperature are very important parameters for the industrial applications of polymeric materials. They define, respectively, the upper and lower temperature limits for many applications, especially for semicrystalline polymers. The glass transition temperature, moreover, defines the upper limit temperature for glassy amorphous materials. In addition, T_m and T_g also influence the fabrication and processing of polymers and polymer matrix composites. The temperatures at which the phenomena occur for polymer are determined by a plot where the specific volume (the reciprocal of the density) is a function of temperature (Fig.2.3).

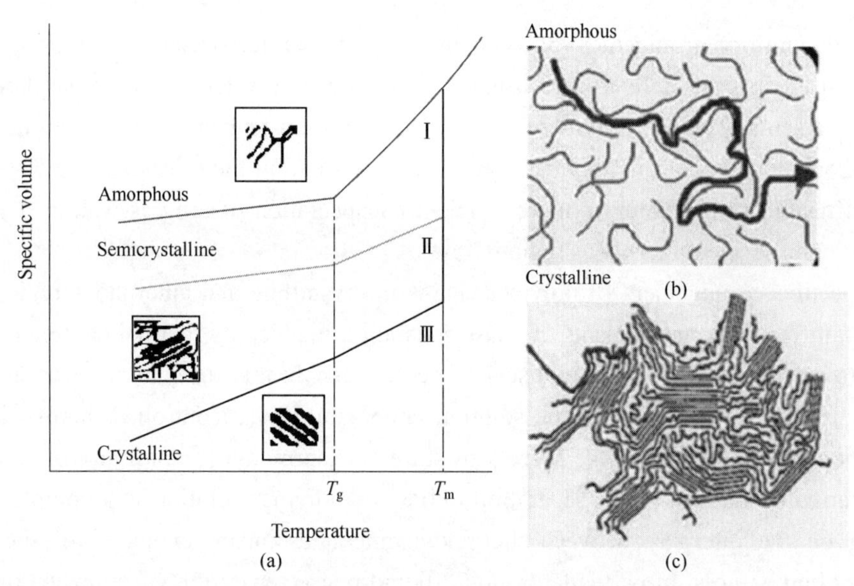

Fig.2.3 Specific volume of polymers as a function of temperature for totally amorphous (a), semicrystalline (b), and crystalline (c) polymers and schematic representation of diffusion mechanisms of small molecules amorphous and crystalline domains.

For crystalline polymers, there is a discontinuous change in specific volume at the melting temperature. The curve for the totally amorphous material is continuous but with a slight decrease in slope at the glass transition temperature. In the case of a semicrystalline polymer, there is the coexistence of crystalline and amorphous phases. The exact temperatures at which the polymer chains undergo these transitions depend on the structure of the polymer, which means that molecular chemistry and structure are factors able to influence melting and glass transition temperatures.

Both the melting and the glass transition temperatures are influenced from the molecular structure and the inter chain bonds of polymers. During the melting process, in fact, there is a rearrangement of the molecules in the transformation from ordered to disordered molecular states. Molecular chemistry and structure influence the ability of polymer chain molecules to make these rearrangements affecting the melting temperature. Chain stiffness also has a pronounced effect on melting temperature because of double bonds in the polymer backbone, in fact, rigid chains have higher T_m. Moreover, the melting temperature and glass transition temperature also depend on molecular mass. At low molecular masses T_m decrease, since low polymer molecular mass have shorter chain length comparing to higher polymer molecular mass. Less significantly, even an increase of molecular mass generally is coupled to the increase of T_g.

The control of amorphous and crystalline domains by thermal process conditions into a polymer system is extremely important to properly design the final properties of devices for biomedical use. For instance, polymer based carriers for drug delivery applications are generally designed by using different polymers in terms of degradability and crystallinity, which allows different drug and growth loading as a function of the relative crystalline/amorphous phase ratio. This is due to the ability of a crystalline polymer to degrade slowly than an amorphous one because of the uniform arrangement of its chains within the lattice structure. Therefore, polymers commonly used in drug delivery are usually a mixture of crystalline and amorphous forms. Indeed, into a semicrystalline polymer such as PCL, the water tends to easily penetrate only into the amorphous domains of the polymer matrix, thus facilitating the release of the water soluble drug by diffusion, so promoting a more efficient encapsulation of drug as well as a controlled release through polymer amorphous regions.

The peculiar arrangement of polymer chains in crystalline and amorphous regions is also relevant on influencing mechanisms of phase separation induced by thermal cooling of polymer solutions to design scaffolds with multiscale pore network. In this case, the mechanism of phase separation is induced by lowering the solution temperature by a controlled thermal history or adding a non-solvent to extract the solvent by different thermodynamic affinity of components. Indeed, thermally induced phase separation (TIPS) of polymer solution is a complex process, depending on the interplay between thermodynamic and kinetic evolution of the polymer solution cooling process. In particular, a liquid-liquid phase separation occurs when the applied temperature is higher than the solvent crystallization temperature or higher than the freezing point, while a solid-liquid phase separation takes place when the solvent crystallization temperature exceeds the coolant temperature. The strict control of physical transition of polymer component by applying appropriate thermal histories can significantly influence the final aspect of the scaffold in terms of morphological and mechanical properties. For example, the cooling kinetics can affect the degree of crystallinity of the polymer matrix, promoting the formation of scaffolds with high mechanical properties in comparison with products obtained by conventional techniques. During liquid-liquid phase separation, the presence of semicrystalline polymers in solution (e.g., polylactide and polycaprolactone) may frequently initiate a gelation process. This leads to the formation of small crystallites, which act as physical crosslinks; these can stabilize the 3D polymer network and may also enhance the mechanical properties. The study of the behavior of binary systems becomes more complex in the presence of a solid-liquid separation mechanism. This is because of the formation of microcellular domains, which grow preferentially during cooling of the polymer solution until the solvent crystallization temperature is reached. Moreover, solvent removal by freeze drying promotes the formation of macropores as consequence of the solvent sublimation, preserving the crystal shape previously formed.

2.5 Mechanical Properties of Polymers

2.5.1 Basic Aspects: Toughness and Viscoelasticity

The mechanical response of polymers is the result of several force contributions: the properties of single polymer chains mediated by intramolecular forces and the properties of polymer chains assembly in the bulk system governed by intermolecular forces. However, it is really complex to identify separately the contribution of different forces which often mutually concur to the final mechanical response offering a multiplicity of different behavior as a function to the specific force activity. From macroscopic point of view, the mechanical behavior of polymers is generally classified into three categories: brittle, ductile, and rubbery (Fig.2.4). Brittle polymers show a high modulus and high ultimate tensile strength but low ductility and toughness and are generally characterized by a glass transition T_g that is much higher than room temperature, (i.e., Polymethylmetacrilate, PMMA). Ductile polymers are semicrystalline polymers such as polyethylene or polytetraluoroethylene (PTFE) that have a T_g below room temperature for the amorphous polymer content. The intrinsic crystalline regions confer the strength to the polymer while the amorphous regions determine the capability of energy adsorption before failure (toughness). These polymers have generally lower strength and modulus but greater toughness than brittle polymers. Finally, rubbery polymers show low moduli since they have a T_g well below room temperature, but they can return to their original shape following high extensions since cross-links prevent significant polymer chain translations.

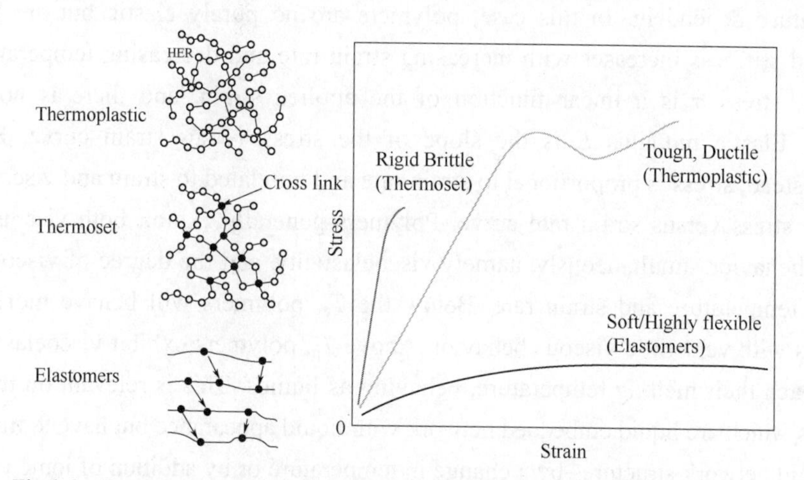

Fig.2.4 Stress versus strain curves of three different polymer classes characterized by different molecular interactions.

According to this classification, it is possible to recognize three different groups of polymers namely thermoplastic, thermoset and elastomers, respectively (Fig.2.4) which may be classified as a function of peculiar intra and inter molecular interactions. Thermoplastic polymers are made of individual polymer chains that are held together by relatively weak van der Waals and dipole-dipole forces. They are generally processed into products by melt processing such as injection molding and extrusion, but may be also dissolved in solvents to form films and other devices by casting technique. Thermosetting polymers contain cross-links between polymer chains. Cross-links are covalent bonds between chains and can be formed using monomers with at least two free-functionalities during synthesis (i.e., polyurethanes, silicones). Depending on the cross-link density, thermosets can swell in certain solvents, whereas may degrade prior to reach the theoretical melting point. Sometimes, cross-links may be created after the polymer is formed (i.e., Vulcanization) with the support of cross-linking agent (i.e., sulfurs) but in this case they are named rubbers or elastomers. Their mechanical properties are strongly influenced by the peculiar organization/conformation of chains and their interactions at atomic, microscopic, and macroscopic scales, thus allowing to variously resist against stretching (i.e., tensile strength), compression (i.e., compressive strength), bending (i.e., flexural strength), sudden stress (i.e., impact strength), and dynamic loading (fatigue). As function of molecular mass and the level of intramolecular forces, polymers may display different mechanical properties under an applied stress. For instance, flexible polymers work better under stretching rather than a rigid polymer working better under compression. In the case of thermoset polymers, as cross-link density increases, the slope of the stress/strain line become steeper, corresponding to higher elastic moduli, and ultimately, to higher polymer strength or stiffness. In the case of thermoplastic polymers as melt or viscous solutions may be handled with more difficulty as their molecular mass increases due to the occurrence of topological interactions called entanglement, which limit the sliding of adjacent polymer chains.

The most part of polymers shows mechanical properties which are also highly strain rate and temperature dependent. In this case, polymers are no purely elastic but are viscoelastic materials and stiffness increases with increasing strain rate and decreasing temperature. For an elastic solid, stress σ is a linear function of the applied strain, and there is no strain-rate dependence. Elastic modulus E is the slope of the stress versus strain curve. For viscous materials, instead, stress is proportional to strain rate and unrelated to strain and viscosity η is the slope of the stress versus strain rate curve. Polymers generally exhibit both viscous and solid mechanical behavior simultaneously, namely viscoelasticity, and the degree of viscous behavior depends on temperature and strain rate. Below the T_g, polymers will behave more or less as elastic solids with very little viscous behavior. Above T_g, polymers exhibit viscoelastic behavior until they reach their melting temperature, behaving as liquids. This is relevant on the design of polymer gels which are liquid embedded network with liquid appearance but have to transform into relatively rigid network structures by a change in temperature or by addition of ionic cross-linking agents. They are largely used for cell interaction studies due to their excellent biocompatibility but mainly to fabricate injectable systems able to release molecular species in vivo.

Moreover, many polymers may deform differently with time when they are under a continuous load (creep mechanism). Ideal elastic solids do not show any creep since strain is proportional to stress, and there is no time dependence. Viscous materials (liquids) deform at a constant rate with a constant applied stress. Viscoelastic polymers may show both mechanisms so variously behave in terms of modulus, ductility, and strength as a function of the strain rate. This is the basic working mechanisms of hydrogels such as hyaluronic acid which are largely used as dumping systems in the place of damaged articular cartilage, for nucleus substitution in spine surgery or ocular therapies. For example, the physical properties of hyaluronic acid (HA) and its derivate compounds, including viscosity, elasticity and highly negative charge, make it useful in various therapeutic uses including tissue engineering and visco-supplementation for osteoarthritis treatment. HA's structural characteristics hinge on this random coiled structure in solution. At very low concentrations, chains entangle each other, leading to a mild viscosity which is strictly dependent upon the molecular mass of polymer. On the other hand, HA solutions at higher concentrations are very highly viscous due to greater HA chain entanglement which confer to the polymer a shear-dependent mechanical behavior. Indeed, the most distinctive property of HA is its peculiar viscoelasticity in the hydrated state as a function of shear rate stimulation. For example, the viscosity of a 1% solution of HA, having a molar mass of 3×10^6~4×10^6 Da, is about 500,000 times that of water at low shear rate, but can drop 1,000-fold when forced through a fine needle. As a consequence, high shear rates promote a reduction of HA viscosity, which is reflected in the force required to overcome internal friction. This also increases the elasticity, namely stored energy, thus permitting the polymer chain recovery after deformation. In consideration of these peculiar viscosity and elasticity properties, HA have been largely used in various surgical or medical preparations to replace the vitreous body after surgery, to manipulate the retina in retinal detachment surgery or protect the corneal endothelium in corneal transplantation (viscoprotection). Indeed, HA solutions may be intraocularly implanted by different surgical practices as a function of viscoelastic properties; high viscosity at low shear rate allows to maintain space and manipulate tissues; moderate viscosity at medium shear rates allows easily for the manipulation of surgical instruments and intraocular lenses within the polymer solution. Very low viscosity at high shear rates allows to minimize the pressure needed to expel the solution through a thin cannula. Meanwhile, high degree of elasticity is fundamental to protect adjacent intraocular tissues, especially the endothelial cells of the cornea, from contact with surgical instruments. For example, highly viscoelastic HA solutions (i.e., Healon®) are commercially available for therapeutic use in alleviating discomfort in "dry eye syndrome". Although HA is not present in tears, in many aspects sodium hyaluronate is similar to mucin, a major component of tears. Mucin with a mean molar mass of about 2 MDa shows, similarly to HA, typical viscoelastic and shear-thinning behavior. This glycoprotein plays an important role in the lubricating, cleansing, and water-retaining properties of tears. In this case, the usefulness of an HA solution as a tear substitute resides in its water-entrapment capacity (hydration) and its function as a viscoelastic barrier between the corneal and conjunctival epithelia. Indeed, the HA eye drops are elastically deformed during eye blinking, but not removed from the surface of the eye with blinking movements.

HAs have been also successfully used to support the normal activity of pathological articular cartilage when the synovial fluid in osteoarthritis joints lacks sufficient shock absorption and lubrication properties mostly due to the not adequate viscosity of fluid at the interface (i.e., visco-supplementation). In particular, a large series of HA injectable preparations are commercially available to modify the local properties of the joint fluid with different viscoelasticity in order to stimulate the production of endogenous HA, to inhibit the effects of inflammatory mediators, and to decrease cartilage degradation also promoting cartilage matrix synthesis.

Referring to structural applications, viscoelastic properties of polymers are historically used to reproduce the mechanical properties of natural ligament into a synthetic prosthesis which have to develop the knee function after injury. Indeed, natural ligaments exhibit significant time and history dependent viscoelastic properties with different strain-rate sensitivity as a function of the proximity to the bone tissue joint, which is crucial to preserve it from uncontrolled failure under different loading conditions. The composite structure of native tissue also contributes to mediate the viscoelastic properties of natural ligaments, which regulate important biomechanical functions of the knee during dynamic loading. In particular, the viscous dissipation evident at low strain and/or low loading frequency is presumably due to the mechanical interaction (friction) between collagen fibers and the hydrated matrix of proteoglycans and glycoproteins as suggested by the dependence of the loss modulus upon the static stress. Current devices for ligament augmentation characterized by fibrous in woven non woven tissues made of monolithic polymers are suitable to replace many of its biomechanical functions due to the intrinsic viscoelastic properties of constituent materials. However, its simplified architecture does not reproduce completely the mechanical features of the natural tissue. Hence recently enabled to develop soft composite materials fabricated by reinforcing hydrogel-like matrixes with rigid polymeric fibers have been proposed alternatively. In this case, the mechanical response is strictly controlled by the structural arrangement of the reinforcing fibers and by the properties of the components so that the static and dynamic mechanical behavior of natural ligaments can be reproduced. Noteworthy, the viscoelastic response of device is also affected by intermolecular forces which concur to the macroscopic mechanical response including fiber hydration, fiber-matrix interaction, and fiber-fiber interaction, which influence the mechanical parameters (i.e., static, storage, loss moduli) of the composite materials and preserve the same constitutive dependency upon the kinematics variables (i.e., strain, frequency) as natural tissues. More recently degradable polymers typically used for tissue engineering approaches allow to design ligament devices where the viscoelastic response progressively changes as the polymer begin to degrade in vivo, opening new advantages in terms of biocompatibility for in vivo ACL reconstruction.

2.5.2 Mechanical Properties Mediated by Intermolecular Forces

Mechanical properties are among the most fundamental properties of polymeric materials in biomedical applications. In traditional applications, polymers firstly have to meet the basic mechanical criteria such as strength (modulus), energy-dissipating capacity (toughness) and elasticity. With the tremendous progress made in polymer science in the last century, a wide range of synthetic polymers with excellent mechanical properties have been developed for various applications including plastic, fiber, and elastomers. Whereas synthetic polymers can be prepared to meet particular mechanical parameter one at a time, it is still challenging to design advanced polymeric materials that show peculiar mechanical properties as a function of the specific application.

In this context, the approach based on the learning for nature allows to design synthetic polymers mimicking the natural biopolymers on natural tissues that combine important mechanical properties including strength, toughness, and elasticity. For example, cell adhesion proteins and connective proteins existing in both soft and hard tissues such as muscle, and bone exhibit a remarkable combination of high strength, toughness, and sometimes elasticity as well—three properties that are rarely found in one synthetic polymer. Starting from this approach, recent advanced structural analysis and single-molecule nanomechanical studies revealed that the combination of these mechanical properties in natural materials originates from their unique molecular and nanoscopic structures. These mechanistic understandings at molecular level provide inspiration to materials scientists for designing biomimetic polymers that have a balance of advanced mechanical properties.

One important strategy nature adopted to enhance mechanical properties is the use of non-covalent weak forces in addition to covalent bonds. By programming weak forces, such as hydrogen bonds, hydrophobic interactions, and ionic interactions, etc., biopolymers can achieve combined mechanical performances that synthetic polymers still cannot rival. Among the weak forces, intermolecular interactions among biopolymers chains are particularly interesting to enhance mechanical properties. For instance, the exceptional mechanical properties of silk produced by spiders have recently attracted much attention from scientists in various disciplines to investigate the molecular origin of its mechanical properties mainly ascribable to intermolecular forces. For example, natural silk produced by spiders shows an exceptional strength (tensile strength of 1.5 GPa) and toughness (150 MJ•m^{-1}), which makes it stronger than steel compared on a weight basis and has a tensile strength similar to Kevlar®. Moreover, it is also characterized by high elasticity and exceptionally high toughness values never attained in synthetic high-performance fibers which are directly ascribable to the peculiar polymer chain organization. Indeed, spider silk is a semicrystalline material made of amorphous segments reinforced by strong and stiff crystalline domains. Molecular studies indicate that the crystalline domains are made of hydrophobic poly(alanine) and poly(alanine-glycine) repeat motifs whereas the amorphous segments are composed of glycine-rich peptides formed both inter- and intramolecularly (Fig.2.5). This peculiar microscopic organization explain the species-specific silk mechanical response well represented by a characteristic nonlinear stress-strain (σ-ε) curve which show four distinct regimes characteristic of silk: (a) stiff initial response governed by homogeneous stretching; (b) entropic unfolding of semi-amorphous protein domains; (c) stiffening regime as molecules align and load is transferred to the β-sheet crystals; and (d) stick-slip deformation of β-sheet crystals until failure.

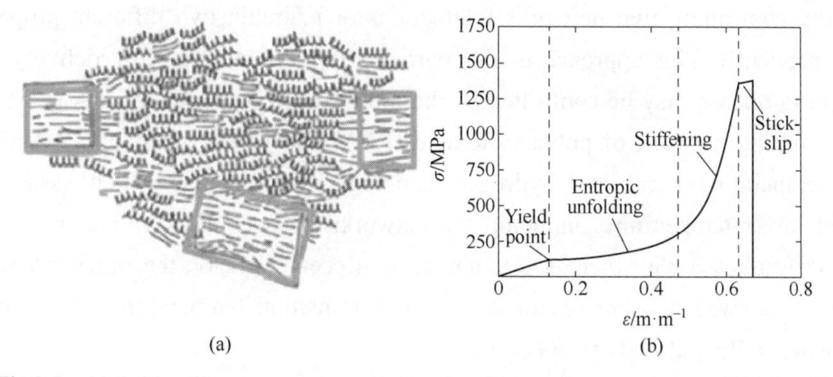

Fig.2.5 (a) Spider silk microscopic model to reproduce chain arrangement of spider silk; (b) Stress-strain curve to describe the mechanical behavior of spider silk.

Based on X-ray evidence, spider silk is characterized by a large number of nanocrystallites separated by amorphous regions made of flexible chains. This allows to underline the important role of crystallites, which act as multifunctional crosslinks playing an analogous role of carbon black in carbon-reinforced elastomers, creating inside the amorphous regions a thin layer with modulus higher than in the bulk. This is an excellent example of using a combination of inter and intramolecular weak forces (hydrogen bonds) to assemble polymers into more ordered structures for enhanced mechanical properties and have inspired many researchers to design biomimetic fibers that hopefully can mimic both its structure and mechanical properties. Sogah and coworkers have proposed a synthesis of peptide and synthetic modules to imitate the crystalline and amorphous silk structures based on the assembly of phenoxathi in-templated parallel β-sheet peptide building blocks which are linked into polymers through polymerization. However, the design of a synthetic polymer that imitate both the structure and mechanical properties of spider dragline silk is still far due to the difficulty to effectively balance of various molecular parameters that control polymer secondary structures to achieve optimal mechanical properties.

In this direction, stimuli responsive polymers are emerging as a versatile solution to adapt the properties of materials to the biological context, by properly imparting external stimuli. Their recent success rises from the low costs of materials combined with the ease to tailor, better than tradition materials classes specific functionalities, by the responsive behavior to pH, temperature, ionic strength, electric or magnetic field, light and/or other chemical or biological stimuli. Indeed, this offers the opportunity to design highly customizable materials with a large variety of properties to be fruitfully used in biomedical applications from drug delivery to tissue engineering.

Among them, crosslinked hydrogels based on interpenetrating networks (IPN) are extremely interesting for their peculiar mechanical properties. They consist of two covalently linked polymer networks which are bound together by physical entanglement as opposed to covalent bonds. This is possible by the polymerization of both networks simultaneously and results in two intermixed networks that can only be separated by breaking bonds. As a consequence, this confers to the polymer the ability to improve the intrinsic mechanical properties by controlling two networks interactions or maintain two different properties when acting independently. This approach is also particularly interesting in drug delivery application where the drug release may be controlled by thermosensitive properties of IPN gel. Specifically, an interpenetrating network of polyacrylic acid (PAA) and polyacrylamide (PAAm) above the transition temperature break the hydrogen bonding among chains which assure the water retention at lower temperature, inducing the networks to swell thus giving the possibility of increasing drug release with increased temperature. Recent work on the same IPN with grafted β-cyclodextrin showed a faster response for lower transition temperature (35℃) promoting a more efficient swelling at body temperature.

The basic mechanisms at molecular level used in nature to design natural materials are increasingly inspiring chemists and materials scientists to design biomimetic polymers to imitate both the structures and properties of their natural counterparts. Due to the complexity and subtlety of biopolymer structures, it remains a major challenge but the rapid improvement of advanced analytical techniques open new insights to develop truly effective synthetic polymers that can rival the biomimetic performance of their natural analogs. Despite many techniques are available to synthesize and process polymers for biomedical applications, there is still a need to either develop novel methodologies or expand their employment to more polymers, in the light of knowledge of chemical and physical interactions and recent discovery of effects of micro environmental stimuli, thus giving the opportunity to design a greater variety of modified polymers that can be used as innovative systems in pharmaceutics and medicine.

2.6 Pharmaceutical Polymers: Principles, Structures, and Applications of Pharmaceutical Delivery Systems

This section presents a general overview of pharmaceutically used polymers with respect to their physico-chemical characteristics and factors affecting drug delivery abilities. Pharmaceutical polymers, chemical structure, and properties are discussed for their applications in controlled drug release systems. An additional focus is on new polymers (dendrimers, hyperbranched polymers), considering their chemical versatility, uniqueness, and future implications. Problems associated with controlled drug release systems are also highlighted. Finally, applications of FDA-approved polymers used for oral drug delivery systems are outlined.

2.6.1 Introduction

Conventional drug delivery systems often only reach concentrations within the therapeutically effective range (therapeutic window) when taken several times a day (Fig.2.6).

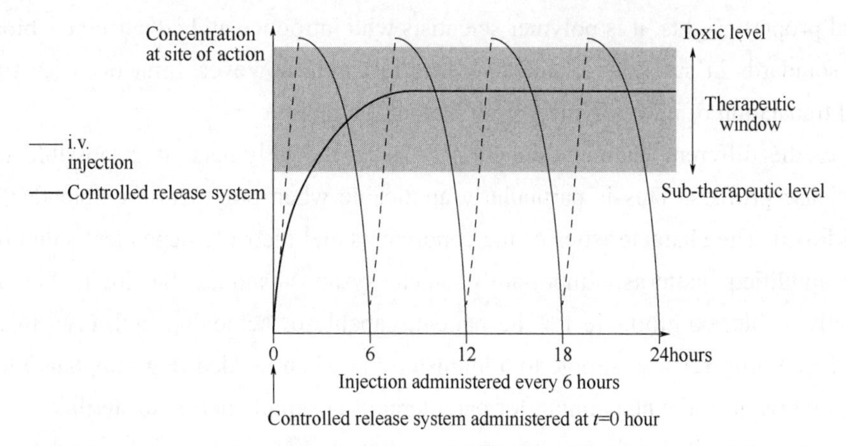

Fig.2.6 Therapeutic window of an API: application by a controlled drug delivery system compared with injection.

In contrast, controlled release systems (CRS) are tailored to sustain the active pharmaceutical ingredient (API) release at a specific rate over a defined period of time to maintain plasma concentration within the therapeutic range. Thus the release profile of a drug delivery system can allow for a reduction in dosing frequency and thus improve compliance and overall treatment effectiveness. Furthermore, toxic peak plasma levels above the therapeutic window can be avoided and undesired side effects minimized. In particular, implants have the potential for drug delivery for several years which makes them valuable tools in long-term medication.

Constant plasma levels realized by zero-order API release, however, are not always the best profile. In fact, circadian variations in the severity of symptoms may require adaptive API plasma levels. For example, exacerbations of bronchial asthma predominantly occur in the early morning. Therefore efficient bronchodilation can be initiated later in the night. Moreover, bactericidal antibiotics are most efficient when exposing the microorganism to fluctuating drug levels by consumption of several doses a day. However, compliance is best when drug intake is limited to a single daily dose at a convenient time. Therefore, pulsatile release systems are currently being introduced into therapy based on polymer blends.

Moreover, polymers can be used to design delivery systems for parenteral use which allow Ehrlich's principle of the "Magic Bullet" to come into reality. Passive targeting to solid tumors via the Enhanced Permeability and Retention (EPR) effect as well as active targeting by antibodies attached to the particle surface can improve clinical efficacy in, for example, cancer therapy. Therefore, the use of CRS and pharmaceutical polymers should increase in the near future. Although many promising CRS have evolved, challenging physico-chemical problems need to be addressed particularly in the areas of polymeric bulk diffusion, membrane permeability, osmotic effect, colloidal aggregation, and polymer dynamics. Although many polymers have found their way into the clinics, there is still a great demand for new polymers. Since the existing polymers do not have sufficient versatility, the physico-chemical properties do not fully cover the required spectrum. Moreover, as most of the polymer candidates are under intellectual property rights, it is polymer scientists who introduce and advance new biomaterials. The high standards in safety assessment in drug research, however, limit or at least retard the successful translation of new polymers from laboratory to clinic.

Due to the different chemical functionalities of the polymers, it is possible to achieve defined release profiles. This is particularly applicable when polymers are "blended" (used in combined form). The characteristics of the copolymers and their pH-dependent solubility impart new and modified patterns. Importantly, such systems should be inert, biocompatible, mechanically stable, comfortable for the patient, capable of achieving high drug loading, safe from accidental drug release, simple to administer and when needed (e.g., implants) to remove, and easy to fabricate and when aiming for parenteral use or implantation to sterilize.

Therefore, controlled delivery systems are often more expensive compared to traditional pharmaceutical formulations. The main disadvantage of CRS, however, is a possible toxicity or non-biocompatibility arising from the polymeric materials used. There could also be undesirable by-products from the polymers due to degradation. Therefore, the production process of the polymer has to be strictly controlled; by-products have to be limited, analytically quantified and subjected to toxicological testing if present in relevant amounts. Moreover, in some instances, surgery may be required to implant or remove the system, which may cause discomfort to the patient.

2.6.2 Pharmaceutical Polymers and Biomaterials

To adjust the release profile of the API from the formulation, polymer and other excipients are either used in a single or in blended form. In general, the selection of the polymer is based on the release-defining characteristics:

① Chemical nature and charge on polymer: anionic, cationic, and neutral;
② Hydrophilicity/hydrophobicity of the polymer;
③ Cross-linking ratios and swelling/de-swelling capacity;
④ pH-dependency/independency;
⑤ Route of administration;
⑥ Release pattern and erosion mechanism;
⑦ Targeted site for absorption (non-parental application).

For economic reasons, drug delivery research often makes use of a platform technology. As shown in Fig.2.7 and Fig.2.8, the total time line from project concept to regulatory approval which is mandatory to launch a delivery product requires an average of 5~8 years.

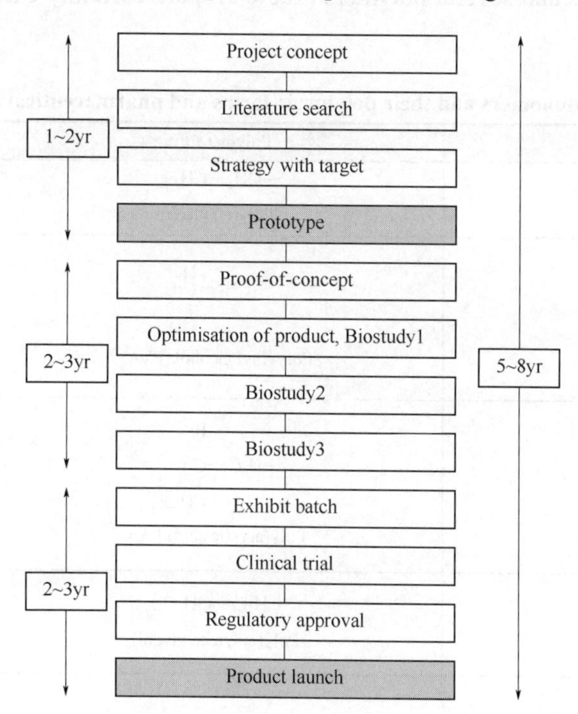

Fig.2.7 Drug delivery systems manufacturing pathway leading to estimation of drug bioavailability and product launch.

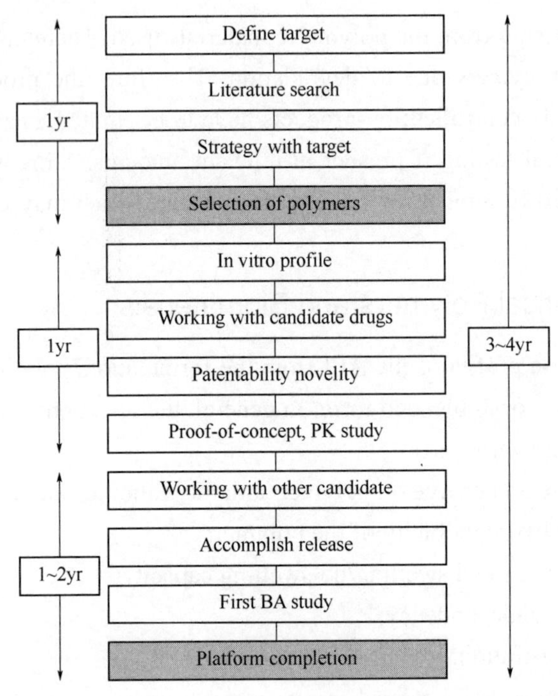

Fig.2.8 Major steps in development of platform drug delivery system (BA= Bioavailability).

2.6.2.1 Classification of Pharmaceutical Polymers and Biomaterials

A range of pharmaceutical polymers and biomaterials is used to control the release of APIs and other active agents and several polymers (Table 2.1) are currently extensively used for drug delivery.

Table 2.1 Monomers and their polymeric forms and pharmaceutical applications

Monomer	Polymer form	Drug delivery applications
1. $H_2C{=}CH_2$ Ethylene	$\text{+}CH_2-CH_2\text{+}_n$ Poly(ethylene)	Controlled delivery systems
2. $H_2C{=}CH-O-C(=O)CH_3$ Vinyl acetate	$\text{+}CH_2-CH(OH)\text{+}_n$ Polyvinyl alcohol (PVA)	Copolymerization with vinyl acetate is used in microspheres and bio-adhesive hydrogels
3. $H_2C{=}C(R)COOH$ R=H (acrylic acid), R=CH_3 (methacrylic acid)	$\text{+}H_2C-C(R)(COOH)\text{+}_n$ Poly(acrylic acid) PAA	Stimuli-sensitive polymer, pH-sensitive. Reversible swelling hydrogels
4. H_2C-CH_2 (epoxide, O) Ethylene oxide	$\text{+}H_2C-CH_2O\text{+}_n$ Poly(ethylene glycol)	Polymer prodrugs to enhance aqueous solubility
5. (cyclic lactone) Caprolactone	$\text{+}O-(CH_2)_5-C(=O)\text{+}$ Polycaprolactone	Used in controlled drug delivery, local delivery, micelles, etc.

Monomer	Polymer form	Drug delivery applications
6. 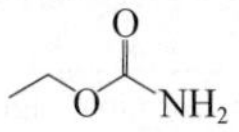 Urethane	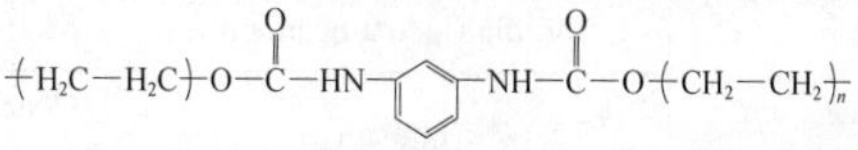 $-[H_2C-H_2C]-O-C(=O)-HN-C_6H_4-NH-C(=O)-O-[CH_2-CH_2]_n-$ Polyurethane	Used for temporal controlled release
7. 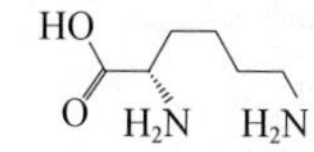 HO, O, H_2N, H_2N Amino acid e.g., lysine	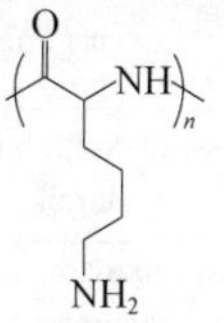 O, NH, n, NH_2 Poly(lysine)	Used in drug delivery systems in various forms including copolymers with polylactic acid (PLA)
8. 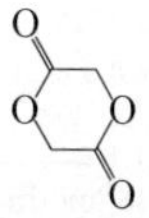 Glycolide; 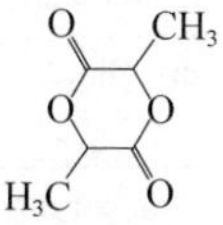 Lactide (CH_3, H_3C)	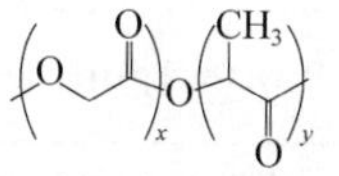 CH_3, x, y Poly(lactic-co-glycolic acid) (PLGA) copolymer	Nanoparticles and microparticles drug, protein, and peptide delivery
9. 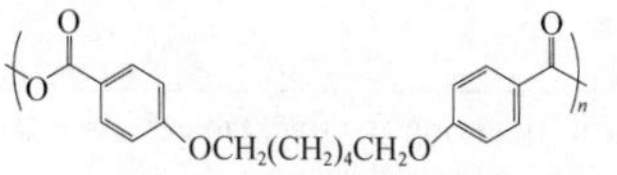 $OCH_2(CH_2)_4CH_2O$, n *p*-(carboxyphenoxy)hexane	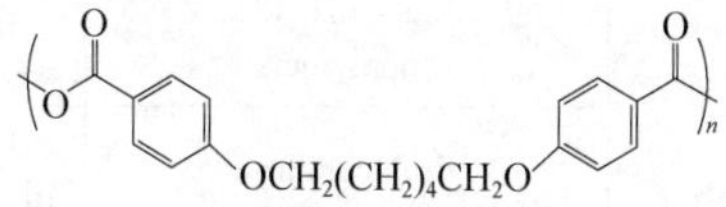 $OCH_2(CH_2)_4CH_2O$, n Poly anhydrides *p*-(Carboxyphenoxy)hexane	Used in oral controlled drug delivery systems

Note: Polymers 7, 8, and 9 degrade within the physiological environment and are being used in drug delivery (systemic route of application)

Table 2.2 gives an overview of the various polymers in pharmaceutical use.

Table 2.2 Polymers in pharmaceutical use

Polymers			Use
Hydrophilic Polymers	Cellulosic	Methyl cellulose	Coating agent, emulsifying agent, suspending agent, tablet and capsule disintegrant, tablet binder, viscosity increasing agent
		Hypromellose (hydroxypropylmethyl) Cellulose	Coating agent, film forming agent, rate controlling polymer for sustained release, stabilizing agent, suspending agent, tablet binder, viscosity increasing agent
		Hydroxypropyl Cellulose	Coating agent, emulsifying agent, suspending agent, stabilizing agent, thickening agent, tablet binder, viscosity-increasing agent
		Hydroxyethyl Cellulose	Coating agent, suspending agent, thickening agent, tablet binder, viscosity-increasing agent
		Sodium carboxymethyl cellulose	Coating agent, suspending agent, tablet and capsule disintegrant, tablet binder, viscosity-increasing agent, stabilizing agent, water absorbing agent
	Noncellulosic: gums/ polysaccharides	Sodium Alginate	Stabilizing agent, suspending agent, tablet and capsule disintegrant, tablet binder, viscosity- increasing agent
		Xanthan gum	Stabilizing agent, suspending agent, viscosity-increasing agent
		Carrageenan	Gel base, suspending agent, sustainedrelease tablet matrix

续表

Polymers			Use
Hydrophobic Polymers	Noncellulosic: gums/ polysaccharides	Ceratonia (locust bean gum)	Matrix, binder
		Chitosan	Coating agent, disintegrant, film-forming agent, mucoadhesive, tablet binder, viscosity increasing agent
		Guar gum	Suspending agent, tablet binder, tablet disintegrant, viscosity-increasing agent
		Cross-linked high amylose starch	Gelling agent, thickening agent
	Noncellulosic: others	Poly(ethylene oxide)	Mucoadhesive, tablet binder, thickening agent
		Poly(ethylene glycol)	Ointment base, plasticizer, solvent, suppository base, tablet and capsule lubricant
		Poly(vinyl pyrrolidone)	Disintegrant, dissolution aid, suspending agent, tablet binder
	Water-insoluble	Homopolymers and copolymers of acrylic acid (Carbomer)	Bioadhesive, emulsifying agent, release modifying agent, suspending agent, tablet binder, viscosity-increasing agent
		Gantrez® maleic acid copolymers	Bioadhesive
		Ethylcellulose	Coating agent, flavoring fixative, tablet binder, tablet filler, viscosityincreasing agent
		Cellulose acetate	Coating agent, extended release agent, tablet and capsule diluents
		Cellulose acetate propionate	Membrane
		Cellulose acetate phthalate	Enteric coating agent
		Hydroxypropylmethyl cellulose phthalate	Enteric coating polymer
		Poly(vinyl acetate) phthalate	Viscosity modifying agent with enteric coating
		Gantrez® AN (copolymers of methyl vinyl ether and maleic anhydride)	Mucosal adhesive resins
		Methacrylic acid copolymers	Film forming agent, tablet binder, tablet diluents
		Eudragit® acrylate copolymers	The wide range of immediate, enteric and sustained release polymers allow the design of any number of combinations to match targeted release profile

2.6.2.2 Orally Applied Polymers: Properties and Applications

(1) Starch

Starch is a complex carbohydrate abundantly found in nature: for example, corn sorghum, wheat, potato, and rice. Starch comprises a mixture of two polysaccharides, namely, amylose (a linear polysaccharide), and amylopectin (a branched polysaccharide with a molecular mass of up to 2×10^8). Amylose is widely used as a binding agent in tablet manufacturing since it exhibits strong binding and swelling properties. The natural polymer also forms a stable viscous dispersion.

(2) Hydroxylpropyl Methylcellulose

The 2-hydroxylpropyl ether of methyl cellulose (HPMC) (Fig.2.9) is derived from alkali-treated cellulose which is reacted with methyl chloride and propylene oxide. HPMC swells and dissolves slowly in cold water to produce a viscous colloidal solution. It is also soluble in most polar solvents. Aqueous solutions are surface active, form films upon drying, and undergo reversible transformation from sol to gel upon heating and cooling.

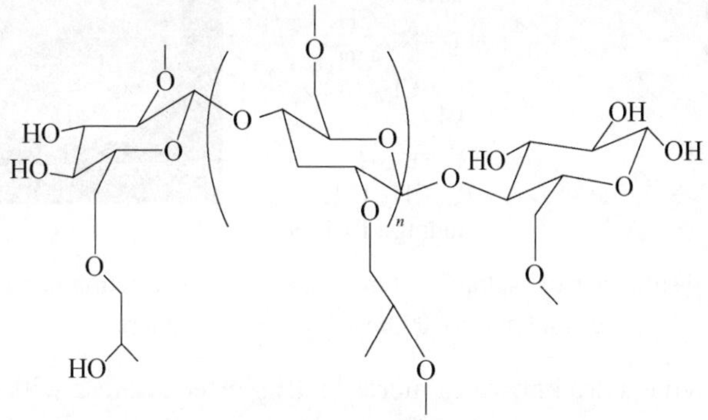

Fig.2.9 Chemical structure of hydroxylpropyl methylcellulose (HPMC) polymer.

Besides the use as a tabletting agent and as an emulsifier in ointments, HPMC is extensively used in oral controlled drug delivery systems. Moreover, the polymer has been widely explored as a matrix-forming agent in the design of patches.

Propranolol hydrochloride release from HPMC matrices was fast yet slowed when coated with a protective layer. A swellable, floating triple-layer tablet was designed to prolong the gastric residence time of anti-infectives in Helicobacter pylori eradication therapy. HPMC and poly(ethylene oxide) (PEO) were used as the rate-controlling polymeric membrane excipients. Tetracycline and metronidazole were incorporated into the core layer of the triple-layer matrix for controlled delivery, while bismuth salt was included in an outer layer for instant release. Floatation was accomplished by incorporating a gas-generating layer consisting of sodium bicarbonate:calcium carbonate (1 : 2 ratio) along with the polymers. In vitro testing revealed sustained release of tetracycline and metronidazole over 6~8 h while the tablets remained afloat.

(3) Eudragit

Many enteric coating formulations are based on anionic polymers of methacrylic acid and methacrylates (Fig.2.10). The polymers are considered unique and versatile since they dissolve at a wide range of pH 5.5~7. In the early 1930s, Roehm & Haas GmbH (Darmstadt, Germany) initiated the systematic research on the first forms of synthetic acrylates and methylacrylates. With the breakthrough development of Eudragit® in 1953, the first pharmaceutical coating drug formulation was introduced into the market. Since then, these coatings have opened a new phase in pharmaceutical research and development (Fig.2.10).

Eudragit in enteric coatings allows pH-dependent drug release. Eudragit polymers, used in mono or in blend form, have a wide range of pharmaceutical applications: as binding agents in tablets, to increase viscosity, to mask the taste of drugs, and as flow-controlling agents in liquids, suspensions, and emulsions. The polymers can also enhance drug stability and modify drug release characteristics.

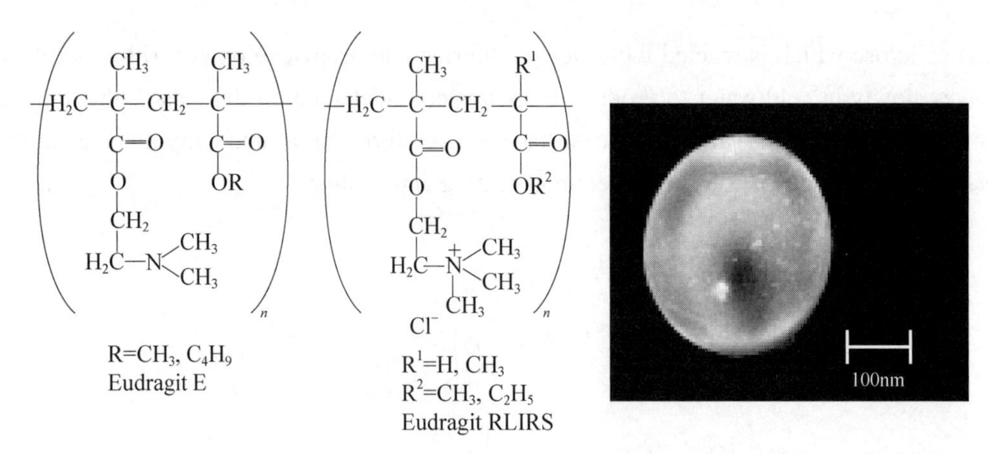

Fig.2.10 Chemical structures of Eudragit polymers and scanning electron microscopy photograph of Eudragit S-100-coated core microsphere.

Overall, these polymers are known to increase drug effectiveness with good storage stability. Their preferred use is in gastrointestinal and colon targeting formulations. Eudragit S-100 is used for uniform coating of microspheres.

Composed of polypropylene foam, Eudragit S, ethyl cellulose, and polymethylmethacrylate (PMMA), floating microparticles were prepared by solvent evaporation, and encapsulation efficiency was found to be high. Good floating properties were observed as more than 83% of microparticles were seen to be floating for at least 8 h. In vitro drug release was dependent upon the type of polymer used. At similar drug loading the release rates increased in the following order: PMMA < ethyl cellulose < Eudragit S. This was attributed to the different permeabilities of the drug in the polymers and the drug distribution within the system.

(4) Carbopol Polymers

Carbopol® polymers celebrate 70 years of timeless innovation and inspiration. These polymers of acrylic acid cross-linked with polyalkenyl ethers or divinyl glycol are produced from primary polymer particles of about 0.2~6.0 mm average diameter. The Carbopol polymers are very mild acids and readily react with alkali to form salts. Aqueous dispersions of Carbopol polymers have an approximate pH range of 2.8~3.2 depending on polymer concentration. Since the Carbopol polymers are readily water-swellable, they find applications in a wide range of pharmaceutical products. Carbopol polymers offer a consistent performance over a wide range of desired parameters (from pH-derived semi-enteric release to near zero-order drug dissolution kinetics) at lower concentrations than competitive systems. The polymers are used for oral drug delivery systems. Carbopol is used as a standard bioadhesive agent for buccal, intestinal, nasal, vaginal, rectal, and ophthalmic applications (e.g., Pilopine HS1 pilocarpine hydrochloride 4% ophthalmic gel has carbopol 940 to impart a high viscosity, Alcon, Freiburg, Germany). Moreover, since Carbopol polymers exhibit high swelling and thickening properties at very low concentrations (less than 1%), they are also used in topical lotions, creams and gels, oral suspensions, and in transdermal gel reservoirs.

(5) Pluronic Block Copolymers

Synthetic copolymers of ethylene oxide and propylene oxide (Pluronic™ block copolymers,

BASF, Ludwigshafen, Germany) (Fig.2.11) are used as antifoaming agents, wetting agents, dispersants, thickeners, and emulsifiers. Biodegradable, biocompatible matrices for drug delivery, including films and microspheres, are formed by blending polymers degrading by hydrolysis such as poly(lactic acid) an Pluronic block copolymers.

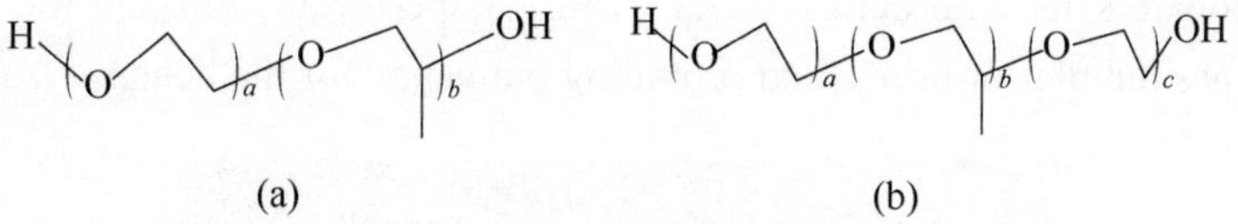

(a) (b)

Fig.2.11 Chemical structure of Pluronic polymer: (a) poly (ethyleneoxide) diblock (b) poly(propyleneoxide) triblock copolymer.

API incorporation into the core of the Pluronic micelles can increase solubility, metabolic stability, and circulation time for the agent. Alakhov et al. studied Pluronic-based megestrol acetate formulations in vitro with respect to drug release and in a rat model for bioavailability of after oral administration of 1~7 mg•kg^{-1}.

An aqueous, micellar formulation comprising a mixture of a hydrophobic (L61) and a hydrophilic (F127) Pluronic copolymer, significantly enhanced the megestrol acetate bioavailability. Moreover, interactions of Pluronic unimers with multidrug-resistant cancer cells can result in sensitization of these cells with respect to various anticancer agents. Animal studies with Pluronic-poly(acrylic acid) (PAA) copolymers have shown that the agents are not absorbed when administered orally. At physiological pH level, the copolymer self-assemble into intra- and intermolecular micelles with hydrophobic cores of dehydrated poly(propylene oxide) and multilayered coronas of hydrophilic PEO and partially ionized PAA segments. These micelles can efficiently solubilize hydrophobic APIs (paclitaxel and steroids), and protect APIs (camptothecin) from the hydrolytic reactions.

(6) Alginates

Alginates (E400-E404) produced by brown seaweeds (mainly Laminaria) are linear unbranched polycarbohydrates containing β-(1→4)-linked D-mannuronic acid (M) and α-(1→4)-linked L-guluronic acid (G) residues (Fig.2.12).

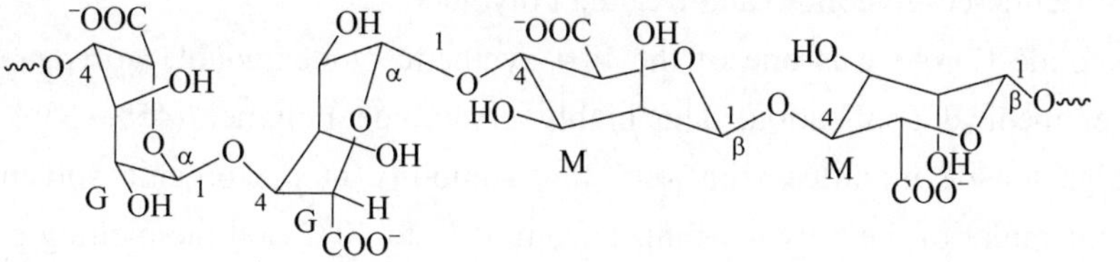

Fig.2.12 Chemical structure of alginate.

Alginates form thermally stable gels in the presence of calcium ions. The formation of gel is at far lower concentrations than any other natural polymer including gelatin. However, the solubility and water-holding capacity of alginate depend on pH (precipitates below pH 3.5), molecular mass, ionic strength and the nature of the ions present.

Alginate beads (Fig.2.13) can encapsulate hydrophobic APIs which allows protection of non-stable APIs and offers potential for drug delivery. Due to enhanced buoyancy and sustained release properties $CaCO_3$-containing beads appear to be excellent candidates for floating systems as shown when incorporating riboflavin and 5-flurouracil. Calcium alginate beads form as alginate undergoes ionotropic gelation by calcium ions and carbon dioxide develops from the reaction of carbonate salts with acid. The evolving gas permeates through the alginate matrix, thereby leaving gas bubbles or pores, and providing buoyancy for the beads.

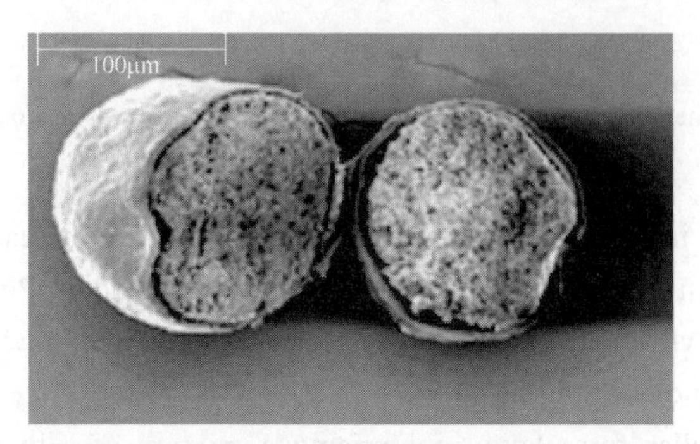

Fig.2.13 Alginate beads provide encapsulation protection and delivery options for hydrophobic drugs, which may come in oily phase.

2.6.2.3 Systemically Applied Polymers

Material scientists continuously wish for new materials and ways to manipulate existing ones in order to fulfill unmet needs. Therapeutic agents linked to polymers such as poly(ethylene glycol) (PEG), poly(lactide-co-glycolide) (PLGA), and *N*-(2-hydroxypropyl) methacrylamide (HPMA) copolymers and polyamines (Table 2.2) are in clinical trails. Dendritic polymers are instrumental in the development of new trends in targeted drug delivery.

Most of the systemic polymers discussed below are used either to form nanoparticles including dendrimers or for prodrug formation of, for example, peptides and proteins for imaging purpose. The polymer architecture and size often affect both pharmacodynamic and pharmacokinetic properties of such a prodrug. Therefore, it is critical to design a prodrug which would release the API at the site of action to avoid adverse effects.

(1) Poly(lactide-co-glycolide) and Related Polymers

Polyglycolide (PGA) was one of the first synthetic biodegradable polymers to be investigated for biomedical application. The highly crystalline polymer (45%~55% crystallinity) exhibits a high tensile modulus with very low solubility in the organic solvents. The glass transition temperature of the polymer ranges from 35℃ to 40℃ and the melting point is greater than 200℃.

Extensive research has been performed to develop a full range of poly(lactideco-glycolide) polymers (PLGA) (Table 2.2). Copolymer composition in the range of 25%~75%, poly(L-lactide-co-glycolide) forms amorphous polymer. 50/50 poly(*dl*-lactide-co-glycolide) degrades in approximately 1~2 months, 75/25 in 4~5 months, and 85/15 in 5~6 months. In addition, poly(lactide) (PLA) and PLGA microspheres are some of the most important components of biopolymers used to design and develop biodegradable microspheres containing bioactive agents for therapeutic application.

Biodegradation and biocompatibility of PLA and PLGA devices. Over the past decade, extensive efforts have been made in the development of poly(*dl*-lactic acid), poly(L-lactic acid) and poly(lactide-co-glycolide) copolymer microsphere CRS. PuraSorb1PLG is a semicrystalline bioabsorbable co-polymer of L-lactide and glycolide with a monomer ratio of 80L:20G .

A co-polymer containing 90% glycolic acid and 10% L-lactic acid was initially used for the development of the multifilament suture Vicryl1. A modified version of the suture, Vicryl Rapid1, which is an irradiated version of Vicryl1 to increase the rate of degradation, is currently in the market. Panacryl1 is another suture from the co-polymer with a higher lactic acid: glycolic acid ratio in order to decrease the rate of degradation.

Microparticles prepared using PLGA are smaller than 10 mm in diameter and are therefore available for uptake by leukocytes, monocytes, macrophages, and other cells of the reticular endothelium system (RES), i.e., liver, spleen, etc. APIs are entrapped into microparticles and nanoparticles by various methods. Efforts have been directed towards understanding plasma protein adsorption of nanoparticles which may facilitate or inhibit phagocytosis by cells of the RES.

(2) PEG Polymers

Poly(ethylene glycol) (PEG) is a linear polyether and is synthesized by an anionic ring opening polymerization of ethylene oxide initiated by nucleophilic attack of a hydroxide ion on the epoxide ring. Polyethylene glycol monomethyl ether (mPEG-OH)-PEGs are now available in many forms including bis functional derivatives (Fig.2.14). The polymer is soluble in both aqueous as well as organic solvents, and is therefore the most preferred candidate in prodrug conjugates. The beginning of PEG chemistry was in 1977 with the findings of Abuchowski, who is considered "the father of PEGylation," and colleagues. They later foresaw the potential of the conjugation of PEG to proteins and small molecules. Abuchowski founded a company, Enzon Inc., which brought three PEGylated drugs to market (the first one contained PEGylated adenosine deaminase). Over the past 20 years the area of PEGylated proteins has expanded dramatically. Higher molecular mass PEGs (M_w > 20,000, especially 40,000) result in increased plasma residence time. The systemic toxicity of an anticancer drug can be reduced by tumor-specific targeting of PEGylated conjugates, and PEGylated adenosine deaminase (Adagen1, Enzon, Bridgewater, NJ) and L-asparaginase (Oncaspar1, Medac, Wedel, Germany) were introduced into the market in the early 1990s. Moreover PEGylated interferon a and G-CSF are approved drugs and PEG-camptothecin (Prothecan) has been subjected to clinical phase II testing.

(3) Dextran

Dextrans are natural macromolecules and consist of linear units with covalently linked $(1\rightarrow6')$ glycopyranose which are branched at α-$(1\rightarrow4')$ position (Fig.2.15). The polyglucose bio-polymer is characterized by α-$(1\rightarrow6')$ linkage with hydroxylated cyclohexyl units and is generally produced by enzymes from certain strains of Leuconostoc or Streptococcus. Dextran contains multiple hydroxyl groups for bioconjugation, and is available in different molecular mass ranges. Dextran has the most compact structure of all polymers, and is another ideal candidate for prodrug formation due to the following properties: water soluble, nontoxic, highly stable glycosidic bonds. Dextran is FDA-approved as a plasma expander; it is also being investigated as a carrier for the passive targeting of tumors and inflamed areas according to an EPR effect. A methylprednisolone prodrug has been prepared by dextran conjugation using succinic acid as a spacer and dextran-conjugated glucocorticoids have been evaluated for colon-specific delivery.

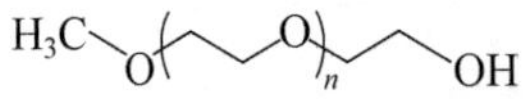

Fig.2.14 Chemical structure of mPEG.

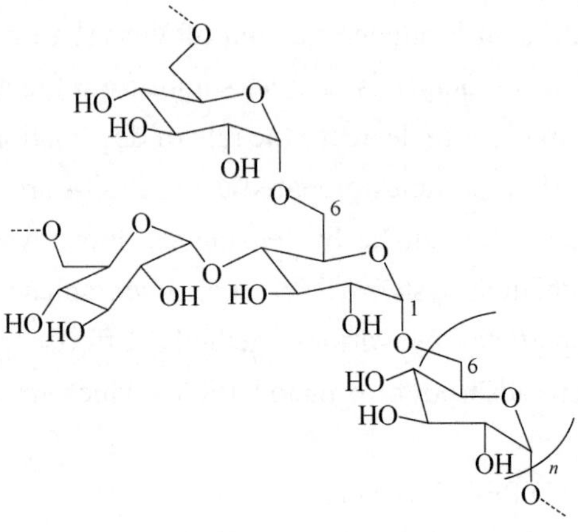

Fig.2.15 Chemical structure of Dextran.

(4) HPMA Copolymers

N-(2-Hydroxypropyl)methacrylamide (HPMA) homopolymer was designed in the 1980s as a plasma expander. The hydrophilic HPMA copolymers increase water solubility of APIs and have proven to be biocompatible, non-immunogenic, and nontoxic. HPMA distribution in the body is well characterized. It has shown great potential as a carrier for targeted delivery of, e.g., antisense oligonucleotides and the controlled release of small molecules.

The HPMA-doxorubicin conjugates (Fig.2.16) were less toxic than the free drug and could accumulate inside solid tumors. The conjugates were synthesized as follows: for subsequent conjugation with a modified antibody, pyridyldisulfanyl groups were incorporated into poly(HPMA) hydrazide using *N*-succinimidyl 3-(2-pyridyldisulfanyl) propanoate. Then the anticancer drug was covalently linked to the remaining hydrazide groups via an acid-labile hydrazone bond. Finally, human immunoglobulin G was modified with 2-iminothiolane by conjugating it to the HPMA polymer through substituting the 2-pyridylsulfanyl groups of the polymer with —SH groups of the antibody.

(5) Poly(anhydrides)

Poly(anhydrides) have been specifically designed and developed for drug delivery applications. These polymers are aromatic poly(anhydrides) based on monomers of *p*-(carboxyphenoxy)propane and *p*-(carboxyphenoxy)hexane as well as aliphatic poly(anhydrides) based on sebacic acid. Poly (anhydrides) are prepared by melt-condensation polymerization, dicarboxylic acid as a starting material, and a prepolymer of a mixed anhydride formed with acetic anhydride.

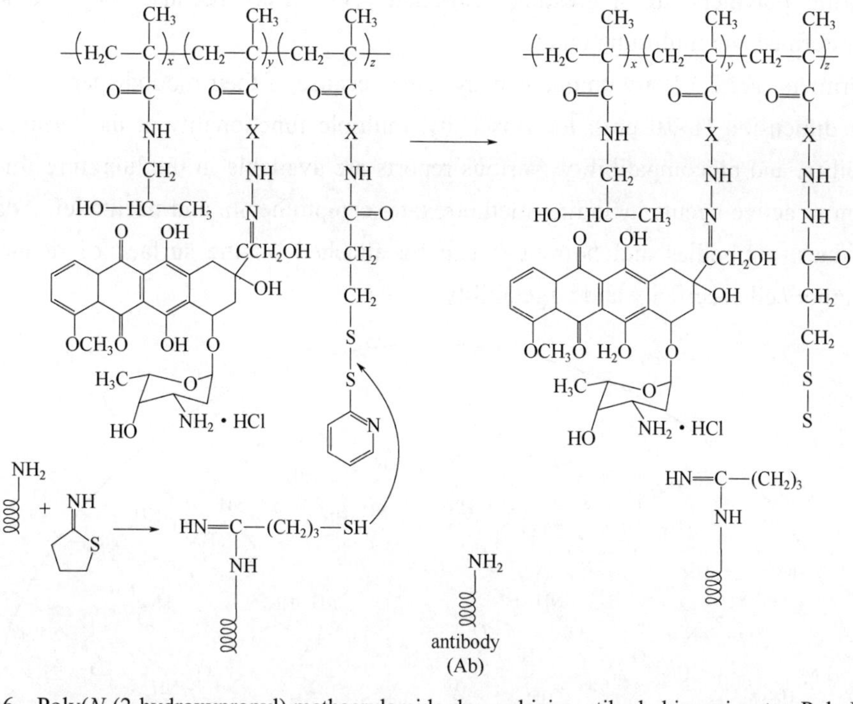

Fig.2.16 Poly(*N*-(2-hydroxypropyl) methacrylamide-doxorubicin-antibody bioconjugates. Poly HPMA hydrazides are modified with *N*-succinimidyl 3-(2-pyridyldisulfanyl) propanoate to introduce the pyridyldisulfanyl groups and for subsequent conjugation with modified antibody (Ab).

Poly(anhydrides) undergo hydrolytic degradation forming water-soluble degradation products which dissolve in an aqueous environment. They also undergo surface erosion due to the high water lability of the anhydride bonds on the surface and the hydrophobicity which prevents water penetration into the bulk. Due to matrix degradation and erosion, the polymers have been widely used for the incorporation of small molecules and proteins, such as insulin, enzymes, and growth factors. An intracranial device of sebacic acid/*p*-(carboxyphenoxy) propane copolymer improves the therapeutic efficacy of a nitrosourea antitumor agent in patients suffering from lethal brain cancer. Poly(anhydrides) possess satisfactory biocompatibility.

2.6.2.4 Current Developments: Dendritic Polymers

In the early 1980s, polymer science research introduced versatile nanosized dendritic polymers, e.g., dendrimers and hyperbranched polymers. The latter are highly branched macromolecules (with 50%~75% branching) with polydispersity index (PDI) typically in the range of 1.5~2.0, but possess a defined chemical structure.

Dendrimers, which are named after the Greek word "dendron" for tree, can be chemically designed and synthesized to possess precise structural characteristics. These versatile polymers are synthesized from monomeric units with new branches being added in steps until a uniform tree-like structure is formed.

Dendritic polymers are interesting drug delivery carriers because they are nanosized, chemically defined and multifunctional.

Dendrimers (Fig.2.17) are unique nanosystems because of their monodispersity (PDI ~1.0), nanometer dimension (1~10 nm), low viscosity, multiple functionality at the terminal groups, high solubility, and biocompatibility. Various reports are available in the literature for dendritic conjugation to active agents including methotrexate, camptothecin, and paclitaxel. Additionally, folate residues, antibodies and hormones can be attached on the surface of dendrimers for potential tumor cell specificity and targetability.

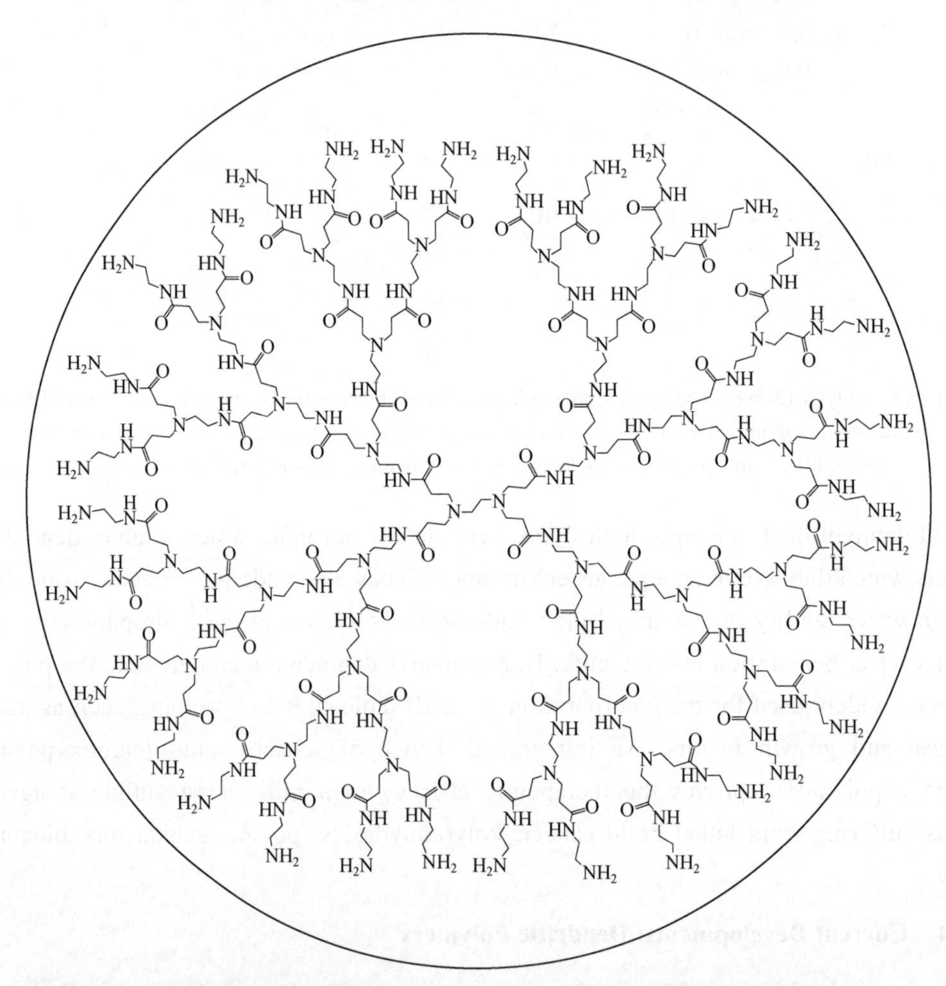

Fig.2.17 PAMAM dendrimer (generation 4) used in biological applications.

Dendritic polyglycerols are aliphatic polyethers and possess multiple functional hydroxyl end groups (Fig.2.18). They can either be prepared as perfect dendrimer by stepwise synthesis or as hyperbranched polymers in kilogram scale. Since dendritic polyglycerols are synthesized in a controlled manner to obtain definite molecular mass and narrow molecular polydispersity (~1.7) , they have been evaluated for a variety of biomedical applications as delivery vehicles for nonpolar APIs , heparin analogs, and multivalent selectin ligands.

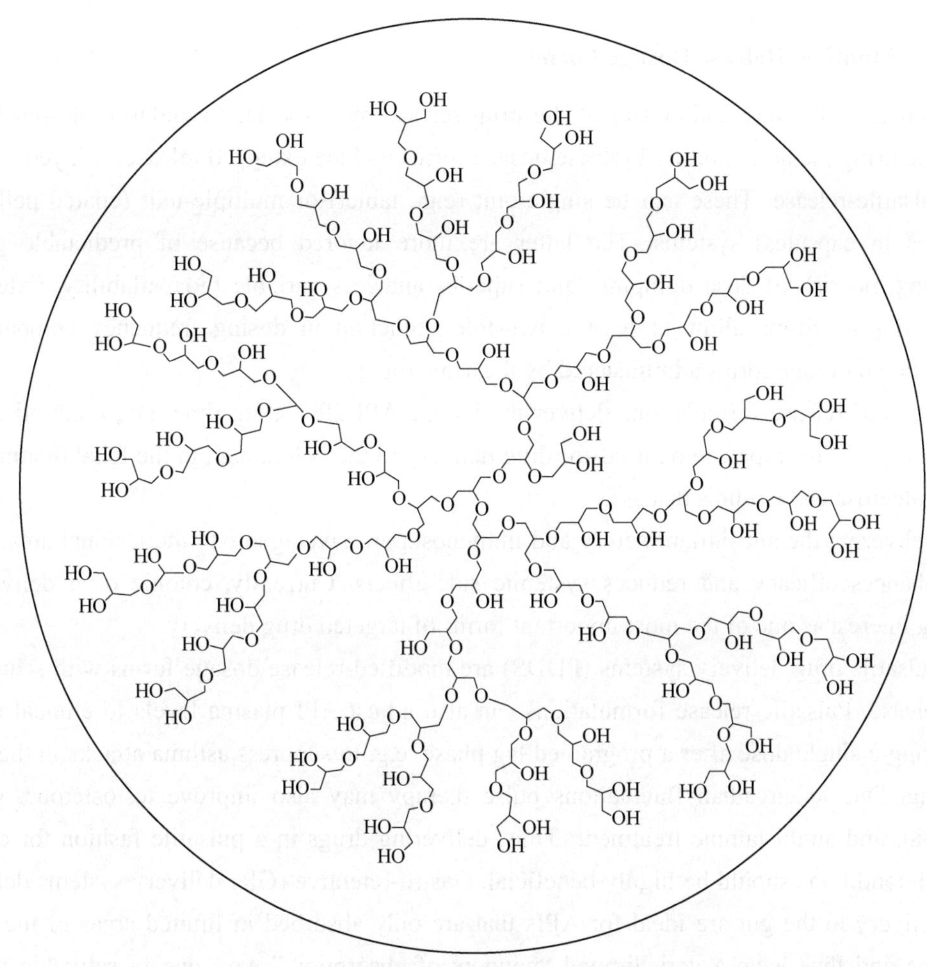

Fig.2.18 Dendritic polyglycerols as new highly biocompatible polymeric scaffolds for delivery and multivalency.

2.6.3 Specific Aspects of Polymers in Oral Drug Delivery

In the following section we describe specific delivery systems for oral use. A challenging aspect in CRS development is to provide prolonged zero-order kinetics for the total drug release with no time lag or burst effect. Such release profiles can optimize the dosage frequency in drug therapy.

With oral treatment, coated pellets or particles in capsules and compressed tablets are cost-effective; however, the advanced matrix systems, reservoir systems and osmotic pumps systems are currently proving to be the most efficient. Moreover gastro-enteric-coated forms are designed to protect the API from degradation by gastric pH and enzymes and to prevent nausea and vomiting caused by an API irritating the gastric mucosa. Many APIs also require protective coatings to improve patient compliance by masking unpleasant taste and odor. It has to be noted, however, that pharmaceutical dosage forms can irritate the gastrointestinal mucosa due to the chemical properties of the polymers in the form of external coating.

2.6.3.1 Modified Release Dosage Forms

Imparting the rate and/or site of the drug release by a special formulation design and/or manufacturing method, modified release dosage forms include extended release, delayed-release, and pulsatile-release. These can be single-unit (e.g., tablet) or multiple-unit (coated pellets or particles in capsules) systems. The latter are more favored because of predictable gastric emptying, no risk of dose dumping, and superior and less variable bioavailability. Extended-release dosage forms allow at least a two-fold reduction in dosing frequency compared to conventional dosage forms administered by the same route.

Delayed release formulations deliver the loaded API after a lag time, i.e., a period of "no drug release." This type can be used for drug delivery to the colon, e.g., in the local treatment of colitis ulcerosa and Crohn's disease.

Delivering the anti-inflammatory and immunosuppressive agents to the lesional area of the gut enhances efficacy and reduces systemic side-effects. Currently, colonic drug delivery is gaining interest as one of the most important forms of targeted drug delivery.

Pulsatile drug delivery systems (PDDS) are modified-release dosage forms with sequential API release. Pulsatile release formulations can also adjust API plasma levels to clinical needs, delivering a single dose after a programed lag phase, e.g. to suppress asthma attacks in the early morning. Due to circadian fluctuations pulse therapy may also improve testosterone, gluco-corticoid, and antihistamine treatment. Thus, delivering drugs in a pulsatile fashion for certain clinical conditions should be highly beneficial. Gastro-retentive (GR) delivery systems delaying API delivery to the gut are ideal for APIs that are only absorbed in limited areas of the small intestine and thus have a very limited "window of absorption," e.g., due to saturable uptake mechanism.

2.6.3.2 Gastro-enteric Coatings

Advantages of enteric coating polymers are good storage stability of the API, pH-dependent drug release, protection of actives sensitive to gastric fluid and subsequent increase in drug effectiveness, and protection of gastric mucosa from aggressive actives. Gastrointestinal and colon targeting are possible. Table 2.3 summarizes drug delivery sites for absorption of actives at pH 5.5~7.0 referring to Eudragit polymers. Prozac Weekly™ capsules (Eli Lilly, Bad Homburg,

Table 2.3 Drug delivery sites in gastrointestinal tract, polymer grades

Drug delivery site	Polymer dissolution at pH	Eudragit®, grade and form
Duodenum	Above pH 5.5	L30 (powder) D-55 (30% aqueous dispersion)
Jejunum	Above pH 6.0	L100 L (powder) S12,5 (12.5% organic solution)
Ileum	Above pH 7.0	S100 (powder) S12,5 (12.5% organic solution) FS30D (30% aqueous dispersion)
Colon delivery	Above pH 7.0	FS30D (30% aqueous dispersion)

Germany), a delayed release formulation, contain enteric-coated pellets of fluoxetine hydrochloride (90 mg fluoxetine). Bioavailability following a single 40 mg oral dose achieves peak plasma concentration of fluoxetine 15~55 ng•mL^{-1} after 6~8 h (FDA 2006a).

2.6.3.3 Matrix Systems

Matrix systems, as the most common controlled delivery system, are tablets and granules. The system is preferred because of its effectiveness as well as ease of manufacturing and thus low cost. It is typically formulated into a once-daily dosage form which contains homogeneously dissolved or dispersed API.

A solidifying matrix dissolvable or dispersible in solvents allows for a controlled release. This applies also to APIs with high aqueous solubility that need to be applied in higher doses. With conventional delivery systems, these agents become almost instantaneously available and toxic plasma levels early after intake are soon followed by sub-therapeutic ones, given that the API is eliminated rapidly from the organism. In contrast, matrix systems can maintain constant plasma levels. Importantly, matrix systems can also be formulated to avoid interactions of API and food components and the advanced systems can be taken irrespective of the meal pattern.

Delayed release is achievable by the use of an erodible monolith, which delivers the actives in the lower gastrointestinal tract (FDA 2006b). Controlled release becomes possible by swelling and eroding of the monolith, and then the API is continuously released throughout the gastrointestinal tract.

Multilayered matrix tablets are developed which can provide multiple release kinetics of a single API or a combination of two (or more) APIs having identical or different physicochemical properties. Burst release by immediate disintegrating of monolith layer delivers the loading dose required to achieve active plasma concentration. Then a second layer slowly releases the API at a controlled rate as described above. By physical separation a physical/chemical barrier can also avoid the incompatibility between two actives, the excipients and active-excipients interactions. A well-known example of such an interaction is the Millard reaction that occurs during tablet compression.

With the commercially available multiple layer matrix system Cipro XR1 (500~1,000 mg daily; Bayer Pharmaceuticals, Leverkusen, Germany) extended ciprofloxacin release from the tablets becomes possible by a barrier layer which separates the core and the dissolution medium. Approximately 35% of the dose is present within an immediate-release component, while the remaining 65% is embedded in a slowrelease matrix. In the case of Cipro XR, maximum plasma ciprofloxacin concentrations are attained between 1 h and 4 h after dosing, bioavailability is close to availability following ciprofloxacin immediate release when applying identical doses (FDA 2004). Multilayered matrix tablets can also be used to generate repeat-action products.

One layer of the tablet or the outer layer of the compressed coated tablet provides the initial dose by rapid disintegration in the stomach. The inner tablet is formulated with components that are insoluble in gastric media but release API in the intestinal environment. The release profile can be further improved by a degradable barrier layer, a commercially available multilayered matrix system is PAXIL CR1 (GlaxoSmithKline, Munich, Germany) containing the psychotropic agent paroxetine hydrochloride.

2.6.3.4 Reservoir Systems

With matrix systems, an unwanted burst effect which leads to higher initial drug delivery and reduction of the effective lifetime of the device may not be eliminated with all APIs, necessitating more advanced systems. In polymeric reservoir systems offering a defined resistance to drug diffusion from the reservoir to sink, the driving force is the concentration gradient of the API molecules between the reservoir and sink. Such systems are capable of presenting a linear release pattern. A reservoir tablet system consists of a semi-permeable barrier which is involved in API release from a core site within the tablet. Commonly used methods include coating of tablets or multiparticulates, microencapsulation of drug particles, and press coating of tablets.

There are several advantages and disadvantages associated with reservoir systems. The advantages include linear release of the API which is independent of the solubility of the active agent in the medium as well as pH level and ion strength of the medium. The disadvantages include the complex manufacturing process, rapid transit of drug levels from the reservoir, low capacity for drug loading and eventual incomplete release of the actives. Complexity of design, manufacturing processes, and the need for specialized equipment may all result in an unbalanced cost–benefit ratio for these systems.

2.6.3.5 Osmotic Pump Systems

Osmotic pump systems consist of a tablet sealed by a semi-permeable membrane with an orifice. The elementary osmotic pump generally has a single-layer core containing the API (typically water-soluble) enclosed in a semi-permeable membrane with one or more laser-drilled orifices. This is the one-chamber elementary osmotic pump system.

In the gastrointestinal tract, water is drawn in through the membrane at a controlled rate, gradually dissolving the active ingredient. By the increased pressure, the resulting API solution flows out through the orifice at the same rate that water is flowing in through the membrane. Examples of products utilizing this system are Acutrim (phenylpropanolamine hydrochloride, Norvartis, Nuernberg, Germany), Efidac (pseudoephedrine hydrochloride, Hogil, White Plains, NY, USA), and Volmax (albuterol sulfate, Merck, Darmstadt, Germany; Santus and Baker 1995).

There is also a second type of osmotic pump systems, the two-chamber osmotic pump system, also called the push and pull system, which is typically used when the API has limited aqueous solubility. In this system, the semi-permeable membrane encloses a two-layer core, one layer containing the active ingredient and the second layer containing a water-swellable osmotic agent (Fig.2.19). As water flows into the core through the rate-controlling membrane, the osmotic agent expands which causes the dissolved API to be pushed out through the laser-drilled orifice(s).

Examples of products utilizing the push-pull system are Procardia XL (nifedipine, Pfizer, Berlin, Germany) and Minipress XL (prazosin hydrochloride, Pfizer, Berlin, Germany).

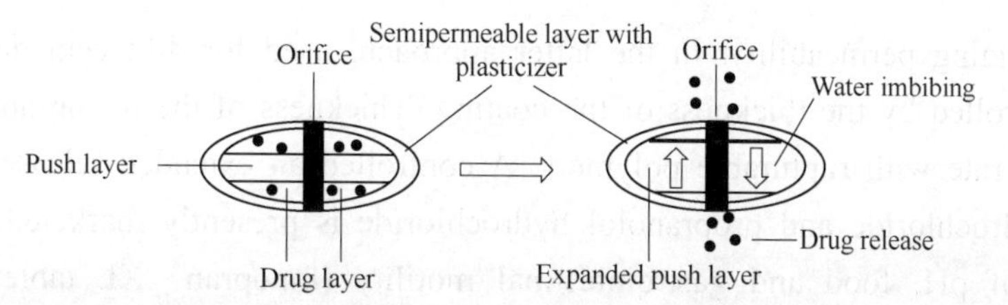

Fig.2.19 Multilayered matrix push layered tablets consisting of polymeric plasticizers.

The weight of the semi-permeable membrane (increased membrane thickness, as measured by weight) slows release rate of the API. According to the FDA, it is important for an applicant developing a gastrointestinal tract dosage form to determine and carefully control the weight of the semi-permeable membrane during the coating operation. This consideration is important in preapproval inspections (FDA 2008a).

Among various pulsatile delivery systems, single-unit pulsatile systems are popular and consist of sub-classes:

① Capsule-based systems (Pulsincap 1, Scherer International Corporation, Michigan, US);

② Osmotic systems (Port 1 System, Therapeutic System Research Laboratory, Ann Arbor, Michigan);

③ Delivery systems with soluble or erodible membranes (chronotropic) consisting of a core containing drug reservoir coated by a hydrophilic polymer (e.g., HPMC);

④ Reservoir systems with a rupturable coating (e.g., soft gelatin or HPMC).

The length of a swellable hydrogel plug and its point of insertion into the capsule controls the lag time of the Pulsincap® system. Making contact with the dissolution fluid, the hydrogel (e.g., HPMC, PMMA, PVA, PEO) swells and the plug sealing the drug content within the capsule is pushed outside, rapidly releasing the API. The length of the plug and its point of insertion into the capsule controls the lag phase. Pulsincap® is reported to be well tolerated in human volunteers.

The osmotic Port® system consists of a capsule coated with a semipermeable membrane. The capsule contains a plug of insoluble yet osmotically active compound and the API. Water entry via the semipermeable membrane increases the pressure within the capsule and after a lag time the plug is expelled. The Port 1 system is set up for methylphenidate dosing in children with attention deficit hyperactivity syndrome to avoid a second dosing during daytime. Alternatively, pulsatile methylphenidate release (three pulses) becomes possible from dosage units forming tablets. Coating is based on erodible polymers.

Pulsys® (MiddleBrook Pharmaceuticals, former Advancis Pharmaceutical Corp.) is a once-a-day delivery system delivering three regular pulses of amoxicillin for an optimal bactericidal effect. Preclinical studies have demonstrated its superior efficacy; publication of results of comparative clinical testing in relation to standard therapy is to be awaited. The product is commercially available (Moxatag™ tablet).

The most reliable gastric emptying is possible with pellet formulations as these are distributed freely in the gastrointestinal tract. The pellets are coated with rupturable polymers or polymers changing permeability. In the latter approach, used for diltiazem delivery, delivery rates are controlled by the thickness of the coating. Thickness of the membranes also controls drug delivery rate with rupturable polymers. A controlled an extended release formulation of verapamil hydrochloride and propranolol hydrochloride is presently marketed; release rate is independent of pH, food and gastrointestinal motility (Innopran® XL tablet, MiddleBrook Pharmaceuticals, Inc.).

References

[1] H. Handa, T. C. Major, E. J. Brisbois, K. A. Amoako , M. E. Meyerhoff, R. H. Bartlett. (2014) Hemocompatibility comparison of biomedical grade polymers using rabbit thrombogenicity model for preparing nonthrombogenic nitric oxide releasing surfaces. J. Mater. Chem. B. Mater. Biol. Med. 2, 1059~1067.

[2] T. Fujiwara, E. Nagaoka, T. Watanabe, N. Miyagi, T. Kitao, D. Sakota, T. Mamiya, T. Shinshi, H. Arai, S. Takatani. (2013) New generation extracorporeal membrane oxygenation with MedTech Mag-Lev, a single-use, magnetically levitated, centrifugal blood pump: preclinical evaluation in calves. Artif. Organs. 37, 447.

[3] N. Pandis, P. S. Fleming, D. Kloukos, A. Polychronopoulou, C. Katsaros, T. Eliades. (2013) Survival of bonded lingual retainers with chemical or photo polymerization over a 2-year period: a single-center, randomized controlled clinical trial. Am. J. Orthod. Dentofacial. Orthop. 144, 169.

[4] A. Goyal, J. Hurkadle, S. Magegowda, P. Bhatia. (2013) Use of light-curing units in orthodontics. J. Investig. Clin. Dent. 4, 137.

[5] W. K. Wan Abdul Khodir, V. Guarino, M. A. Alvarez-Perez, C. Caiero, L. Ambrosio. (2013) Trapping of tetracycline loaded nanoparticles into PCL fibre networks in periodontal regeneration therapy. J. Bioact. Compat. Polym. 28, 258.

[6] L. R. Pires, V. Guarino, M. J. Oliveira, C. C. Ribeiro, M. A. Barbosa, L. Ambrosio, A. P. Pego. (2013) Loading poly(trimethylene carbonate-co-ε-caprolactone) fibers with ibuprofen towards nerve regeneration. J. Tissue Eng. Reg. Med.

[7] V. Guarino, M. A. Alvarez-Perez, A. Borriello, T. Napolitano, L. Ambrosio. (2013) Conductive PEGDA/pani macroporous hydrogels for nerve regeneration. Adv. Healthc. Mater. 2, 218.

[8] M. P. Lutolf, J. A. Hubbell. (2005) Synthetic biomaterials as instructive extracellular microenvironments for morphogenesis in tissue engineering. Nat. Biotechnol. 23, 47.

[9] S. P. Zustiak, J. B. Leach. (2010) Hydrolytically degradable poly(ethylene glycol) hydrogel scaffolds with tunable degradation and mechanical properties. Biomacromolecules. 11, 1348.

Chapter 3

Polymers as Drugs

Polymeric drugs are defined as polymers that are active pharmaceutical ingredients, i.e., they are neither drug carriers nor prodrugs. In general, the underlying concept behind these therapeutic agents is the utilization of high molecular mass and functional characteristics of polymers to selectively recognize, sequester, and remove low molecular mass and macro-molecular disease causing species in the intestinal fluid. The high molecular mass nature of these therapeutically relevant polymers makes them systemically non-absorbed, thus providing a number of advantages including long-term safety profiles over traditional small molecule drug products. Furthermore, multiple functional groups in the polymers incorporate polyvalent binding interactions that can result in pharmaceutical properties not found in small molecule drugs. This chapter summarizes some of the most recent efforts for the discovery and development of polymeric drugs that have proceeded from discovery phase to market place. Examples include sequestration of low molecular mass species such as bile acids, phosphate, and iron ions as well as polyvalent interactions to bind toxins, viruses, and bacteria as well as polymeric enzyme inhibitors and fat binders as anti-obesity agents. Furthermore, use of functional polymers to treat autoimmune disease and sickle cell anemia has been reviewed.

3.1 Introduction

During the last three decades, the role of polymers in biomedicine has seen significant growth. The unique physico-chemical properties offered by polymeric materials have been exploited in a variety of biomedical applications. A wide range of functional polymers with novel structural architectures have been synthesized during this period and have been evaluated in biological environments. The use of polymers materials in drug delivery and as components in artificial organs, tissue engineering, medical devices, and dentistry is well known. However, an increasingly important aspect of the field of biomedical polymers is the recognition of the role of polymers as new and novel chemical entities for therapeutic application. For this purpose, the polymer may be intrinsically bioactive, or can be utilized as a carrier for site specific and sustained delivery of chemo- and biotherapeutic agents.

Following the original work of Ringsdorf presenting the concept of site targeted polymeric drugs, a large body of scientific literature has appeared, affirming the role of functional polymers as vehicles for therapeutic agents against a variety of diseases. These polymer-drug conjugate systems have enabled the delivery of small molecule drugs and biotherapeutic agents at a controlled rate, and have achieved the targeted delivery of chemotherapeutic agents to specific sites (e.g. to minimize dose-dependent toxicity and enhance selectivity for anti-neoplastic agents). More recently, several functional polymers have been evaluated as non-viral vectors for the delivery of genetic materials for gene therapy applications. The research efforts in the area of polymer-based drug delivery systems, including polymer-drug conjugates, has brought about considerable progress in this area including clinically approved products. Numerous high-quality research papers and excellent review articles dealing with this aspect of biomedical polymers have been published over the last two decades. Since several books and review articles pertaining to polymeric drug delivery systems and polymer-drug conjugates have been published, this article has been limited to the review of biomedical polymers that act as active pharmaceutical ingredients.

While functional polymers as carriers for therapeutic agents have been extensively studied, examples of polymers acting as active pharmaceutical ingredients are relatively scant. Early studies pertaining to the use of polymers as therapeutic agents include the use of poly(ethylenesulfonate) as a topical anticoagulant and as an antitumor agent. Although this polymer had shown anti-tumor activity against a broad selection of tumor cell lines, it lacked an acceptable therapeutic index. Almost three decades ago a relatively simple class of polymers, maleic anhydride-divinyl ether copolymer, was studied for their effect on tumor cell lines. It was demonstrated that these anionic polymers are able to modulate the immune system by stimulating T-cell activity. Although the therapeutic effects of anionic polymers were found to be modest, they certainly foretold the interesting adaptations of polymers to such complicated disease states as cancer. In spite of these early promises, effective research effort to develop intrinsically bioactive polymers as therapeutic agents is a relatively recent phenomenon. As potential therapeutic agents, the high molecular mass characteristics of polymers would appear to offer several advantages over classical small molecule drug candidates. Although polymers do not fit most “drug-like” definitions and inherently violate “Lipinski’s Rules”, a re-evaluation of the attributes of polymers highlights many possible benefits of polymeric pharmaceuticals that are unattainable with traditional small molecule drugs. These benefits may include: lower toxicity, greater specificity of action, and enhanced activity due to multiple interactions with disease targets (polyvalency). In spite of these potential benefits, the concept of polymeric drugs has been a subject of considerable skepticism among drug discovery and development scientists. As a class of potential pharmacophores for drug development, synthetic polymers have been thought to be uninteresting by medicinal chemists and regulatory authorities. Some of the underlying concerns attributed to polymers as new chemical entities for therapeutic applications include the issue of broad molecular mass distribution (polydispersity) and compositional and structural (microstructure including stereochemistry) heterogeneity. These shortcomings were considered to impede drug development and regulatory approval. Furthermore, the high molecular mass characteristics of polymers would potentially undermine their systemic absorption through oral administration (oral bioavailability). However, these seemingly detrimental pharmacological features of polymers can indeed be exploited to design and develop novel therapeutic agents for disease conditions where low molecular mass drugs have either failed or exhibited inadequate therapeutic benefits.

The systemic non-absorption characteristics of high molecular mass polymers taken through the oral route may offer therapeutic benefits where it is desirable to minimize (even stop) systemic exposure of drug substances. For example, by combining this non-systemic bioavailability characteristic with the ability of macromolecular sorbents to selectively recognize and sequester molecular and macromolecular components in gastrointestinal fluids, it has become possible to develop a powerful new class of therapeutic agents that can selectively bind and remove detrimental species (attributed to several disease indications) from the gastrointestinal (GI) tract. As with any pharmaceutical agents, sequestration of each target molecule or pathogen requires a unique strategy and this strategy depends not only on the chemical nature of the target, but the location, concentration, and quantity of the target to be removed at a therapeutically acceptable dose.

Some of the most recent efforts in the area of discovery and development of polymers as therapeutic agents are reviewed in the present article. The case studies presented in this article concentrate on the development of polymeric drugs for the sequestration of low molecular mass species such as bile acids, phosphate, and iron as well as polyvalent interactions to bind toxins, viruses, and bacteria. Furthermore, the use of polymers as immunomodulating agents to treat autoimmune diseases by inhibiting specific biological antigen binding events is highlighted. Recent developments in these areas of polymeric drugs including examples of approved and marketed pharmaceutical products are presented.

3.2 Polymers for Molecular Sequestration

A number of potentially detrimental substances implicated for various disease states, are present or circulate in the GI tract. These species can either enter the body with food, or from the environment (exogenic), or they may be produced in the human body as a result of metabolism (endogenic). Effective removal of these detrimental species from the GI tract in a selective manner offers a promising approach to treat a number of diseases. Thus, biologically appropriate and non-toxic polymeric sorbents represent an ideal class of therapeutic agents for this purpose. The non-absorption of these polymeric resins through the intestinal wall would deter systemic exposure and should result in minimal toxicity. By incorporating appropriate functional groups and by modulating the physicochemical characteristics of polymers, a wide range of possibilities to tailor-make polymers with high selectivity and capacity for targeted species are conceivable. The following sections provide an overview of a number of polymeric sequestrants that have been discovered and developed during the last few years as therapeutic agents to treat human diseases.

3.2.1 Polymeric Drugs for the Sequestration of Inorganic Ions

The process of electrolyte homeostasis is the key to critical physiological functions such as myocardial and neurological functions, fluid balance, oxygen delivery, and acid-base balance. Perturbation of this delicate electrolyte balance by either excessive ingestion or impairment of the elimination process due to dysfunctional metabolism can bring about detrimental pathological consequences. Although certain non-renal tissues like muscle and liver contribute to maintaining the electrolyte balance, the kidney is the primary organ responsible for maintaining electrolyte homeostasis. As a result, impairment of renal function is the main factor responsible for electrolyte imbalance and can result in life-threatening metabolic disorders. Thus, the use of non-absorbed polymeric sequestrants to bind these excessive ions (implicated in various pathologic conditions) selectively in the GI tract is, in principle, a novel approach to treat or prevent disease conditions associated with electrolyte imbalance.

3.2.1.1 Polymeric Sequestrants for Potassium Ions: A Treatment for Hyperkalemia

Potassium ions play a critical role in regulating the transmembrane potential for cellular functions. Therefore, maintenance of the critical ratio of potassium ions between intra- and extracellular fluids is important to all living cells. Hyperkalemia (elevated level of serum potassium, usually greater than 5.0 $mEq•L^{-1}$) can result from burn and crush muscle injuries, acidosis, or through the use of anti-hypertension drugs like angiotensin-converting enzyme inhibitors. A rise in serum potassium can manifest moderate to serious health problems such as paresthesias, areflexia, respiratory failure, and bradycardia. Since the kidney is the primary route to remove excessive body potassium, patients with impaired renal function are incapable of maintaining potassium homeostasis. Traditional approaches to treat hyperkalemia include the use of insulin, glucose, sodium bicarbonate, and calcium chloride. However, these treatments have their own shortcomings. For example, excess intake of calcium leads to hypercalcemia, which in turn leads to other complications like myocardial infarction and kidney stones.

Use of an insoluble anionically charged polymer resin to sequester excess potassium ions in the GI tract and their subsequent elimination in the feces is a simple approach that is being used to treat this disease. This approach enables patients to remove excess potassium from their bodies in spite of impaired kidney function. A number of low molecular mass ligands that complex potassium ions are known in the literature. Utilizing this principle of potassium ion binding by organic ligands, a cation-exchange resin based on sodium polystyrene sulfonate (**Scheme 1**) was developed to sequester potassium ion in the GI tract. This polymer, marketed under the brand name Kayexalate has been approved in the United States for the treatment of hyperkalemia since 1975. However, a potential problem associated with the use of Kayexalate is induction of hypernatremia (elevated serum sodium), since the polymer resin exchanges 1.0 equivalent of potassium for 1.5 equivalents of sodium. Furthermore, cases of intestinal necrosis have been attributed to Kayexalate. The adverse effects associated with this polymeric potassium sequestrant provide an opportunity to develop new generations of potassium sequestering polymers. Unfortunately, no new efforts have been made to develop second generation sequestration therapies to treat hyperkalemia.

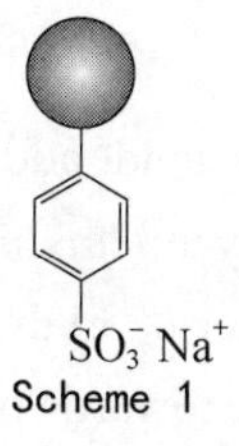

Scheme 1

3.2.1.2 Sequestration of Phosphate Ions: Polymeric Drugs for Chronic Renal Failure

The management and control of elevated levels of serum phosphate (hyperphosphatemia) is a critical health concern for patients suffering from chronic and end-stage renal disease. Consequences of inadequate control of serum phosphate level can manifest a number of pathologies of clinical significance.

These include soft tissue calcification (leading to cardiac calcification and cardiac-related complication), renal bone disease leading to reduced bone density, and secondary hyperparathyroidism. These detrimental factors make hyperphosphatemia a major risk factor for mortality among patients suffering from end-stage renal disease (e.g., dialysis patients).

The kidney is the primary route for the excretion of phosphate from a healthy human body. Therefore, patients with impaired renal function accumulate phosphate as a result of an imbalance between ingestion and excretion of dietary phosphate. Phosphate binder therapy has been the mainstay for the treatment of hyperphosphatemia. The traditional phosphate binders have been calcium- and aluminum-based agents. These inorganic cations remove phosphate through the formation of corresponding insoluble phosphate salts in the GI tract. Since aluminum and calcium salts have the propensity for systemic absorption through the intestinal mucosal layer, prolonged treatment involving these agents carry the liability of bringing about undesirable toxic and metabolic side effects (such as neurological disorders, cardiac calcifications, etc.) in renal-compromised patients. These shortcomings of calcium- and aluminum-based binders limit their long-term use as phosphate sequestrants.

Non-absorbed cationic polymers as sequestrants for phosphate ions offer an effective and safe approach to treat hyperphosphatemia in renal failure patients. The binding of phosphate ions by polycationic species is a well studied phenomenon in molecular recognition and supramolecular chemistry research . Towards this end, a great deal of research efforts have been made over the past several years to design and synthesize novel compounds such as oligomeric/macrocyclic amines, ammonium salts, and guanidinium compounds (e.g., **Schemes 2-4**) as synthetic receptors for phosphate and related anionic guests (viz. phosphate, pyrophosphate, and phosphonate anions). Electrostatic interaction is the primary driving force for complexation of phosphate-based anions with these organic cationic hosts.

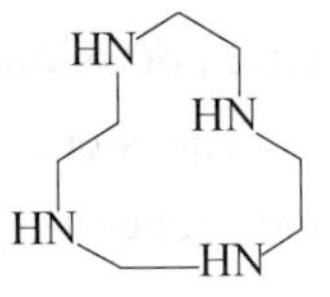

Scheme 2 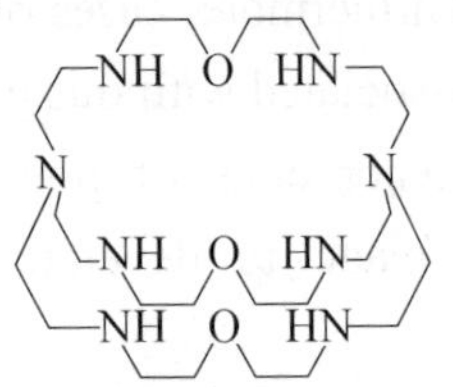 Scheme 3 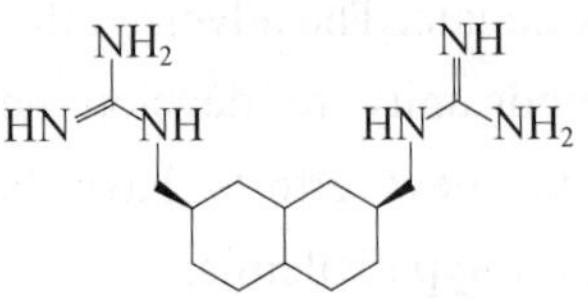 Scheme 4 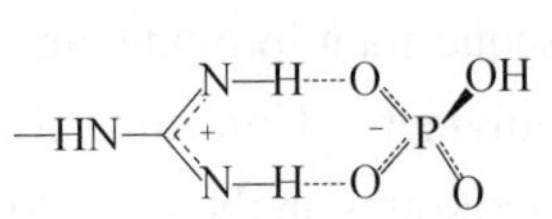 Scheme 5

Hydrogen bonding is considered to lend additional binding strength. The underlying principle of the physical organic chemistry of this anion recognition process was recruited to discover non-absorbed, cationic, polymeric hydrogels (such as polymeric amines and guanidinium compounds) that show affinity towards phosphate ions derived from dietary sources. Being non-absorbed, these polymeric sequestrants are confined to the GI tract. Therefore, they act as effective therapeutic agents (free from the side effects associated with calcium and related metal salt-based phosphate binders) to treat hyperphosphatemia.

Binding of phosphate ions to guanidinium groups of arginine residues of proteins (involving two electrostatic bonds and two stereochemically favorable hydrogen bonds, structure Scheme 5) is a well-known and important phenomenon in biological systems. This biological principle of phosphate recognition was utilized by Hider and Rodriguez in designing and synthesizing a series of insoluble polymers containing guanidinium groups as sequestrants for phosphate anions . The synthetic procedure employed to prepare these guanidinium-functionalized polymer resins is illustrated in Fig.3.1. Phosphate binding studies involving these polymeric guanidinium salts under in vitro conditions has shown that these polymers bind phosphate selectively in the presence of other competing biologically important anions such as chloride, bicarbonate, etc.

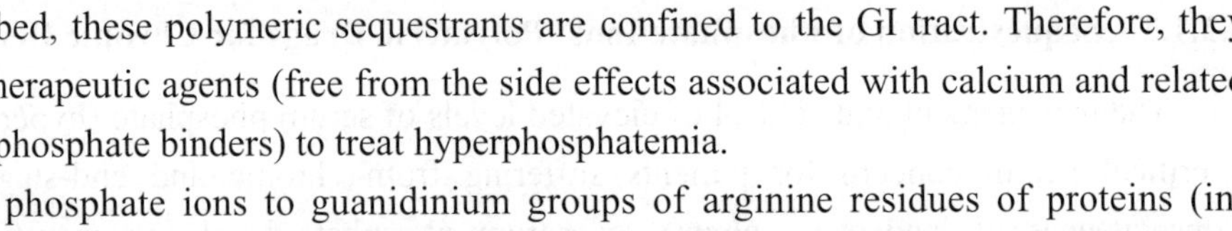

Fig.3.1 Synthesis of polymeric guanidinium salts through chemical modification of poly(acrylonitrile) resin.

A series of amine containing, polymeric phosphate sequestrants were discovered and systematically investigated. This work finally led to a marketed drug for the treatment of hyperphosphatemia (vide infra). Using the knowledge of phosphate binding properties of macrocyclic polyamines and related compounds as anion receptors (described above), a series of amine containing functional polymers were synthesized. These amine-functionalized polymeric hydrogels were prepared either by postpolymerization cross-linking of polymeric amines, or by the crosslinking copolymerization of appropriate amine containing vinyl monomers. By these processes a variety of polymer structures bearing primary, secondary, tertiary amine groups, as well as quaternary ammonium groups were obtained. A general procedure for the preparation of polymeric amine-based hydrogels by the crosslinking of soluble polymers is illustrated in Fig.3.2.

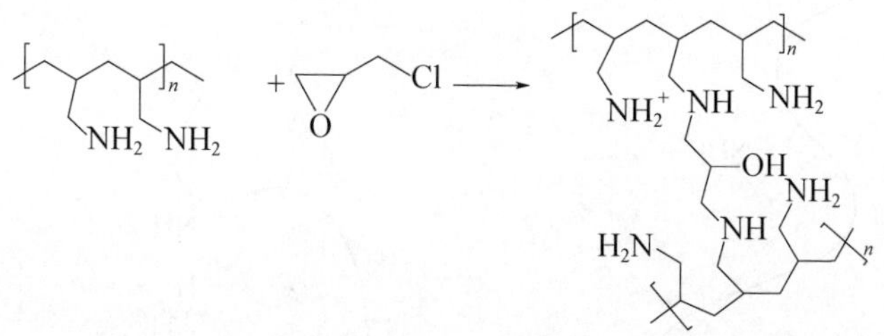

Fig.3.2 The general synthetic procedure to prepare amine-functionalized hydrogels by a post-polymerization crosslinking reaction.

The main structural repeat units of a selection of representative polymeric amines used are summarized in Table 3.1. The in vitro phosphate binding properties of some of these polymeric amine-based hydrogels are given in Table 3.2. A systematic structure-activity relationship (SAR) study enabled us to identify epichlorohydrin crosslinked polyallylamine (**Scheme 6**) as the lead, which was advanced to preclinical and clinical development. These polymers are believed to bind phosphate ions through electrostatic and possibly through hydrogen bonding interactions (See Fig.3.3).

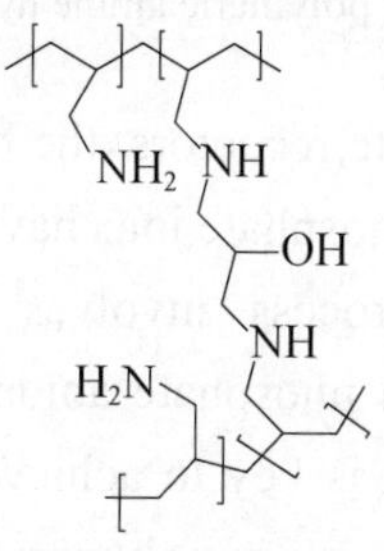

Scheme 6

Table 3.1 Structural repeat units of representative polymeric amine precursors used to prepare phosphate sequestrants

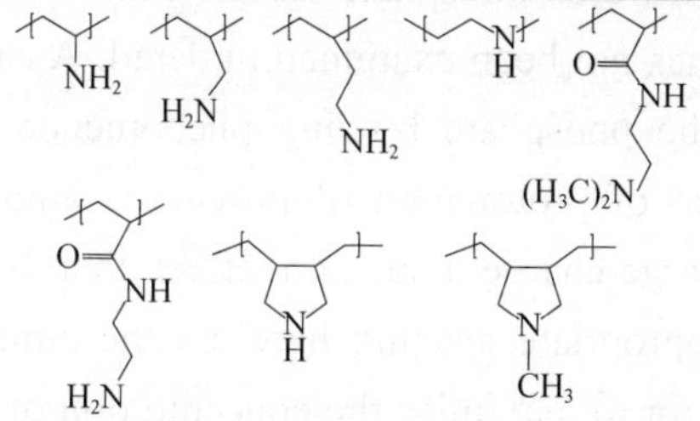

Table 3.2 In vitro equilibrium phosphate-binding properties of amine-functional hydrogels at 5 mmol•L^{-1} Phosphate

Nature of polymeric amine	Phosphate bound to polymer/meqv•g^{-1}
Polyallylamine/epichlorohydrin	3
Polyethyleneimine/acryloyl chloride	1.2
Diethylenetriamine/epichlorohydrin	1.5
Poly(dimethylaminoethylacrylamide)	0.8
Poly(4-trimethylammoniummethyl)styrene chloride	0.7

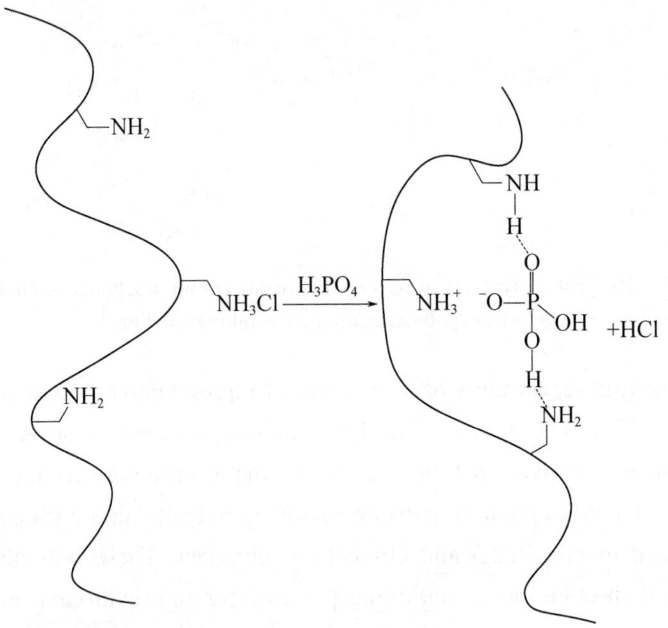

Fig.3.3 Binding interactions between polymeric amine hydrochloride gels and phosphate anions.

Like low molecular mass phosphate receptors, the binding strengths and capacities of these polymeric phosphate sequestrants for phosphate ions have been found to depend on the pH of the medium. Since the complexation process involves electrostatic interaction between the polymer-bound ammonium groups and phosphate anions, the optimum level of protonation of amine groups along the polymer chain is key to achieving maximum binding capacity. Due to higher local concentrations of amine groups in polymeric systems, and the divalent (and possibly trivalent) nature of phosphate anions, polymeric amines also exhibit stronger affinity towards phosphate compared to their corresponding small molecule receptor analogs. This enhanced binding affinity of polymers towards phosphate anions can be attributed to a chelating effect. This phenomenon, however, has not been examined in detail. A carefully study of the pH effect may shed further light on the phosphate binding phenomenon in these polymeric systems. Furthermore, since the degree of protonation of polymeric amines is dependent on polymer architectures (due to the charge-charge repulsion effect that is prevalent in polyelectrolyte systems), incorporation of appropriate spacing between the amine groups along the polymer chain is another important factor to maximize the concentration of cationic groups, which would in turn influence the phosphate binding capacities and strengths of these polymeric sequestrants. Finally, polymers containing primary amine groups were found to be better phosphate sequestrants than polymers bearing secondary and tertiary amines, while polymers containing quaternary amines were found to be the poorest sequestrants of phosphate anions. As mentioned above, SAR studies revealed that crosslinked polyallylamine gels are very potent phosphate-binding polymers and possess properties suitable for pharmaceutical applications. This class of polymers exhibited maximum phosphate binding in the pH range encountered in the milieu of the GI tract.The polymer gels have been found to be non-toxic and are essentially nonabsorbed. These polymer gels were very well tolerated in multiple clinical trials with end-stage renal failure patients undergoing hemodialysis. After successful multiphase clinical trials as the first metal-free phosphate sequestrant for the treatment of hyper phosphatemia, polyallylamine (**Scheme 6**) was approved in the United States by the FDA in 1998 under the generic name of sevelamer hydrochloride. It has been marketed since under the brand name Renagel by Genzyme Corporation. In subsequent years, sevelamer hydrochloride has been approved in Europe, Japan, and a number of other countries. Since its regulatory approval, Renagel has demonstrated effective, long-term control of serum phosphate levels and has shown several clinical benefits over and above traditional inorganic binders for the management of hyperphosphatemia in renal failure patients. The ability for improved control of serum phosphate without increasing the exposure to toxic metal ions like aluminum and eliminating the intake of additional calcium offers a number of clinically relevant benefits. For example, without increasing the calcium load or promoting arterial calcification, Renagel may help prevent cardio-vascular complications in patients suffering from end-stage renal disease. Furthermore, Renagel has been found to reduce serum parathyroid hormone and also reduce total and low-density lipoprotein cholesterol (LDLc) inhemodialysis patients. Since cardiovascular events are the most common causes of mortality among dialysis patients, Renagel thus offers a very effective treatment in the management of renal failure. Renagel, represents one of the first tailor-made, polymeric drugs that exhibits prophylactic and therapeutic properties through the selective sequestration and removal of unwanted dietary components in the GI tract without presenting any systemic side effects. These clinically proven benefits of Renagel provide an opportunity to impact patient survival and morbidity as well as to reduce overall health care costs associated with the treatment of end-stage renal disease.

3.2.1.3 Sequestration of Iron: Polymeric Drugs for the Treatment of Iron Overload Disorders

Iron is an important metal ion in biological systems. While it is essential for the proper functioning of all living mammalian cells, the presence of excess iron in the body leads to toxic effects. Through the well-known Fenton reaction (Fig.3.4) excess iron catalyzes the transformation of molecular oxygen to oxygen-derived free radicals such as hydroxyl radicals. These oxygen derived radicals in turn cause damage to many vital biological molecules. This peroxidative tissue or organ damage forms the basis for several pathological conditions including neuro degenerative diseases. Under normal physiological conditions, iron metabolism is tightly conserved with the majority of the iron being recycled within the body. Normal physiology does not provide a mechanism for iron loss, i.e., iron is not normally excreted in urine, feces, or bile. Some loss of iron from the body occurs only through bleeding or normal sloughing of epithelial cells. Certain genetic disorders can, however, lead to increased absorption of dietary iron (as in the case of hemochromatosis) or transfusion-induced iron over load (as in the case of β-thalassaemia and sickle cell anemia).

$$2O_2^{\bullet} + 2H^+ = HO^- + O_2(\text{SOD catalyzed})$$
$$H_2O_2 + H^+ + Fe^{2+} = HO + Fe^{3+} + H_2O$$
$$O_2^{-\bullet} + Fe^{3+} = O_2 + Fe^{2+}$$

Fig.3.4 Production of reactive oxygen species through an iron-catalyzed Fenton reaction.

In the case of hemochromatosis, excess iron can be removed from patients' bodies by venesection. On the other hand, removal of iron using an iron chelator is the only effective way to relieve iron overload in patients with β-thalasemia or sickle cell anemia. Neither of these therapies are optimum for the treatment of these diseases. The current standard of care for iron chelation therapy is desferrioxamine (**Scheme 7**), which is the only approved iron chelator for the treatment of iron overload conditions in the USA. Desferrioxamine has reportedly been associated with several drawbacks. It has a narrow therapeutic window and due to lack of oral bioavailability, it requires administration for 8~12 h per day by parenteral infusions. Thus, there is a clear need for the discovery and development of a new generation of orally active iron-chelating agents for the treatment of iron overload conditions.

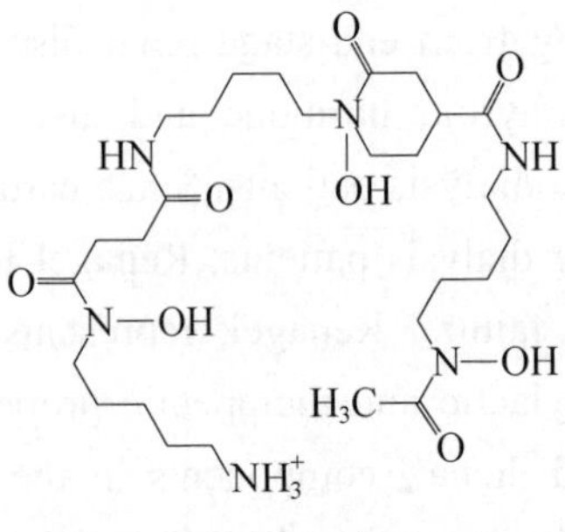

Scheme 7 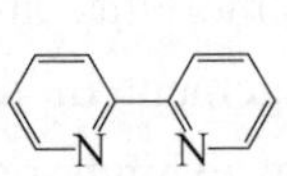 Scheme 8 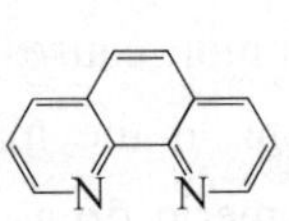 Scheme 9

Non-absorbed polymeric ligands that selectively sequester and remove dietary iron from the GI tract would offer an attractive method for the treatment of certain iron overload conditions. Clinically useful polymeric iron chelators require several features. Because of the importance of other metal ions for normal human physiology, the polymeric ligand must possess high affinity, capacity, and selectivity towards iron. Furthermore, the chelator should be biocompatible, and should not be absorbed from the GI tract. This is particularly important for hemochromatosis. However, binding dietary iron in the GI tract alone may not be sufficient to treat transfusion related iron overload such as beta thalassemia. For clinical applications, the important properties of chelators include metal ion selectivity and high stability constant for the ligand-metal complex. Since iron exists in two oxidation states (Fe^{2+} and Fe^{3+}), chelators could be designed to sequester both forms of iron. Thus, the design of pharmaceutically relevant polymeric iron chelators has been based on the knowledge of low molecular mass chelators.

For Fe(Ⅱ), soft donor atoms (e.g., nitrogen-containing ligands such as bipyridine, **Scheme 8** and phenanthroline, **Scheme 9**) can be employed. Although these ligands are selective for Fe(Ⅱ), they also possess affinity towards other biologically important divalent metal ions such as Zn(Ⅱ) and Cu(Ⅱ). On the other hand, oxyanions like hydroxamates and catecholates are selective towards Fe(Ⅲ) and these ligands in general show higher selectivity towards trivalent metal ions over divalent metal ions. Nature offers a precedent: natural iron chelators like siderophores such as desferrioxamine (**Scheme 7**) and enterobactin (**Scheme 10**) contain hydroxamate and catechol groups respectively, and they exhibit selectivity towards Fe(Ⅲ).

On the basis of the above criteria, crosslinked polymeric hydrogels containing hydroxamic acid and catechol moieties (**Schemes 11** and **12**) as well as crosslinked polymeric amines were prepared and were evaluated as iron chelators. Under in vitro conditions, all of these polymers sequester iron at high pH. At lower pH, the polymers containing hydroxamic acids maintained their iron binding properties, while other polymers showed poor iron binding properties. The iron binding isotherms for a hydroxamic acid containing hydrogel at different pH values are shown in Fig.3.5. In vivo studies using rodents have shown that the use of a hydroxamic acid-based hydrogel has arrested intestinal absorption of dietary iron. The polymers were well tolerated by the test animals, indicating their overall biocompatibility.

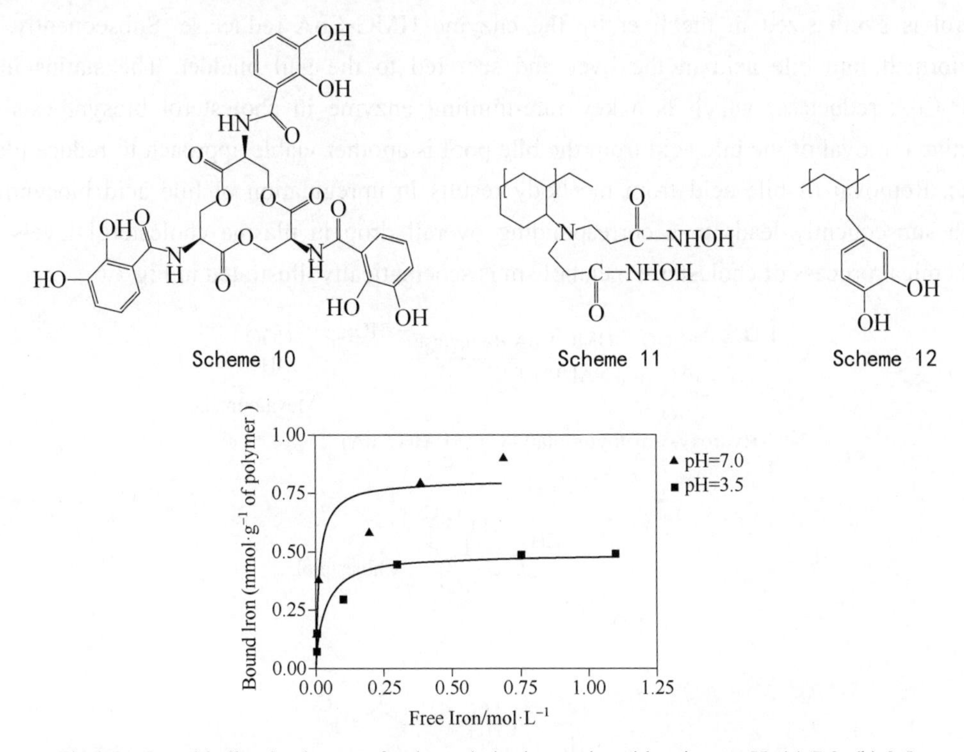

Fig.3.5 Iron-binding isotherms of polymeric hydroxamic acid resins at pH: (a) 7.0; (b) 3.5.

3.2.2 Bile Acid Sequestrants: Polymeric Cholesterol-Lowering Drugs

Increased plasma total cholesterol and low-density lipoprotein cholesterol (LDLc) are established risk factors for atherosclerosis, which is the underlying cause of coronary heart disease and most strokes. The reduction of elevated LDLc is one of the most common therapeutic approaches to treat this disease. The majority of individuals at risk of cardiovascular disease require only a modest (20% to 30%) reduction in LDLc level to minimize this serious and often life threatening health risk. HMG-CoA reductase inhibitors (more commonly known as statins) are the most widely used drugs for reducing plasma LDLc and have been shown to significantly reduce the risk of coronary events and strokes. These findings have led to recent guidelines for expanding the use of cholesterol lowering drug therapies to more patients at risk of cardiovascular disease.

Despite the clinical success of statins, there is still a need for alternative therapies to reduce blood LDLc. For example, statins are not recommended for pregnant women. They are also not recommended for pediatric use or for patients suffering from liver disease. Furthermore, some patients do not achieve the LDLc goal with statin therapy alone. There are also long-term potential safety issues associated with statins such as liver dysfunction and musculoskeletal symptoms. Since cholesterol lowering therapies are generally life-long, these safety factors are significant for such an extended treatment. This has been evident from the recent withdrawal of a statin, cerevastatin (Baycol), from the market as a result of several cases of rhabdomyolysis leading to death.

The molecular regulation of cellular cholesterol metabolism has been elucidated. Cholesterol is synthesized in the liver by the enzyme HMG-CoA reductase. Subsequently, it is transformed into bile acid in the liver and secreted to the gall bladder. The statins inhibit HMG-CoA reductase, which is a key rate-limiting enzyme in cholesterol biosynthesis. The effective removal of the bile acid from the bile pool is another viable approach to reduce plasma LDLc. Removal of bile acid from the body results in upregulation of bile acid biosynthesis, which subsequently leads to a corresponding overall drop in plasma cholesterol levels. The biochemical process of cholesterol metabolism is schematically illustrated in Fig.3.6.

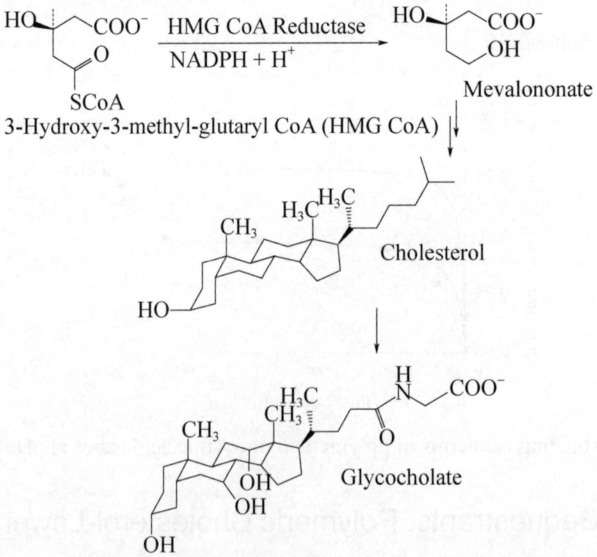

Fig.3.6 Biosynthesis of cholesterol and its metabolism into bile acid.

Bile acid sequestrants (BAS) are crosslinked polymeric cationic gels that bind anionic bile acids in the GI tract and subsequently eliminate them from the body along with the feces. The use of these polymeric gels for sequestering bile acids is indeed an established approach for treating patients with elevated plasma cholesterol. Being non-absorbed, these polymeric drugs do not exhibit the systemic side effects that are associated with statins, and the BAS have over 30 years of clinical experience with a good safety record.

Until recently, two cationic polymers, namely cholestyramine (**Scheme 13**) and colestipol (**Scheme 14**) have been the only approved bile acid sequestrants on the market. Despite the appeal of their safety profiles, these two first generation bile acid sequestrants have seen decreased use since their introduction to the market. While these two BAS exhibit high in vitro capacity, they have shown low clinical potency. For example, the doses required for a 20% cholesterol reduction with cholestyramine and colestipol are typically 16 to 24 g/day. The requirement of this high daily dose led to reduced patient compliance and hence limited use of these first generation BAS in clinical settings.

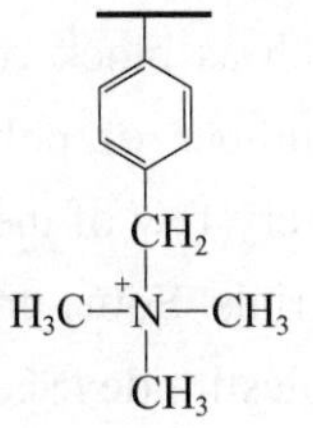

Scheme 13

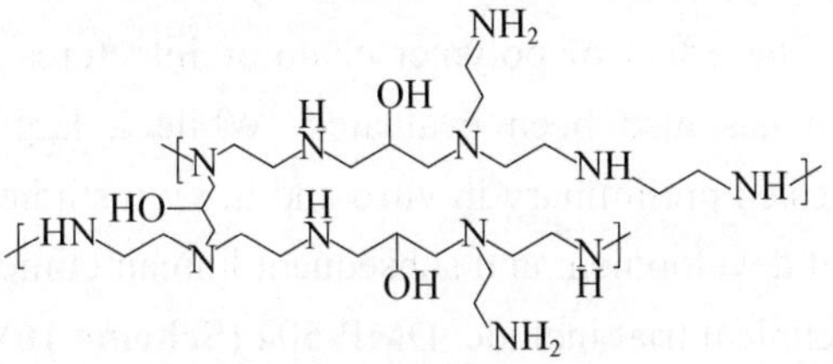

Scheme 14

The low in vivo efficacy of BAS has been ascribed to competing forces of the active bile acid transporter system of the GI tract, which the BAS has to encounter to strongly hold on to the bile acid from systemic reabsorption. It appears that, for a polymer to be a potent BAS, it must have high binding capacity, strong binding strength, and selectivity towards bile acids (over other anionic and amphiphilic species in the GI tract) in the presence of competing desorbing forces of the GI tract. Therefore, a potent BAS needs to exhibit slow off-rates of bound bile acids from the polymer resin to effectively overcome the active transport of bile acids from the GI tract. The design rules for making potent BAS need to take into consideration the physicochemical features of bile acids associated with their structures (**Scheme 15**). A typical bile acid possesses an anionic group and a hydrophobic core, which are responsible for their biological detergent properties. Therefore, while electrostatic interaction is the primary force required for bile acids to complex with cationic polymers, a second attractive force that needs to be considered is the hydrophobic interaction between the sequestrant and the bile acid. Moreover, favorable swelling characteristics of these cationic hydrogels in physiological environments are required for attaining high capacity (that would make use of maximum binding sites in the polymers). Thus, a balanced combination of hydrophilicity (high capacity) and hydrophobicity (to slow down the rate of desorption), along with an optimum density of cationic groups would constitute key features of potent BAS.

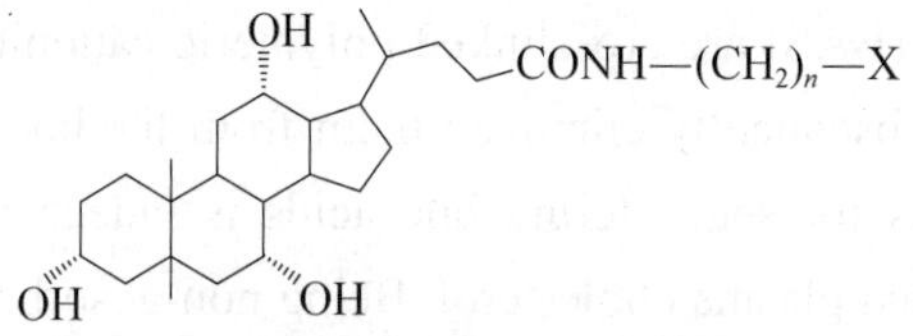

15a: n=1 X=COOH; 15b: n=2, X=SO_3H

Scheme 15

Careful consideration of these desired features has led to the discovery of a number of new generation bile acid sequestrants over the last decade from our laboratories as well as from other groups. Thus, a large body of literature (that has been published over the years) is now available. Structural features of some of the representative new bile acid sequestrants are summarized in Table 3.3. In general, the common features of these cationic hydrogels are the presence of amine/ammonium groups containing a whole range of substituents around the nitrogen atom. Additional structural features of these polymers include the presence of hydrophobic chains. Different kinds of polymer backbones including vinyl and allyl amine polymers, (meth)acrylates, (meth)acrylamide, styrene, carbohydrates, polyethers, and other condensation polymers have been considered. The effect of polymer chain architectures (such as block copolymer) on bile acid sequestration has also been evaluated. While a large number of polymers have been synthesized and tested preliminary in vitro and in vivo studies, very few of these polymers have entered preclinical development and subsequent human clinical trials. Some promising BAS that have entered the clinical trial include: DMP-504 (**Scheme 16**), colestimide (**Scheme 17**), SK & F 97426-A (**Scheme 18**), and colesevelam hydrochloride (**Scheme 19**). The key structural features of these polymers are an optimum combination of charge density, hydrophobic tails, and water swelling properties. From this new generation of bile acid sequestrants only two compounds have been approved by the regulatory agency for marketing. Colestimide has been approved for marketing in Japan and is sold under the trade name Cholebine. Colesevelam hydrochloride has been approved for marketing in the United States and is being sold under the trade name WelChol. Both polymers exhibit lower rates of side effects and are better tolerated than previously marketed BAS. These BAS have been recommended for use as monotherapy or in combination therapy by co-administration with other cholesterol lowering drugs, such as statins.

Table 3.3 Chemical structures of some representative bile acid sequestrants

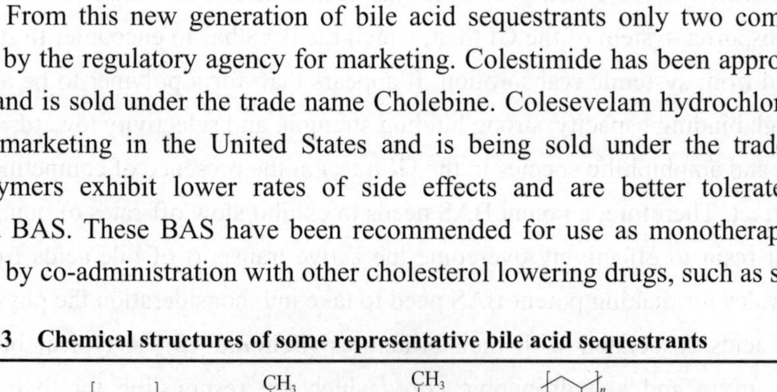

Scheme 16

Scheme 17

Scheme 18

Scheme 19

3.3 Polyvalent Interactions and Anti-Infective Polymeric Drugs

Polyvalent interactions are characterized by the simultaneous binding interaction between multiple ligands on one molecular entity and multiple receptors on another (cells, viruses, proteins, etc). These phenomena are frequently encountered in biological systems. Due to a multitude of chelating effects, polyvalent interactions can be significantly stronger than the corresponding monovalent interactions. Polyvalent interactions form the initiation steps for a large variety of key biological processes such as cell-surface or receptorligand recognition events. They can provide the basis for mechanisms of both agonizing and antagonizing biological interactions that are fundamentally different from those available in monovalent systems. A schematic illustration of the principle of polyvalency is presented in Fig.3.7.

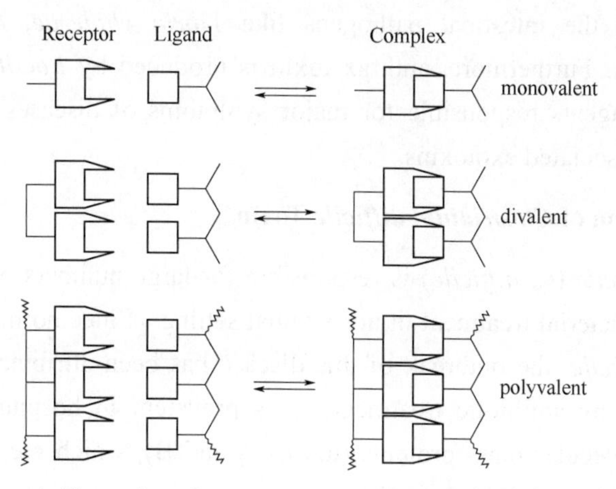

Fig.3.7 Schematic illustration of monovalent, divalent, and polyvalent receptors, ligands, and formation of their complex.

Recognition of this important interaction in biological systems is the basis of a new paradigm for the design and discovery of human therapeutics. This approach is particularly relevant when the interaction between a monovalent ligand and a polyvalent receptor is weak, since polyvalency can significantly enhance the binding strength leading to highly potent drug candidates. Thus, a polyvalent agent bearing two or more chemically bound ligands can interact at two or more receptor sites on a target pathogen leading to increased binding strengths resulting in effective inhibition and/or sequestration of the target pathogen of interest. Polymeric systems, wherein a collection of similar or different ligands can be covalently linked together on a single polymer chain provide a unique scaffold to elaborate this concept. In principle, the inherently polyvalent nature of polymeric materials can translate into longer-lasting and more potent therapeutics. Utilization of this new concept of polyvalent ligand-substrate interaction has led in recent years to the discovery of a number of therapeutically relevant polymeric species that have been found to sequester/inhibit pathogenic toxins, viruses, and bacteria. Some of these polymeric drugs have advanced into human clinical trials, further supporting the validity of this novel concept in drug discovery (vide infra).

3.3.1 Polymeric Sequestrants of Toxins

Toxins are produced by pathogenic microorganisms in the host body. They are classified into two categories: exotoxins and endotoxins. Exotoxins are generally proteinous materials released by these microorganisms. Endotoxins are lipopolysaccharides and consist of polysaccharide segments and glycolipid segments that constitute the outer cell membranes of all Gramnegative bacteria. Upon secretion from microorganisms, these toxins can travel within host organism and can cause damage in organs far from the initial site of infection.

The pathogenic effects of bacterial toxins can include diarrhea, hemolysis, destruction of leucocytes, paralysis, diarrhea, and septic shock. The outcomes of these events can be life threatening. Some of the microorganisms that release life-threatening toxins in the GI tract include food poisoning organisms like *Staphylococcus aureus*, *Clostridium perfringens*, and *Bacillus cereus*, and the intestinal pathogens like *Vibrio cholerae*, *Escherichia coli*, and *Salmonella enteritidis*. Furthermore, anthrax toxin is produced by *Bacillus anthracis*. In most cases, the causative agents responsible for major symptoms of diseases associated with these pathogens are their associated exotoxins.

3.3.1.1 Sequestration of *Clostridium difficile* Toxin

Clostridium difficile (*C. difficile*) is responsible for large numbers of episodes of diarrhea that arise from anti-bacterial treatment in nosocomial settings. Since normal colonic flora inhibit the growth of *C. difficile*, the outbreak of this disease has been attributed to the disruption of normal colonic flora by antibiotic treatment, as is prevalent in hospital settings. *C. difficile* releases two high molecular mass proteins (toxins A and B), which are the primary causes of diarrhea in patients with *C. difficile* infection.

The traditional approach to treat *C. difficile*-associated diarrhea has been the use of one of two antibiotics: metronidazole or vancomycin. Although the use of antibiotics is often effective in eliminating *C. difficile* infection initially, each repeated use sterilizes the gut, thereby increasing the chance for further infection by *C. difficile*. In many cases, repeated cycles of antibiotic treatment, followed by re-infection by *C. difficile* leads to long hospital stays and patient morbidity. Furthermore, there has been an increasing body of documentation pointing to growing resistance of *C. difficile* to antibiotics.

Since disease symptoms are attributed to the toxins released by *C. difficile*, and re-colonization of *C. difficile* is attributed to the inhibition of normal microbial flora growth by antibiotic treatment, selective sequestration and neutralization of these toxins appears to be an attractive and safe antibiotic-free therapy to treat this disease. The proposed approach would treat the infection without disrupting normal bacterial growth in the GI tract. Being proteinaceous substances, bacterial toxins possess multiple binding sites (derived from amino acid side chains). The binding sites can be used as anchoring sites to effectively interact with multifunctional polymers (carrying complementary binding sites) through polyvalent interactions. Some of the early studies to use polymeric ligands to sequester *C. difficile* toxins utilized anion exchange resins like cholestryramine and colestipol (the bile acid sequestrants described above). However, the effectiveness of these polymeric ion-exchangers as sequestrants to bind and remove *C. difficile* toxins was found to be modest at best.

Systematic investigations have led to the discovery of polymeric multivalent ligands that effectively bind and neutralize *C. difficile* toxins. From the various polymers evaluated, a series of high molecular mass, water-soluble anionic polymers (**Schemes 20~22**) have been found to be the class of polyvalent ligands that are particularly effective in sequestering and neutralizing *C. difficile* toxins. Two important characteristics of these polymers are the presence of sulfonic acid groups and high molecular mass. Careful in vitro studies using pulsed ultrafiltration binding experiments and fluorescence polarization spectroscopy revealed that the binding constants of complexes between one of these polymers and toxins A and B were 133 $nmol•L^{-1}$ and 8.7 $\mu mol•L^{-1}$, respectively. The ability to bind these toxins is not a general feature of all anionic polymers. For example, while the sodium salt of poly(styrene sulfonic acid) (**Scheme 20**) effectively binds both toxins, poly(sodium 2-acrylamido-2-methyl-1-propanesulfonate) (**Scheme 22**), a high molecular mass polyanion of similar charge density does not bind either of the toxins to any measurable extent. This implies that the chemical characteristics of monomer units that make up these polyvalent ligands influence the toxin-binding properties of the polymer. Furthermore, the binding strength was found to be dependent on molecular mass, with lower molecular mass polymers exhibiting very low binding strength. From these observations it is evident that sequestration of *C. difficile* toxins by polyanions is not purely electrostatic in origin. On the other hand, it appears that the sodium salt of poly (styrene sulfonic acid) and related anionic polymers interact with toxins A and B through a multitude of weak interactions. Amplification of these individual interactions, as a result of polyvalency, translates into a substrate-binding event of very high binding strength. From fluorescence polarization data, it was estimated that one molecule of toxin A interacts with about 800 monomer units present in a polymer chain. Thus, a single polymer chain of molecular mass of 300 kDa can wrap around a toxin molecule about 3~4 times. This suggests that for effective toxin binding of therapeutic relevance, large polyvalent interactions between the polymer and protein surface is needed .

Scheme 20 Scheme 21 Scheme 22

A schematic presentation of the process of multiple contacts between the polymer and toxin molecule is illustrated in Fig.3.8. The in vitro binding activity of these high molecular mass polyanions correlate well with the in vivo biological activities. Thus, the lead polymer prevented the mortality of 80% of hamsters suffering from severe *C. difficile* colitis.

Fig.3.8 Schematic representation of the sequestration of *C. difficile* toxin by the polyvalent ligand.

These polymers are non-antimicrobial and do not interfere with the activities of standard antibiotics. High molecular mass sodium salt of poly(styrene sulfonic acid) derivative was selected as the clinical candidate for treating *C. difficile* infection. This compound, under the generic name of Tolevamer, has been successful in both phase Ⅰ and phase Ⅱ human clinical trials. Besides the above anionic polymer, another class of polymers bearing pendant oligosaccharide groups has also been investigated as possible sequestrants for *C. difficile* toxins that have progressed to human clinical trials. The underlying principle behind this polymeric sugar agent for the treatment of *C. difficile* infection is that toxin A has shown a lectin-like activity, which allows it to bind to an oligosaccharide receptor on epithelial cells.

Furthermore, toxin B has been found to bind erythrocytes. These findings suggest that the cell invasion/binding of *C. difficile* may be mediated by interactions with cell surface carbohydrate receptors. Therefore, polymers bearing pendant sugar residues may compete with human cells towards *C. difficile* toxin. After identifying oligosaccharide sequences, appropriate oligosccharide molecules that are specific for both toxins were conjugated to different polymer backbone (**Scheme 23**). Lengths of tethering arms linking the polymer backbones with oligosaccharide moieties were appropriately optimized to maximize the binding strengths between polymeric ligands and toxins.

Scheme 23

The oligosaccharide sequences that were found to improve toxin binding include maltose, cellobiose, isomaltotriose, and chitobiose. Polymer carriers examined in this study include substituted polystyrene and other polyolefin backbones. Inorganic carriers such as biogenic silica and kaolinite were also used. These polymeric toxin binders (under the trade name of SYNSORB) were found to be effective in neutralizing *C. difficile* toxins and in controlling diarrhea in animal models. One of these polymeric carbohydrate agents had also entered human trials. However, due to lack of therapeutic efficacy, it was withdrawn after a phase Ⅱ clinical trial.

3.3.1.2 Sequestration of Anthrax Toxin

Anthrax toxin is produced by the Gram-positive bacteria *Bacillus anthracis* and is the primary cause for the major symptoms of the disease. In general, the occurrence of anthrax in clinical settings is rare. However, growing concern over bioterrorism and biological warfare involving *B. anthracis* in recent years has put the effort to discover and develop anti-anthrax agents on a high-priority list. Although vaccines against anthrax exist, several factors make mass vaccination difficult. While treatment with antibiotics could eradicate bacteria from the host, the continuing action of the toxin in inducing damage after symptoms have become evident makes antibiotic treatment of limited clinical benefit. Therefore, development of agents that can sequester anthrax toxin (thereby inhibiting its action) is an attractive adjunct to antibiotic therapy for treating anthrax infection. Like most intracellularly acting toxins (such as ricin and botulinum neurotoxin), anthrax toxin consists of two subunits: an activating region (subunit A) and a promoter region (subunit B). The subunit B interacts with the cell surface receptor that is specific for the toxin. However, the bacterium secretes three separate proteins: a single receptor-binding moiety termed protective antigen (PA), and two enzymatic moieties, termed edema factor (EF) and lethal factor (LF). Upon release from bacteria as non-toxic monomers, these three proteins combine and undergo a cascade of processes through the formation of a cell bound heptameric fragment called PA63, which finally attacks the macrophages. The heptameric PA63 moiety binds to host cells via a polyvalent interaction of very high binding constant ($K_d \approx 1$ nmol•L^{-1}).

The biological pathway leading to anthrax infection (as described above) suggest that development of a polyvalent ligand that would compete with PA63 could be a potential agent to inhibit binding of anthrax toxin to the surface of the host cell. While the overall structural requirements for binding of PA63 heptamer are yet to be investigated, some structural studies have shown the importance of hydrophobic interactions involving Tyr and Trp residues as well as

H-bond donor/acceptor sites. By utilizing phage-display library screening, a dodecameric peptide sequence (P1) was identified that binds to PA63 and thus interferes with its interaction with host cells. In a cell culture assay, this peptide exhibited modest potency ($IC_{50} \approx 150\ \mu mol \cdot L^{-1}$). By attaching multiple copies of this peptide P1 to a polyacrylamide-based polymer backbone, the corresponding polyvalent ligand was obtained. This polymeric ligand contained, on average 22 P1 units and ~ 900 acrylamide units. The IC_{50} value of this polymer in inhibiting the binding of PA63 to host cells was found to be 20 $nmol \cdot L^{-1}$. This value corresponds to a nearly 7500-fold increase in potency on a per-peptide basis relative to the free P1. Optimization of this class of polyvalent ligands by increasing the P1 concentration in the polymer backbone and incorporation of additional hydrophobic groups and H-bond donor/acceptor sites are reportedly under investigation to further enhance the potency of this class of polymers to treat anthrax infection.

One of these polymeric agents was evaluated in vivo in Fisher 344 rat models that were intoxicated with anthrax lethal toxin. The polymer, when dosed intravenously, delayed the symptoms and eliminated the toxicity at a dose of 12 nmol equivalent of peptide. No obvious toxicity associated with i.v. administration of the polymer was observed in the test animals during the week-long treatment. The efficacy of this polyvalent ligand in blocking the action of anthrax toxin in vivo suggests that this approach could be useful in developing therapeutically relevant agents to combat possible future risks of bioterrorism involving anthrax toxin.

3.3.2 Polyvalent Ligands as Antiviral Agents

Viral infection is initiated by the attachment of viruses to specific cellular receptors. This virus-cell receptor attachment process is mediated by viral attachment proteins (VAPs). It has been postulated that the presence of synthetic analogs of cellular receptors would compete with VAPs towards forming the receptor-binding epitopes of VAPs. It has been calculated that in general, attachment of viruses to cell surfaces involves the interaction of a virus species with at least 10 monomeric cellular receptor units organized into polymeric cellular receptor units. The characteristic feature of such polyvalent cooperative binding is the amplification of multiple low affinity individual contacts to a binding event of very high association constant. This suggests that by presenting polymeric ligands bearing multiple copies of the cellular receptor analogs, competitive inhibitors for viruses can be developed. This design principle has been utilized to develop prototype antiviral agents against viruses such as influenza and rotavirus.

3.3.2.1 Inhibition of Influenza Virus

Influenza virus A is the primary causative agent responsible for serious cases of human influenza. The influenza infection is initiated by attachment of the virus to the mammalian cell membrane through a process known as hemagglutination. The hemagglutination process is a multivalent interaction between trimers of hemagglutinin (a carbohydrate binding protein present on the viral surface) with multiple sialic acid groups present on the surface of the mammalian epithelial cell. These sialic acid residues are parts cell-surface glycoproteins. This biological process of viral invasion of mammalian host cells is schematically shown in Fig.3.9. One of the potential strategies to treat influenza virus infection is to block the binding of the virus to mammalian cells by presenting polymers bearing several sialic acid groups as competitors for cell surface ligands. Although individual viral surface hemagglutinin (HA) binds individual cell-surface sialic acids (SA) weakly, the virus attaches to the cellular surface through multiple interactions between clusters of HA and SA residues. While an individual HA-SA interaction has a K_d of ~ 2.5×10^{-3} mol•L^{-1}, the interaction between the virus and an erythrocyte has a K_d of ~ 10^{-12} mol•L^{-1}. In other words, the polyvalent interaction of the latter system is about a billion-fold stronger than the corresponding individual monovalent interaction. In principle, appropriately designed polymeric SA derivatives containing multiple copies of SA may be able to compete with and disrupt the strong virus-cell binding. Some research groups investigated this phenomenon quite systematically by designing and synthesizing an array of polymers containing side chain SA groups. Some of these polymers have shown very impressive affinity enhancement towards influenza virus. The chemical structures of some of the representative polymers are summarized in Table 3.4. The most effective inhibitor among these polymers is a linear polyacrylamide derivative containing, on a side chain, the C-glycoside of SA (**Scheme 24**). This polymeric ligand prevents hemagglutination at a concentration of 35 pmol•L^{-1}, while the corresponding monomeric α-methylsialoside is a very weak inhibitor that inhibits the hemagglutination process at a concentration greater than 2 mmol•L^{-1}.

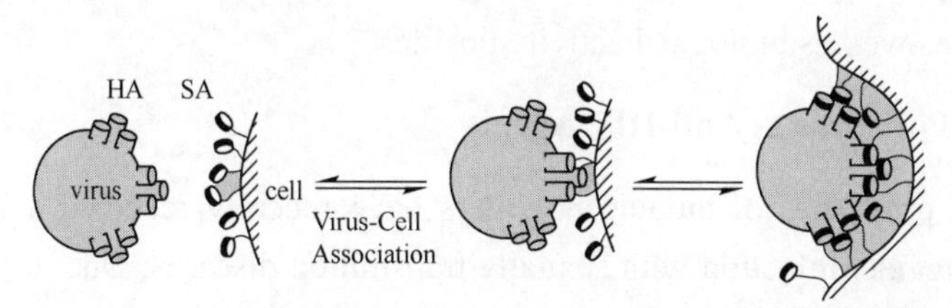

Fig.3.9 Schematic illustration of the biological process involving viral invasion of mammalian host cells.

Table 3.4 Representative examples of polyvalent ligands as influenza virus inhibitors

Scheme 24

This is probably the highest increase in potency for any polyvalent system to inhibit virus-erythrocyte binding. A systematic investigation to elucidate the mechanism of this polyvalent interaction between influenza virus and polymeric ligands has shown that a balanced combination of several factors including high affinity multiple ligand density, steric stabilization, and entropically driven enhanced binding contribute to the overall effectiveness of these polymeric inhibitors to prevent cell surface attachment of viruses. Although a great deal of in vitro studies have been carried out, in vivo assays exhibiting the success of these polyvalent ligands in protecting animals against influenza infection have yet to be demonstrated. The success of this approach to develop antiviral agents for influenza would also depend on their favorable toxicity as well as biological activity profiles.

3.3.2.2 Anionic Polymers as Anti-HIV Agents

The antiviral properties of anionic polymers have recently received a lot of attention as agents to protect against infection with sexually transmitted diseases. Due to the cationic nature of most viruses, several anionic polymers are known to bind viruses. As early as the 1960s, researchers had studied the anti-viral properties of a variety of synthetic polymers. However, not all anionic polymers inactivate viruses. Several classes of anionic polymers have been studied for their ability to inactivate the HIV virus. These polymers include poly(styrene-4-sulfonate), 2-naphthalenesulfonate-formaldehyde polymer, and acrylic acid-based polymers. Certain chemically modified natural polymers (i.e., semisynthetic) such as dextrin/dextran sulfates, cellulose sulfate, carrageenan sulfate, and cellulose acetate phthalate have also been investigated for this purpose. Of a number of such anionic polymers that have shown in-vitro and in-vivo anti-HIV activity, a couple of polymeric drug candidates have proceeded to early stage human clinical trials for the evaluation of safety/tolerability. While most of these have shown the desired tolerability and safety, further clinical trials are necessary to discern the therapeutic benefit and see if anionic polymers will be applicable as anti-HIV therapies.

3.3.3 Polyvalent Antimicrobial Agents

The emergence of microbial pathogens that are resistant to multiple classes of available antimicrobial agents is becoming a major worldwide public health concern. These multidrug resistant bacteria are ubiquitous in both hospital and community settings. The majority of these strains have been found to carry multiple drug resistance factors. At present, the only effective treatment for multiply resistant bacterial infections is vancomycin. Unfortunately, vancomycin resistance itself is becoming a growing problem. For example, methicillin-resistant *Staphylococcus aureus* (MRSA) strains, vancomycin-resistant *enterococci*, and amikacin- and β-lactam-resistant *Kleisiella pneumoniae* are some of the bacterial species that are resistant to vancomycin. This rapid emergence of multidrug resistant bacterial strains and the potential threat they pose to human life means that there is a pressing need to discover and develop novel antibacterial agents to overcome the challenges posed by multidrug resistant pathogens.

The polyvalent ligands as antibacterial agents have been considered to exhibit several potential advantages over monomeric antimicrobial agents. Cluster effects from polyvalent ligands would lead to amplification of weak non-covalent bonding interactions between the bacterial surface receptors and the polymeric ligands. Aggregation and precipitation of bacteria by polyvalent ligands is potentially another favorable feature. Finally, polyvalent ligands utilizing multi-point attachments could enhance lysis of the bacterial cell membrane/wall.

Similar to viral infection, most bacterial infections are initiated by adhesion of microorganisms to the mucosal surfaces of the host, mediated in part by bacterial protein adhesins. These adhesins interact with carbohydrate determinants of host cell glycolipids or glycoproteins. This underlying mechanism for bacterial infection suggests that appropriate polyvalent sugar derivatives could competitively block the attachment of microbial adhesin to the host mucosal surface resulting in protection against infection.

This concept has been explored through the synthesis of a number of polymers bearing acid-functionalized glycoside moieties. Olefinic monomers containing glycoside moieties and acid functional groups such as *O*-sulfo and *O*-carboxymethyl groups were prepared and converted to various copolymers. These polymers were found to be effective in vitro against a number of bacterial targets.

A second approach to design polyvalent ligands as antimicrobial agents based on cationic polymers has also been systematically explored. The mechanism of the antimicrobial action of these polymers has been attributed to their enhanced ability for cell lysis. These polymers were designed as mimetics of certain cationic amphiphilic peptides containing multiple arginine and lysine residues. These cationic peptides, which are known as antimicrobial peptides (or defensins) cause cell lysis. The cell lysis is mediated by the interaction of the positive charges of the peptides with negative phosphate head groups of cell membrane phospholipids . A series of amphiphilic cationic polymers were prepared bearing amine and quaternary ammonium groups as well as hydrophobic tails as defensin analogs.

These polymers were found to exhibit antimicrobial activity against a number of microbes. In particular, some of these polymers (**Schemes 25~27**) were found to be very effective against *Cyptosporadium parvum* (*C. parvum*). Until the advent of therapeutic HIV protease inhibitors, *C. parvum* was a primary target for drug discovery to treat GI tract infections in individuals with HIV infection. In an in vivo study, some lead polymers from this series of polycations were found to be superior to the commonly prescribed antibiotic, paromomycin.

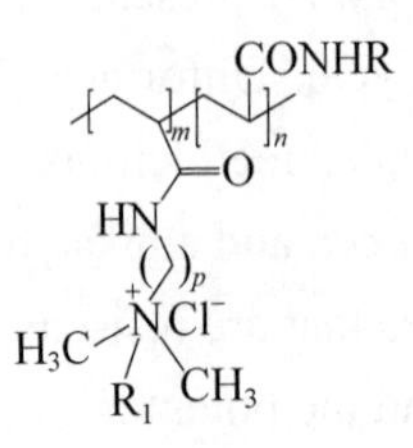

Scheme 25

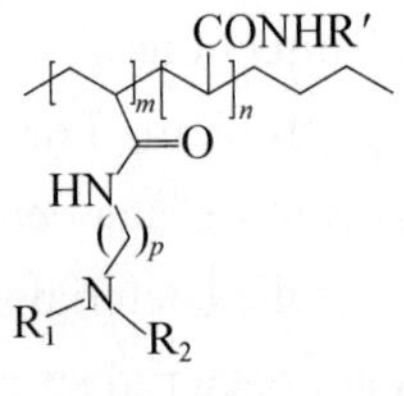

Scheme 26

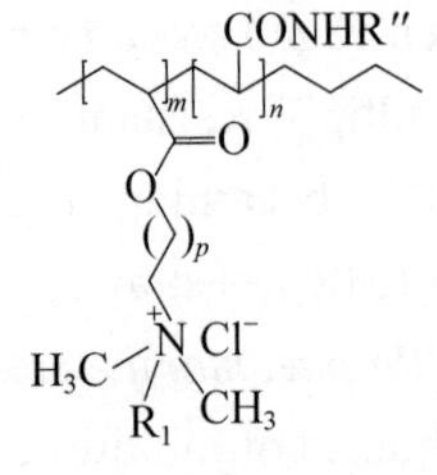

Scheme 27

Synthesis of a multivalent vancomycin derivative exhibiting significantly enhanced antibacterial activity against vancomycin-resistant *enterococi* (VRE) has been recently reported. This work was built upon the finding that dimeric vancomycin showed enhanced affinity towards the L-lysyl-D-alanyl-D-alanline peptidoglycan precursor that is responsible for the growth of the bacterial cell wall. A polymeric vancomycin derivative was obtained by ring opening metathesis polymerization (ROMP) of a functional cyclic olefin monomer containing the vancomycin moiety (Fig.3.10).

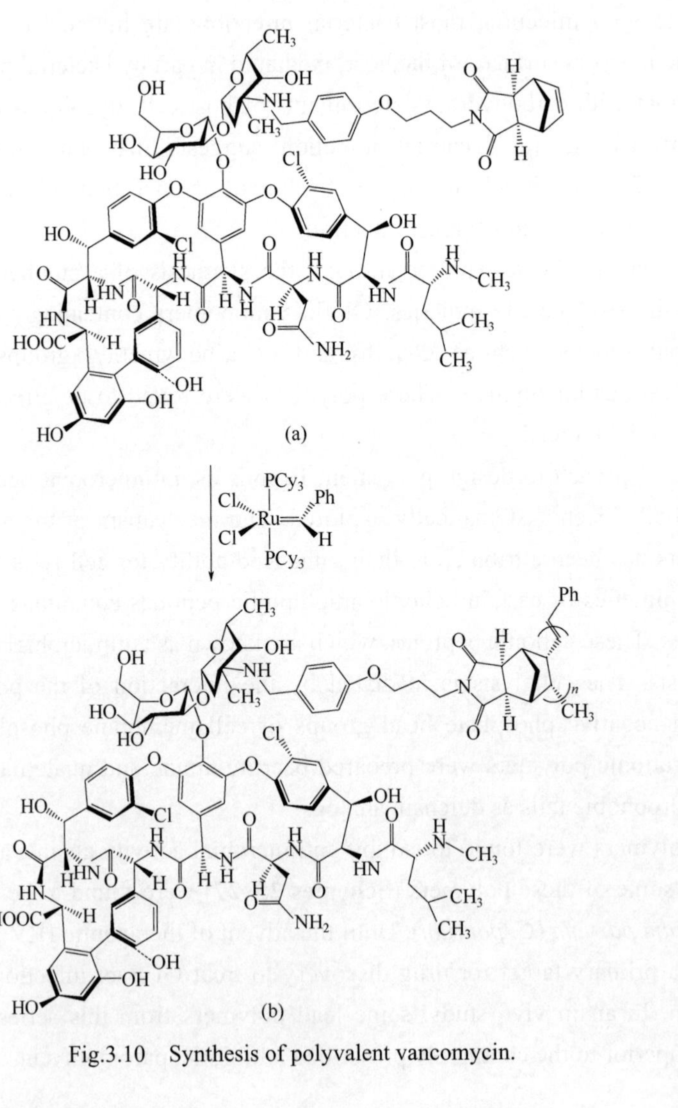

(a)

(b)

Fig.3.10 Synthesis of polyvalent vancomycin.

The antimicrobial property of the monomer [Fig.10 (a)] was similar to native vancomycin. Upon incorporation into the polymer backbone, the activity of the corresponding polymeric vancomycin derivative [Fig.5.10 (b)] was found to have increased by nearly 60 fold, when tested against *S. aureus* and *enterococci* . Although no in vivo data is available to date, this promising in vitro result supports the important role that polyvalency can play in the discovery of a new generation of antimicrobial agents.

3.4 Polymeric Drugs for the Treatment of Autoimmune Diseases

In simplest terms, autoimmune diseases arise when the host immune system mistakenly attacks itself. Ordinarily the immune system uses a number of defense mechanisms to prevent the development of these autoimmune responses by directing T cells (defense cells) to distinguish foreign invaders. Bypassing the protection against autoimmunity leads to inflammation in various parts of the human body. Multiple sclerosis and rheumatoid arthritis (RA) are two of the most common autoimmune diseases.

Multiple sclerosis (MS) is one of the most common inflammatory diseases of the central nervous system associated with immune activity directed against central nervous system antigens. As a result, this disease affects the brain and the spinal cord. Destruction of the regulatory mechanism that guards against autoimmunity leads to inflammation of the central nervous system.

This disabling disease affects over 2.5 million people worldwide, particularly young adults. The pathophysiology of MS has been attributed to the infiltration of autoreactive T cells, degradation of myelin basic protein (demyelination), production of excessive inflammatory cytokines such as tumor necrosis factor alpha (TNF-α), etc. Although the knowledge around the immunopathogenesis of MS has expanded over the years, therapeutic advances to treat this debilitating disease have been modest. The mainstays of therapeutic strategies to treat MS include the use of agents that trigger immunosuppression and immunomodulation such as mitoxantrone, recombinant interferon-β (e.g., Avonex®, Humira®), and glatiramer acetate, etc.

Glatiramer acetate (GA) is an amino acid-derived synthetic copolymer. It acts as an alternative to interferon-β for the treatment of certain forms of MS. The discovery of this polymer represented a significant breakthrough in polymeric drug discovery. GA (**Scheme 28**) is a random copolymer composed of four L-amino acids (alanine, lysine, glutamic acid, and tyrosine in a molar ratio of 4.2 : 3.4 : 1.4 : 1.0). This copolymer is prepared by the ring opening random copolymerization of the corresponding *N*-carboxy-α-amino acid anhydrides. Systematic in vitro (using murine T-cell lines) and in vivo (using experimental allergic encephalomyelitis animal models) studies have indicated the beneficial effects of this polymer in treating MS. The possible mode of therapeutic action of GA has been attributed to its ability to compete with immunodominant myelin basic protein (MBP), which is one of the major autoantigens implicated in the pathogenesis of MS.

MBP is known to sensitize autoreactive T-cells. The autoreactive MBP-specific T-cells migrate into the central nervous system (CNS) and mediate the pathogenesis of MS. The composition of glatiramer acetate is postulated to resemble a portion of MBP (MBP 85~89). This epitope presented by GA competes with MBP for binding with T-cell receptors (through cross-reaction) leading to antigen specific intervention of the autoimmune process. In human clinical trials glatiramer acetate demonstrated a significant decrease in the number of relapses and rate of progression of the disease. Glatiramer acetate has been approved by FDA and European agencies for the treatment of the relapsing-remitting form of MS in the USA and Europe, and is marketed under the brand name Copaxone by Teva Pharmaceutical Industries. Copaxone is the first synthetic polymeric drug approved for systemic treatment.

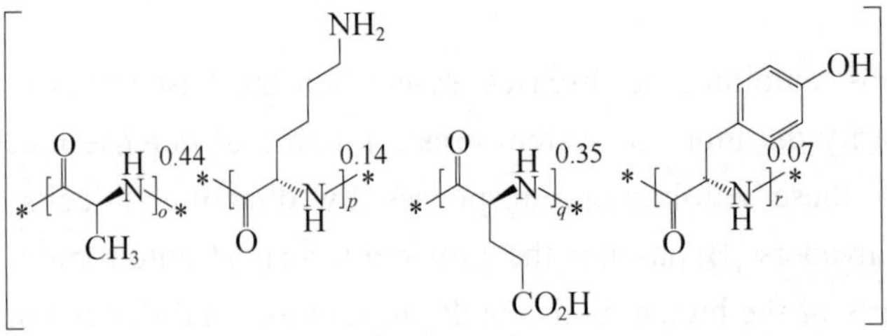

Scheme 28

Although widely used for the treatment of MS, Copaxone does not completely eliminate the frequency of relapse of the disease. Since the mode of action of GA has been considered to be due to its higher binding strength compared to MBP, the design of a new generation of copolymers based on the anchor residues in MBP as well as by tailoring their mode of arrangement (copolymer sequence) along the polymer chain may lead to more effective immuno-modulators. Towards this end, a systematic study has been carried out to discover a new generation of amino acid-derived immuno-modulating copolymers to effectively treat MS. In this investigation, a systematic replacement of tyrosine and glutamic acid with other amino acids were carried out resulting in a series of copolymers with improved binding affinity towards he binding pockets of T-cell receptors. among the various copolymer tested, two copolymers, namely poly(valine-tryptophan- alanine-lysine) and poly(phenylalanine-tyrosine-alanine-lysine) (**Schemes 29** and **30**) were found to exhibit better activities than GA in a number of in vitro and in vivo assays. This new generation of copolymers reportedly produced protective anti-inflammatory cytokines that led to the suppression of the histopathological evidences in EAE animal models. Further development of these promising copolymers for human clinical trials has been undertaken by Peptimmune Corporation.

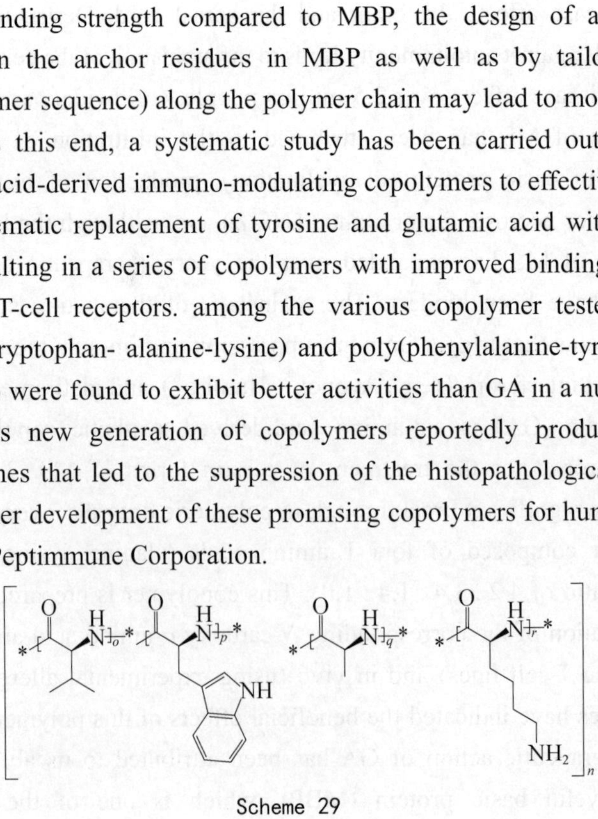

Scheme 29

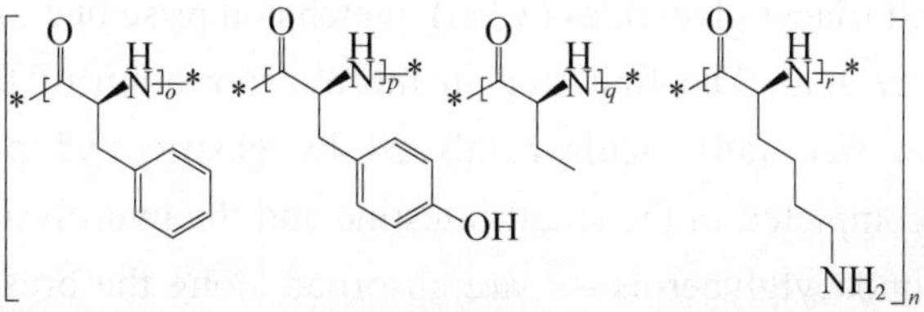

Scheme 30

Since GA has been thought to work by antigen-specific intervention against MS, utility of this class of polymeric drugs can be expanded to treat other autoimmune diseases. It is reported that GA and other related copolymers (containing three amino acids) compete with type Ⅱ collagen peptide 261~273, a candidate autoantigen in rheumatoid arthritis (RA) for binding to RA associated T-cells. This suggests the possible utility of this class of immunomodulating polymers as a potential therapy for the treatment of RA and other autoimmune diseases.

3.5 Polymeric Anti-Obesity Drugs

3.5.1 Obesity and Medical Need

Human obesity has been declared to be one of the most significant health problems in modern times with over 500 million people being overweight. Obesity is associated with an increased risk of developing several serious diseases including hypertension, coronary heart disease, type Ⅱ diabetes, stroke, osteoarthritis, and cancer. Obesity increases the likelihood of mortality by 20% and recently surpassed smoking as the number one cause of death.

There are few medications available for the treatment of obesity. These few anti-obesity agents exhibit modest to minimal efficacy and have been associated with poor side-effect profiles. Thus, there is an urgent need for the discovery and development of new therapeutic agents for the treatment of this significant disease. In general, the strategies to treat/prevent obesity are based on their mechanism of action in maintaining energy balance. When energy intake exceeds expenditure, a state of positive energy balance exists and vice versa. As a result, the role of anti-obesity drugs is to induce negative energy balance until the desired weight loss has been achieved. Besides surgical procedures, the therapeutic approaches to treat obesity fall into four classes:

(i) appetite suppressants that act on the CNS by stimulating anorexigenic signals or by blocking orexigenic signals (e.g., sibutramine);

(ii) inhibitors of fat absorption that act by inhibiting metabolizing enzymes responsible for the digestion of nutrients (e.g., orlistat);

(iii) enhancers of energy expenditure that act by increasing thermogenesis;

(iv) stimulators of fat mobilization that act by decreasing de novo synthesis of triglyceride.

3.5.2 Inhibition of Lipase to Control Digestion and Absorption of Dietary Fat

Consumption of dietary fat is an important contributor to human obesity. Dietary fats are present mainly as mixed triacyl glycerides (TAG), which comprise one molecule of glycerol and three molecules of fatty acid. The first step in the transportation of fat to the circulation is hydrolysis of TAG to free fatty acids catalyzed by gastric and pancreatic lipases. The hydrolysis process is completed in the small intestine and the hydrolysis products——the free fatty acid and sn-2-monoacylglycerols——are absorbed along the brush border membrane of the small intestine (see Fig.3.11). Gastric and pancreatic lipases are the principal lipolytic enzymes in the GI tract that are responsible for the hydrolysis of TAG. Thus, inhibition of lipase enzymes in the GI tract reduces the amount of fat that can be absorbed. Targeting this mechanism to develop anti-obesity drugs appears to offer an inherently safe approach as it involves peripheral targets. Orlistat (also known as tetrahydrolipstatin, **Scheme 31**) is a potent, specific and irreversible inhibitor of pancreatic and gastric lipases and works by inhibiting the action of lipase enzymes in the stomach and small intestine. The mode of action of orlistat involves the formation of a covalent linkage between the serine hydroxyl group in the catalytic triad and the β-lactone ring of orlistat.

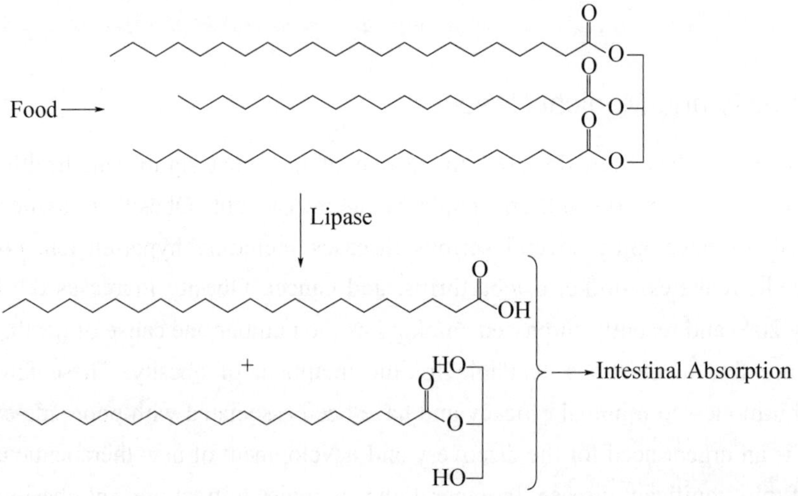

Fig.3.11 Lipase-catalyzed hydrolysis of triglyceride and its subsequent intestinal uptake

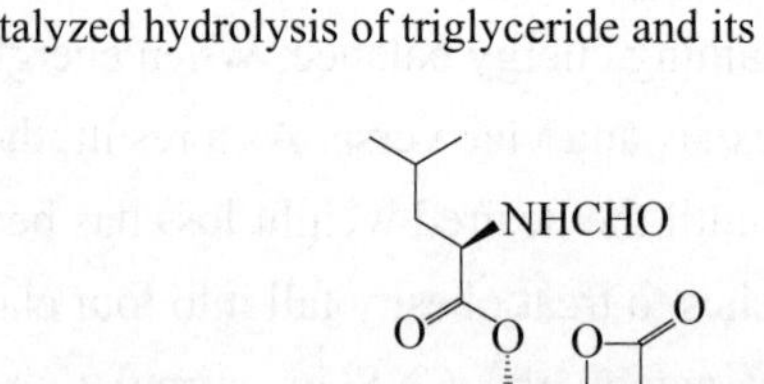

Scheme 31

In human clinical trials orlistat reduced fat absorption by approximately 30%. The undigested fat is almost exclusively excreted with the feces. At the recommended dose it produces a weight loss of ~ 10% after one year of treatment. Orlistat has been approved by regulatory agencies for the treatment of obesity and has been marketed by Roche under the trade name of Xenical. However, since orlistat blocks fat hydrolysis and increases fecal fat loss, the drug is associated with gastrointestinal side effects that are the direct outcome of TAG malabsorption. These side effects include abdominal pain, steatorrhea, increased flatus, diarrhea, etc. As a result of these side effects, there is decreased compliance with orlistat over time.

3.5.3 Polymeric Fat Binder and Dual Acting Polymeric Lipase Inhibitor-Fat Binder

Besides orlistat, a number of other natural and synthetic inhibitors of human pancreatic and gastric lipases have been identified. Some of these have progressed to clinical development. However, no lipase inhibitor can be expected to overcome the above mentioned side effects. This is due to the fact that as these GI side effects are mechanism related (caused by the presence of non-hydrolyzed fat), options are limited for further improvement in patient compliance by targeting lipases alone. However, a therapeutic intervention that could simultaneously inhibit fat hydrolysis and condense unhydrolyzed fat droplets into a less fluid form may provide a novel approach for lipase inhibitor therapy without the aforementioned side effects.

Synthetic functional polymers bearing lipase inhibiting groups and also possessing lipid condensing properties offer a therapeutic opportunity for this purpose. As the first step towards demonstrating the role of synthetic polymers in eliminating the side effects associated with leakage of unabsorbed fat from patients undergoing lipase inhibitor therapy, we adopted a strategy to discover fat-binding polymers that can be co-administrated along with a lipase inhibitor, such as orlistat. Dietary fats are usually present in an emulsified form and even the bulk fat in human diets is quickly emulsified in the GI tract. Typically gastric and pancreatic lipases hydrolyze dietary TG at the oil-water interface of a fine emulsion stabilized by physiological emulsifiers such as bile acids and phospholipids secreted by the gall bladder. After initial hydrolysis of some TG in the stomach, the remaining TG and fatty acid move into the small intestine. Here the free fatty acid facilitates the formation of a fine emulsion with particle sizes in the range of one micron. This emulsified lipid phase is efficiently hydrolyzed by pancreatic lipase. The resulting fatty acid is subsequently absorbed by the enterocytes comprising the intestinal wall. However, in the presence of a lipase inhibitor, unhydrolyzed fat passes unchanged through the GI tract. Eventually the emulsifiers and water are absorbed and the triglyceride collects in the lower intestine leading to the oily stool and associated side effects.

Since the unhydrolzyed fat is emulsified in the presence of anionic emulsifiers such as phospholipids, fatty acids, and bile acids, we considered the possibility that non-absorbed cationic polymers could form polyelectrolytesurfactant complexes with these physiological emulsifiers. The resulting macromolecular network may possess physical properties that could substantially influence the state of the fat emulsion. For example, the polyelectrolytesurfactant network could provide a rigid or waxy matrix able to encapsulate or stabilize the oil droplets, preventing them from coalescing into a bulk oil phase. In this scenario, the polymeric fat binder could effectively reduce or eliminate the presence of fluid TG in the lower intestine, thereby minimizing the side effects observed for patients undergoing lipase inhibitor therapy for obesity. The proposed rationale for fat binding is illustrated schematically in Fig.3.12.

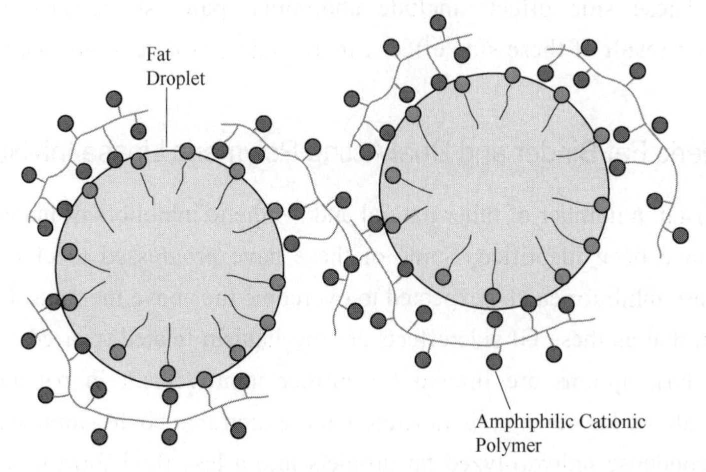

Fig.3.12 Schematic representation of sequestration/stabilization of free fat droplets by fat binding polymers.

Some laboratories prepared a systematic array of polymers of varying hydrophobicity and charge. Then devised tests to measure the fat binding efficacy of these materials. This included an in vivo animal model using rats (fed with lipase inhibitor) to screen polymers for their fat-binding properties. The effect of polymers in overcoming the oily stool side effect (induced by lipase inhibitor treatment) was evaluated with this animal model.

Although non-ionic polymers such as hydrophobically modified polyethylene glycol and polyethyleneglycol-polypropylene glycol block copolymers are known to exhibit emulsifying properties, they performed very poorly in animal model. Similarly, copolymers containing anionic and zwitterionic monomers were found to be ineffective. On the other hand synthetic polycations (based on monomers containing amine and ammonium groups) exhibited strong lipid binding properties. A systematic SAR study led to several potent and non-toxic copolymer compositions that completely eliminated the fluid lipid side effect in vivo. Table 3.5 summarizes a representative list of functional polymer segments that showed favorable fat-binding properties and led to overcoming the lipase inhibitor-induced GI side effect.

Table3.5 Representative examples of fat-binding polymers

In subsequent studies, a novel lipid-binding copolymer was discovered that covalently incorporates a potent and novel lipase inhibitor. This novel copolymer exhibited excellent lipase inhibition in vivo with no fluid-lipid side effects. This polymer has shown safety and efficacy in a number of preclinical studies and has been licensed to Peptimmune, Inc., where it is currently undergoing human clinical trials.

3.6 Polymer Therapy for Sickle Cell Disease

3.6.1 Sickle Cell Disease

Sickle cell disease is caused by a mutation in the gene responsible for the production of hemoglobin. Substitution of a single amino acid (valine for glutamic acid) in the sixth position of the B-chain of the hemoglobin molecule results in a hydrophobic region upon deoxygenation. Because the hydrophobic regions aggregate, the abnormal hemoglobin (called hemoglobin S) polymerizes into strands and forms long, rod-like structures. These elongated hemoglobin fibers distort the blood cell, producing the characteristic crescent or "sickle" shape. Surface molecules are also expressed that promote abnormal adhesion of the defective blood cell. Normal red blood cells are smooth, flexible, and shaped like a donut and can pass easily through small blood vessels. The damaged sickle shaped blood cells are stiff and are unable to pass through the blood vessels leading to reduced blood flow and sometimes blockage. The blockage (vaso-occlusion) results in severe, debilitating painful episodes and ultimately to the damage of tissues, and are one of the most common and difficult problems caused by sickle cell disease.

3.6.2 Non-Ionic Surfactant for Treating Sickle Cell Disease

In 1987, a treatment suggestion for vaso-occlusive event was proposed with the demonstration that pluronic F-68 was able to improve the filterability and rheology of sickle cells. Pluronic F-68 reduces the endothelial adherence and improves the rheology of liganded sickle erythrocytes. Chemically, pluronics are A-B-A type triblock copolymers containing a segment of polypropylene oxide (PPO) that is sandwiched between two polyethylene oxide (PEO) segments (**Scheme 32**). Because PPO is a hydrophobic segment, pluronic is able to adsorb to hydrophobic molecules or surfaces while the hydrophilic PEO segments can extend into aqueous phases, these polymers have surfactant properties. The physical properties of pluronics can be modified by changing the PEO and/or the PPO block size. This strategy enables the synthesis of block copolymers that have found a wide range of applications from drug and gene delivery to surface patterning .

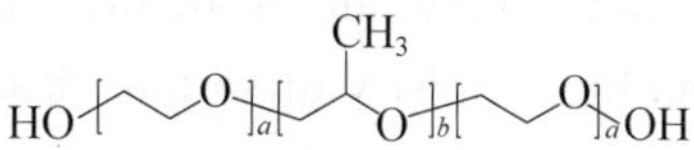

$$HO\!\left[\!\!\diagup\!\!\diagdown\!O\right]_a\!\left[\!\!\diagup\!\!\overset{CH_3}{|}\!\!\diagdown\!O\right]_b\!\left[\!\!\diagup\!\!\diagdown\!O\right]_a\!H$$

Scheme 32

Stemming from the use of pluronics as solubilizers for small molecule drugs, it became apparent that pluronics is more than just a simple delivery system and provides beneficial biological effects both symbiotically with other treatments and alone.

A purified version of pluronic F-68 is being developed by SynthRx under the brand name Flocor as a potential treatment of vaso-occlusive crisis associated with sickle cell disease. After establishing the tolerability of this polymer in clinical studies, further studies of sickle cell patients treated with Flocor showed a reduction in the length of painful episodes and an increase in the number of patients who achieved resolution of the symptoms. The effect observed was significant but small, and the effects seen were more pronounced for children under the age of 15. Further development of Flocor has been reportedly planned, especially for pediatric sickle cell patients.

Although its exact mechanism of action is unknown, Flocor has demonstrated several properties that inherently seem beneficial to sickle cell patients, including: lowering blood viscosity, decreasing red blood cell aggregation, and decreasing friction between red blood cells and vessel walls to increase microvascular blood flow and decrease cell injury. Additional studies would be needed to further clarify Flocor's mechanism(s) of action that may lead to a new generation of polymers for the treatment of sickle cell disease.

3.7 Conclusions and Outlook on Polymers as Drugs

Although polymeric materials have been developed as biomaterials and drug delivery systems, they have not generally been thought of as useful therapeutic agents on their own. In the present article we have attempted to illustrate the extraordinary potential of polymers in the discovery and development of novel human therapeutics. Through appropriate consideration of both disease targets and their mechanisms of action, several functional polymers have been discovered that exhibit useful pharmacological properties. These polymeric drugs capitalize on the unique physicochemical properties of polymer materials and in many cases, exhibit therapeutic properties that cannot be achieved by traditional small molecule drugs. The diversity and activity of these inherently pharmacologically active polymers for the treatment of a number of human diseases that have been either developed or are being evaluated is indeed very impressive. Sequestrants that are confined to the GI tract and carry out their disease-modifying activities are nice examples that support this point. The recognition of polyvalent interactions as a design principle for developing pharmaceutical agents has demonstrated that polymeric drugs can provide a new paradigm for the next generation of pharmaceuticals. Although many have not yet met the goal of exhibiting in vivo biological activity, the successful development of tolevamer suggests that it is not solely an academic curiosity. Finally, our improved understanding of cell biology and biochemistry of various disease targets should further enable us to design polymeric drugs with the desired immunological pharmacological properties for systemic applications. Once these design criteria are identified, exciting and potentially more selective polymer therapeutics will be discovered to treat human diseases, whose medical needs have been either unmet or inadequately met.

References

[1] W. Scheler. (1987) Makromol Chem Macromol Symp 12:1.

[2] R. Langer, D. A. Tirrell. (2004) Nature 428:487.

[3] J. Kohn. (1996) Pharm Res 13:815.

[4] J. Jagur-Grodzinki. (1999) React Funct Polymers 39:99.

[5] H. Ringsdorf. (1975) J Polym Sci Polym Symp 51:135.

[6] Y. Luo, G. D. Prestwish. (2002) Curr Cancer Drug Targets 2:209.

[7] K. Ulbrich, M. Pechar, J. Strohalm, V. Subr. (1997) Makromol Chem Macromol Symp 118:577.

[8] M. C. Garnett. (1999) Crit Rev Ther Drug Carrier Sys 16:147.

[9] Z. H. Israel, A. J. Domb. (1998) Polym Adv Technol 9:799.

[10] L. Gros, H. Ringsdorf, H. Schupp. (1981) Angew Chemie Int Ed Eng 20:305.

[11] R. Langer. (1998) Nature 392:5.

[12] R. Duncan. (2003) Nature Drug Discovery 2:347.

[13] P. T.Kuo, F. T. Hopkins, H. Wurzel. (1958) Circulation Research 6:178.

[14] W. Regelson, J. F. Holland. (1962) Clin Pharmcol Ther 3:730.

[15] D. S. Breslow.(1976) Pure Appl Chem 46:103.

[16] M. A. Chirogos, W. A. Stylos. (1980) Cancer Res 40:1967.

[17] C. V. Uglea, L. Panaitescu. (1997) Curr Trends Polym Sci 2:241.

[18] A. Bentolila, I. Vlodavsky, R. Ishai-Michaeli, O. Kovalchuk, C. Haloun, A. J. Domb. (2000) J Med Chem 43:2591.

[19] P. K. Dhal, S. R. Holmes-Farley, W. H. Mandeville, T. X. Neenan. (2004) Encyl Polym Sci Technol. 3rd Edn. Wiley, New York.

[20] L. R. Johnson. (1994) Physiology of Gastrointestinal Tract. Raven Press, New York.

[21] E. J. Braun. (2003) Comp Biochem Physiol A: Mol & Int Physiol 136A:499.

[22] G. O. Perez, J. R. Oster, R. Pelleya, P. V. Catalis, D. C. Kem. (1984) Nephron 36:270.

[23] T. J. Marrone, K. M. Terz. (1992) J Am Chem Soc 114:7542.

[24] G. W. Gardiner. (1997) Can J Gastroenterology 11:573.

[25] B. B. Gerstman, R. Kirkman, R. Platt. (1992) Am J Kidney Dis 20:159.

[26] C. Hsu. (1997) Am J Kidney Dis 29:641.

[27] S. Ribeiro, A. Ramos , A.Brandao. (1998) Nephrol Dial Transplant 13:2037.

Chapter 4

Pharmaceutical Applications of Natural Polymers

Natural polymers have a very broad range of applications in both the polymer and pharmaceutical industries. The pharmaceutical industry is a very broad field where there is a continued need to consider various applications. It is logical, therefore, to state that understanding the roles of natural polymers in the pharmaceutical industry helps in turn the polymer industry to determine the broader applications of these polymers and incorporate the desired requirements to meet the end applications (e.g. provide various functionalities). Drug delivery methods form a key part of the pharmaceutical applications of polymers. This chapter first discusses portals of drug administration into the human body which gives an overview of the possibilities of applications of natural polymers, and then discusses some specific applications in detail. Transdermal drug delivery, nasal drug delivery, vaginal, ocular, oral drug delivery methods using natural polymers are discussed with some example case studies. As hydrogels play important roles in drug delivery, a separate section is dedicated in discussing the applications of natural polymer-based hydrogels in drug delivery.

4.1 Introduction

Polymers derived from natural resources have been widely researched as biomaterials for a variety of biomedical applications including drug delivery and regenerative medicine. These molecules have biochemical similarity with human ECM components and hence are readily accepted by the body. Additionally these polymers inherit several advantages including natural abundance, relative ease of isolation and room for chemical modification to meet the technological needs. In addition these polymers undergo enzymatic and hydrolytic degradation in the biological environment with body friendly degradation bye products. Natural polymers include the list of polysaccharides and animal derived proteins. Polysaccharides are an important class of biomaterials with significant research interest for a variety of drug delivery and tissue engineering applications due to their assured biocompatibility and bioactivity. Polysaccharides are often isolated and purified from renewable sources including plants, animals, and microorganisms. Essentially these polymers have structural similarities, chemical versatilities and biological performance similar to ECM components, which often mitigate issues associated with biomaterials toxicity and host immune responses. The building block of carbohydrate monosaccharide's are joined together by *o*-glycosidic linkages to form polysaccharide chains. Polysaccharides offer a diverse set of physicochemical properties based on monosaccharide's that constitutes the chain, its composition and source.

The popular list of polysaccharides used for a variety of biomedical applications includes cellulose, chitin/chitosan, starch, alginates, HAs, pullulan, guar gum, xanthan gum, and GAGs. In spite of many merits as biomaterials, these polysaccharides suffer from various drawbacks including variation in the material properties based on the source, microbial contamination, uncontrolled water uptake, poor mechanical strength and unpredictable degradation pattern. These inconsistencies have limited their usage and biomedical applications related technology development.

Numerous synthetic polymers with well defined mechanical and degradation properties have been developed to meet the technological needs in the biomedical applications. However these polymers from the biological standpoint lack much desired bioactivity and biocompatibility and may cause toxicity and immune response.

Polysaccharide structure offers freely available hydroxyl and amine functionalities that make it possible to alter its physicochemical properties by chemically modifying polysaccharide structure. For instance grafting synthetic monomers on the polysaccharide chains offer an easy way to control polymer solubility undesired solvents, water uptake and degradation. These semi synthetic polymers offer best feature of the both natural and synthetic polymers. Various cross linking techniques to restrict the polysaccharide chain movement to control their water uptake, degradation and mechanical properties have also been developed. Polysaccharides based porous scaffolds, fiber matrices, hydrogels, and micro and nanoparticles have been developed for variety of tissue regeneration and drug delivery applications. In the recent years glycochemistry has gained research momentum for understanding carbohydrate biological functions and development of carbohydrate based drugs and vaccines. Engineered carbohydrate based polymeric structures may serve as an alternative material platform for a variety of regenerative medicine and drug delivery applications. A new nonpetroleum based biomaterial platform to meet the versatile needs in biological science and biomedical engineering could be achieved by collaborative efforts between academia, government and industry partnership. The collaborative efforts should include bringing scientist working in different disciplines of chemistry, biology, polymers, materials sciences and engineering to work toward these activities. The collaborative efforts could lead to the development of a methodology for synthesis natural polymer based semi synthetic polymers and provide a greater depth of understanding of carbohydrate biological functions, polymer structure, material properties degradation and mechanical properties. Further the development of modeling tools to predict the structure, property and biological activity of carbohydrates for biomedical applications is a step in this direction. The goal of new initiatives should focus on the development of natural polymer based orthopedic fixation devices, biomedical implants, drug delivery vehicles, carbohydrate based drugs, hydrogels, surfactants, coagulants, and absorbents for a variety of biomedical applications. The research activities in this area could generate commercially available technologies and product from the renewable resources and contribute immensely toward economic development.

4.2 Portals of Drug Administration in the Human Body

The controlled delivery of drug molecules requires either a device or a vehicle for administration into specific localized tissues or systemic distribution via plasma fluid in blood. The human body has several portal entries for drug administration as outlined in Fig.4.1. These portals are intramuscular, percutaneous, intrathecal, subcutaneous, gastrointestinal, ocular, intravenous, nasal, pulmonary, sublingual, baccal, rectal and vaginal. Intravenous, intramuscular, percutaneous, intrathecal, subcutaneous and transdermal are collective terminologies associated with parenteral administration. Pulmonary drug administration through the lungs is the least common portal delivery because of a limited number of excipients, especially natural polymeric excipients with reduced polydispersity of size and ideal particle densities concerning the drug particle formulation.

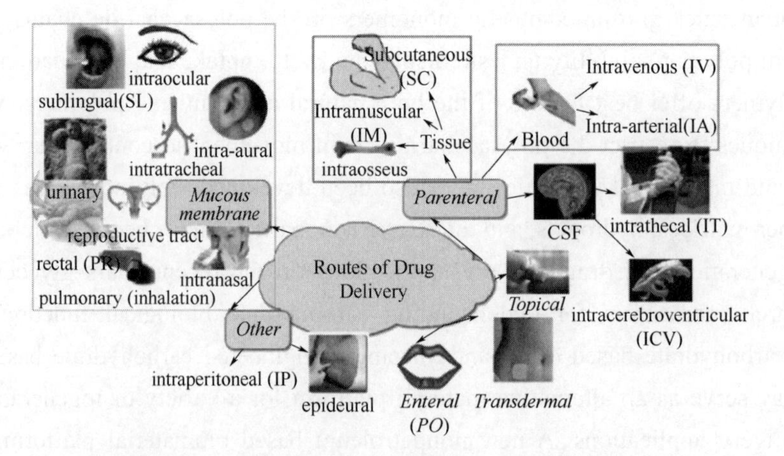

Fig.4.1 An outline of main portals for drug administration.

4.3 Transdermal Drug Delivery Devices

Polymers are used extensively in transdermal drug delivery systems. They control the rate of drug release from the device, act as primary packaging parts, coatings, penetration enhancers and provide ease in drug device handling and structural support to the device in the form of a backing layer. Their unique properties make them ubiquitous component of transdermal patches. As petrochemical-based resources for the production of synthetic polymers become more expensive and in short supply, the production of transdermal drug delivery device components from more readily available natural polymers becomes eminent. This section looks at the application of natural polymers in transdermal drug delivery. The different parts of the transdermal drug delivery system are discussed such that the use of natural polymers in each individual part, namely, the matrix, adhesive layer, rate controlling membrane, backing layer, release liner and penetration enhancer are then discussed. Further areas to be explored are also suggested.

Polymers are used more extensively in transdermal drug delivery (TDD) than any other material as they possess unique properties which are significant to the drug delivery process. They are effective in aiding the control of drug release from carrier formulations. Polymers commonly used in TDD include cellulose derivatives, polyvinylalcohol, chitosan, polyacrylates, polyesters such as PLGA, PELA, and PLA and silicones. Natural polymers are a preferable option in TDD as they are readily available, inexpensive, potentially biodegradable and biocompatible and can undergo various chemical and surface modifications to fit the requirement of the TDD system. A TDD system comprises of a combination of one or more polymers and an embedded drug to be delivered into or through the skin in a controlled and sustained manner.

Polymers used for TDD systems are required to be chemically inert and pure according to high analytical product yields. It should also possess adequate physical properties which correspond with the intended application. The material must not age easily and be suitable for processing. Furthermore, biodegradability and safety are paramount properties in the design of a TDD patch system due to the long-term exposure of the skin in contact with the patch.

4.3.1 Types of Transdermal Drug Delivery Systems

TDD systems are classified into three types, namely, reservoir, matrix and microreservior system. Each of these is described below.

4.3.1.1 Reservoir System

The reservoir system comprises of a reservoir of drug in the form of a suspension, solution or liquid gel embedded between an impervious backing layer and a rate controlling membrane. Suspensions and solutions are two distinct types of liquid mixtures. The definition of a suspension and a solution is well understood. The definition of a liquid gel can sometimes be difficult to formally express. A gel is a semi-solid, colloidal solution consisting of one or more crosslinked polymers dispersed in a liquid medium. A liquid gel is softer, less resilient and easily spreadable colloid gel. The reservoir could also be the drug dispersed within a solid polymer matrix. An adhesive polymer is often placed between the rate controlling membrane and the skin.

4.3.1.2 Matrix System

The matrix system comprises of drug molecules dispersed within a polymer matrix. The matrix system is of two types, the drug in adhesive system and the matrix-dispersed system. In the drug in adhesive system, the drug is dispersed in a polymer adhesive. The drug loaded adhesive polymer is then spread by solvent casting or in the case of hot-melt adhesives, where, it is melted onto an impervious backing layer. Additional layers of adhesive polymer are then applied on top of the reservoir. In the matrix-dispersed system, the drug is dispersed homogeneously in a polymer matrix which is either lipophilic or hydrophilic. The polymer is then placed on a backing layer and above this matrix, an adhesive layer surrounds the matrix perimeter.

4.3.1.3 Microreservoir System

This system combines the reservoir and matrix-dispersed systems. The drug is first suspended in aqueous solution in a water soluble polymer. This is then dispersed homogeneously in a lipophilic polymer, which results in the formation of microscopic spheres of drug reservoirs dispersed within a polymer matrix.

4.3.2 Natural Polymers in Transdermal Drug Delivery

Polymers have been used in transdermal drug delivery as far back as the 1980s. Most transdermal patches contain a matrix of cross-linkage of linear polymer chains from which the drug is to be absorbed into the skin. Polymers used in transdermal drug delivery include cellulose derivatives, polyvinyl alcohol, polyvinylparrolidone, polyacrylates, silicones and chitosan. Both natural and synthetic polymers have been used either as matrices, gelling agents, emulsifiers, penetration enhancers or as adhesives in transdermal delivery systems. For example It has been reported the successful delivery of testosterone into lab rats using a transdermal delivery system with a silicone elastomer as synthetic polymer matrix.

Another group explored the use of pectin hydrogels for the transdermal delivery of insulin. Pectin hydrogels loaded with insulin were administered to diabetic rats with type Ⅱ diabetes mellitus. The results obtained showed that the transdermal patch delivered insulin across the skin in a dose dependent manner with pharmacological effect. More recent studies explored the use of natural polymers such as rubber latex as backing layer adhesive in nicotine patches. There is, therefore, scope for research into the use of natural polymers in transdermal drug delivery.

Although synthetic polymers seem to be more commonly employed in the development of TDD systems, natural polymers from plant and animal sources are emerging as a preferred alternative as they pose the advantage of being biocompatible, biodegradable, degrading into non-toxic monomers and are more readily available. Synthetic polymers derived from petroleum sources and synthetically modified polypeptides are known to have limited pharmaceutical implementations due to toxicity and slow biodegradation rates. The following section discusses the use of natural polymers in the different parts of a transdermal patch system.

4.3.2.1 Controlled Release Systems

Natural polymers in combination with other natural and/or synthetic polymers have been used in hydrogels for pharmaceutical application. A recent study looked at the development of controlled release system based on thermosensitive chitosan-gelatin-glycerol phosphate hydrogels for ocular delivery of latanoprost, a drug used in the treatment of glaucoma. The formulation can be delivered via subconjunctival injection reducing the need for repeated dose administration and possible side effect from conventional treatments of the condition.

4.3.2.2 Matrix

Polymers are attractive for use as matrices in transdermal patches due to certain useful properties which they possess. In addition to being biodegradable and biocompatible, they contain various functional groups that can be modified as required and combined with other materials and tailored for specific applications.

When exposed to biological fluids, biodegradable polymers will degrade releasing the drug that is dissolved or dispersed within them. There are on-going research studies into the application of natural polymers in TDD as polymer matrices. In this area biocompatibility and biosafety are a paramount requirement. Release of APIs (active pharmaceutical ingredients) from a polymer matrix occurs via various mechanisms including polymer erosion, diffusion, swelling followed by diffusion and degradation. The mechanism initiated depends on the type of system.

The use of various natural polymers as matrices has been explored by different research groups. These include natural polymers of chitosan, a polycationic (pH 6.5 or less in solvent) natural polysaccharide which is obtained from one of the most abundant polysaccharides in nature, chitin . Chitin is a natural polymer which forms the shells of crustaceans, some insects, fungi, yeasts and plants. Chitosan is deacetylated chitin with a degree of deacetylation ranging from 60% to 95 %.

The rate of drug delivery from a chitosan matrix can be controlled by varying the manner in which the chains are crosslinked. The most common crosslinkers used for fabrication of chitosan gels are glutaraldehyde, formaldehyde, glyoxal, dialdehyde starch, epoxy compound, diethyl squarate, pyromellitic dianhydride, genipin, quinone and diisocyanate. Preparations of chitosan in the form of beads, microspheres and gels have been shown to deliver drugs such as local anaesthetic drugs, lidocaine hydrochloride and anti-inflammatory drugs, prednisolone . Chitosan has also been used as a matrix for transdermal delivery of large protein molecules such as insulin. It is robustly physicochemically stable and possesses mucoadhesive property which makes a good candidate for TDD.

Pectin is also another natural polymer used in TDD matrices. Pectin is a water soluble polysaccharide composed of different monomers, mainly D-galacturonic acid, sourced from the cell walls of plants which grow on land. Pectin is commercially extracted from fruits and its appearance ranges from a white to light brown powder. Recent studies on application of pectin in TDD have looked at modifying pectin to act as matrix for TDD And pectin was used as a matrix for delivery of chloroquine through the skin; The results showed that pectin was effective as a matrix for TDD delivery of chloroquine resulting in more effective and convenient treatment of malaria. Soybean lecithin has also been used as gel matrices to deliver scopolamine and dcoxatenol transdermally.

Lecithin is a component of cells that is isolated from soya beans or eggs. It is processed into Lecithin organogel (LO) to act as a matrix for topical delivery of many bioactive agents into and through the skin. When purified and combined with water it shows excellent gelating properties in non-polar solvents. LO provides a temperature independent resistant to microbial growth as well as being a viscoelastic, optically transparent and non-birefringent micellar system. LO is a dynamic drug delivery vehicle as it dissolves both lipophilic and hydrophilic drugs. It effectively partitions into the skin thereby acting as an organic medium to enhance permeation of otherwise poorly permeable drugs into the skin.

A combination of more than one polymer can also be used in a TDD matrix and this also applies to natural polymers. For example a matrix comprising of xanthan gum and sodium alginate was developed. In vitro evaluation of the TDD system showed good compatibility and controlled release of the model drug Domperidone following in vitro release in a glass diffusion cell.

Other natural polymers that are being explored for use as matrices in TDD include collagen, gelatin, agarose from seaweed, natural rubber, polyethylene obtained from bioethanol, and polylactide (PLA), a polyester of lactic acid which is produced from starch or cane sugar fermentation by bacteria.

4.3.2.3 Rate Controlling Membrane

Rate controlling membranes are used when the TDD patch is a reservoir type such that the rate at which the drug leaves the device is regulated by the membrane which is either a porous or non-porous membrane. Various natural polymers are being explored for use as rate controlling membranes. These polymers usually have attributes such as good film forming properties and variable film thickness.

An optimised formulation of DamarBatu (DB), a natural gum from the hardwood tree of the Shorea species such as *S. virescens* Parijs, *S. robusta* and *S. guiso* was developed. The optimised formulation was shown to successfully deliver Eudragit RL00, the model drug. Following in vitro drug release, skin permeation studies and other analysis concluded that Eudragit RL100 is a suitable film for TDD. In other studies DB has also been evaluated as a rate controlling membrane for TDD of a model drug diltiazem hydrochloride.

Gum copal, a biological polymer gum has also been tested as a film for TDD. The effect of different plasticizers was tested on the effectiveness of gum copal as a rate controlling membrane.

The effectiveness of the film produced was estimated from tensile strength of the film, uniformity of the thickness, moisture absorption, water vapour transmission, elongation, foldability and drug permeability. PEG400 was found to be the plasticizer which gave the best permeability among those tested. However, a more sustained delivery was achieved in vitro with a formulation containing 30 % *W*/*W* DPB (dibutylphalate).

Another natural polymer with good film forming properties is zein. It is a protein obtained as a by-product from the processing of corn. Zein shows potential as a low cost and effective alternative to synthetic films for TDD.

4.3.2.4 Adhesives

Adhesives are required in TDD systems to ensure the device remains in contact with the skin. For TDD the selected adhesive must meet certain criteria such as skin compatibility, biodegradability and good adhesion over long period due to the long-term contact with the skin and drug formulation.

Pressure sensitive adhesives (PSA) are materials which adhere or stick to the surface following application of normal finger pressure and remains attached exerting a strong holding force. When removed from the attached surfaces, PSAs should ideally leave no residues. Adhesion refers to a liquid-like flow which causes wetting of the skin surface as pressure is applied with the adhesive remaining in place after the removal of the applied pressure. The adhesion is achieved as a result of the elastic energy that has been stored during the breaking of bonds caused by applied pressure. The effectiveness of the PSA is, therefore,an attributable to the relation between viscous flow and stored elastic energy. Synthetic polymers seem to have dominated the adhesives used in TDD. Commonly, used ones include acrylic, polyisobutylene and silicones.

Use of adhesives on skin is an idea that has been around for many decades, one of the earliest applications being in bandages for wound healing by Johnson and Johnson company in 1899. When deciding on what kind of polymer to incorporate as an adhesive, an understanding of the properties of the skin is essential. The surface energy of the skin, which acts as the adherent in the case of TDD, must be greater than or equal to the surface energy of the adhesive. Furthermore the skin properties vary with the factors such as age, gender, race and environmental conditions. Therefore, the effect of properties such as moisture content of skin and the viscometric property of the adhesive should be established.

Adhesives in transdermal patches may exist as a single adhesive layer or a drug in adhesive type, the latter is preferred as the simplest to apply however, it is rather complicated to produce. For drug-in-adhesive type patches, issues which must be addressed include the tendency of the drug or adhesive to crystalize. This will have an effect on the drug delivery rate as it permeates through the adhesive layer.

Pressure sensitive adhesives generally comprise an elastomeric polymer, a resin for tack, a filler, antioxidants, stabilisers and crosslinking agents. Although synthetic polymers seem to be more commonly used as adhesive in TDD systems, the development of adhesives from natural polymers is becoming a rather attractive area of interest. Various sources in nature have been explored for obtain adhesives.

Carbohydrates are readily available polymers of plants. Cellulose, starch and gums are the most common forms that are used in production of adhesives. There are studies which have been focused on the production of adhesives obtained from cellulose recovered from domestic and agricultural waste. These include soy protein, raft lignin and coffee bean shells.

Adhesives formed from carbohydrates include carboxymethyl cellulose (CMC), hydroxyethyl cellulose, ethyl cellulose, methyl cellulose, cellulose acetate and cellulose nitrate. Those formed from starch such as tapioca, sago and potatoes can be more readily converted to adhesives following modification through heating, alkali, acidic or oxidative treatment. The adhesives often require further additives during processing. Recent studies focused on extracting natural polymeric adhesives include that studies acacia mangium bark extracts as a source of natural polymer adhesives.

The dicotyledonous tree bark which is commonly grown in Malaysia as a source of raw material for veneer, pulp and paper showed a promising prospect as an alternative to adhesives produced from petrochemicals. In other works adhesive production from waste materials such as de-inked waste paper has been studied. In a particular study de-inked waste paper from magazines were washed using detergent under stirring. This was then followed by further processing under heat at 150 ℃ and treatment with acid and ethylene glycol. The glycosides which resulted from the breaking down of the cellulose were then transesterified using rice-bran castor and soy oils to convert it to polyols. Polyurathenes are then produced from the polyols. The adhesives produced using the methods described when tested showed strong adhesive properties than the commercial adhesives and also showed significant water resistance. Marine organisms and bacteria have also been shown to be the sources of natural adhesives. The main limitation with these sources is the expensive production process.

4.3.2.5 Penetration Enhancer

Polymers are also used as penetration enhancers to aid the permeation of drugs across skin. Polyethylene glycol solution is an example of such penetration enhancers of prodrugs across skin models. However use of polymers as additives in formulations also carries some limitations such as inhibiting the bioconversion of the drug. Transdermal films incorporating 0.5 % tenoxicam have been developed from varying ratios of glycerol, PEG 200 and PEG 400. Using Fourier transform infrared spectroscopy, it was found that increasing the concentration of PEG enhanced the penetration of tenoxicam into the skin. Polymers are also employed as other formulation additives in the form of viscosity enhancers and as emulsifiers. Chitosan, a natural polymer has been used as a penetration enhancer, which acts by opening up the tight junctions which exists between epithelial cells.

Recently, research studies aimed at fabricating micron-sized penetration enhancers which partially disrupt the stratum corneum layer creating a more permeable pathway for drugs to enter into the skin via natural polymers is emerging. For example dissolving polymer microneedles could be fabricated from silk Fibroins obtained from bombyx mori silk worm. The resulting structures were rapidly dissolving microneedles with adjustable mechanical parameters and were biocompatible with skin. Maltose has also been used to fabricate dissolving polymer microneedle using traditional casting methods as well as using the extrusion drawing method. More recent studies have looked at the application of hydrolyzed collagen extracted from fish scales for production of microneedles as mechanical penetration enhancers.

4.3.2.6 Backing Layer

The backing layer comes in contact with the drug matrix or reservoir therefore the chemical inertness of the material used for the backing layer is required. The backing layer must also be compatible with the excipient formulation. Back-diffusion of the drugs, penetration enhancer or excipient must not occur even over a long period of contact. While maintaining chemical inertness it must also be ensured that the backing layer is flexible enough to allow movement, transmission of moisture vapour and air in order to prevent skin irritation during long-term contact with skin. Adequate transmission of moisture vapour and air also prevents the weakening of the adhesive hold on the skin surfaces. In more modern designs of TDD patches, the backing layer could be solidified with the reservoir to form a single structure such that it serves as a storage space for the reservoir. More recent studies have explored the use of natural polymers as backing layer of nicotine transdermal patches from natural rubber latex.

4.3.2.7 Release Liner

The adhesive side of the transdermal patch is usually covered with a liner which protects the adhesive and the rest of the patch during storage. Although mostly for packaging purpose, the liner is in direct contact with the adhesive layer throughout the storage period. The material used as a release liner should be chemically inert and resistant to the permeation of the drug, penetration enhancer and moisture. The liner should also not cross link with the adhesive such that it becomes difficult to remove. Example of a release liner used in commercial TDD is the ScotchPak™ 1022 and Scotchpak™ 9742 liner which are produced from fluoropolymers by 3M Drug Delivery Systems.

Although currently the use of synthetic polymers seem to dominate that of natural polymers in TDD, there is increasing research interest in incorporating natural polymers in new ways in TDD systems. This is attributable to the desire to produce pharmaceutical products with more desirable environmental impacts, reduce dependency on fast diminishing petrochemical resources and developing more sophisticated TDD systems with better effectiveness and biosafety. However, there is yet to be a transdermal drug delivery system which is developed fully from natural polymers. The dependency on synthetic polymers therefore still persists. Future research efforts directed towards developing novel natural polymers from new biological sources.

Consequently as new polymers emerge, extensive studies will be required to identify the physical and chemical properties of the new biomaterials. Furthermore developing newer processing methods and new combinations of polymers could be optimised leading to improved effectiveness in transdermal drug delivery.

Natural polymers have proven valuable in transdermal drug delivery systems. They have a wide range of applicability and pose several advantages over synthetic polymers in this application. Nature offers an abundant supply of polymers with numerous properties. Understanding these sources and properties allow us to further modify these polymers to suit specific requirements. The area of transdermal drug delivery still faces certain limitations such as skin irritation and limited range of drugs which can be delivered through this means. Exploring new polymers from natural sources could provide new solutions and offer clinical and commercial development in the area of transdermal drug delivery.

4.4 Topical Drug Delivery

Delivery of drugs into the body topically can be employed to treat conditions which exist on or close to the surface of the skin. This could vary from aches and bruises to severe burns and mild and chronic conditions such as eczema and psoriasis. This form of delivery refers to when a drug formulation is applied directly to the external skin surface or surface of the mucous membrane of the vaginal, anal oral, ocular or nasal area for local activity. Topical delivery through the other entry routes (i.e. oral, vaginal, ocular, etc.) is not to be confused with the other forms of delivery.

4.4.1 Advantages and Disadvantages

Topical delivery is relatively convenient and has relatively better patience compliance than, e.g. oral or intravenous injection which could impose adverse impacts such as nausea, low bioavailability due to metabolism of drug in the gastrointestinal tract, needle phobia and general preferences. The specificity of topical delivery is also advantageous as it can be directly applied to the affected area to act locally, similarly the medication can be easily terminated by simply cleaning off the medication. Topical delivery particularly becomes a favourable option where other routes of entry into the body are deemed unnecessary or unsuitable depending on the individual or nature of drug. In cases where for instance oral delivery of the drug could induce adverse effect which could even is more severe than the actual condition being treated. For instance many of the adverse effect associated with antifungal drug fluconazole are gastrointestinal related and could be avoided by applying a topical formulation for effective delivery of the drug.

Main challenges in the area of topical drug delivery alongside skin irritation and allergic reaction include skin penetration into target region especially for drugs with large particle size. In particular, situations such as in fungal infection where the penetration into the stratum corneum is further inhibited as an attack mechanism of the pathogen to prevent shedding of the stratum corneum, penetration enhancement of the topical agent becomes of relative importance in the effectiveness of the drug formulation.

4.4.2 Composition of a Topical Formulation

The main components of a topical formulation includes a vehicle which could be in the aqueous form, mainly water or alcohol, or it could be an oil such as mineral oils, paraffin, castor oil, fish liver oils, cotton seed oil, etc. A vehicle should maintain effective deposition and even distribution of the drug on the skin; it should allow delivery and release to the target site and maintain a pharmacologically effective therapeutic concentration of the drug in the target site. In addition to these properties a suitable vehicle should be well formulated to meet patient's cosmetic acceptability and be well suited for the anatomic site.

Emulsifiers are important to maintain stability and the distribution of the water and oil emulsion throughout the shelf and usage lifespan of the formulation Typical synthetic emulsifiers include polyethylene glycol 40 stearate, sorbitan monooleate (commercial name: Span 80), polyoxyethylene sorbitan monooleate (commercial name Tween 80), stearic acid and sodium stearate. Natural polymers used as emulsifiers include starch, gum acacia, alginates, xanthan gum, irvingia gabonensis mucilage, and tragacanth gum. Gelling agents are also important in increasing the bulk of the drug and thicken the topical formulation according to stable viscoelasticity. Examples include sodium alginate, cellulose in modified forms as sodium carboxymethyl cellulose (NaCMC), hydroxypropylmethyl cellulose (HPMC) and hydroxypropyl cellulose (HPC).

4.4.3 Types of Topical Formulations

Topical drug delivery systems could be in the form of gels, emulgels, emulsions, liposomes, liquids, powders and aerosols. Gels and emulgels are relatively new forms of topical delivery formulations. Gels are formed when a large amount of aqueous or hydro-alcoholic solutions are entrapped within a network of colloidal solid particles or macromolecules, while emulgels are a combination of a gel and emulsion. Emulgels are targeted at addressing the limitation of gels to delivery of hydrophilic compounds by enabling the delivery of hydrophobic compounds better than using gels or emulsions. To create emulgels for hydrophobic drugs, oil-in-water (O/W) emulsions are needed to entrap the hydrophobic drugs followed by addition of a gelling agent to the emulsions, while for hydrophilic drugs; a waterin-oil (W/O) emulsion is used. The desirable features of an emulgel include more effective cutaneous penetration, greaseless, spreadability, extended shelf life compared to gels or emulsions, biofriendly, non-staining, water soluble, moisturising and a generally transparent and pleasing appearance.

4.4.4 Natural Polymers in Topical Delivery Systems

Natural polymers are used in topical drug delivery as gelling agents, emulsifiers, stabilizers, thickeners, etc. Cellulose, alginates, chitosan, albumin, starches and xanthan gum are examples of natural polymers which have been applied in the production of topical formulations. The derivatives of cellulose such as HPMC or CMC are particularly common candidates in topical formulations and they pose a good alternative to the commonly used carbopol, a synthetic polymer. Gels are of interest for topical delivery of pharmaceutical agents as they are easy to apply, spread and remove, thus encouraging patience compliance. Excipients used in topical delivery of psoralen using natural polymers; pectin, xanthan gum, egg albumin, bovine albumin, sodium alginate and guar gum are compared in Fig.4.2. Table 4.1 shows the reagent component concentrations of the respective polymer, humectant, drug, solvent, antioxidant and preservatives in a developmental formulation of Psoralen, labelled F1~F8. Psoralen is a drug used in the treatment of skin conditions such as psoriasis, vitiligo, mycosis fungoides and eczema, but also possesses antitumor, antibacterial and antifungal properties. It belongs to a class of furano-coumarins compounds found in the psoralea corylifolia L. plant.

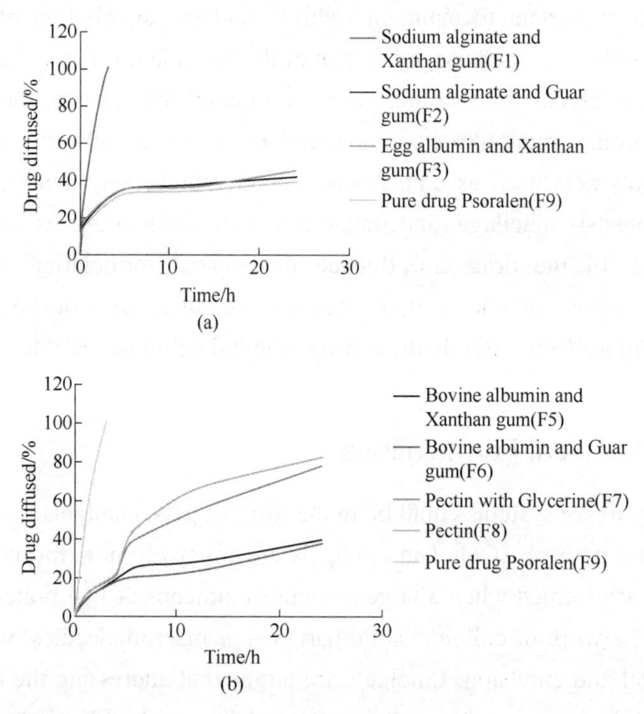

Fig.4.2 Comparing diffusion profiles of topical drug, psoralen using various natural polymerbased excipients (a) and (b).

Table 4.1 Various formulations of Psoralen using natural polymer excipients

Materials	Formulation Code							
	F1	F2	F3	F4	F5	F6	F7	F8
Psolaren/g	0.05	0.05	0.05	0.05	0.05	0.05	0.05	0.05
Sodium alginate/g	0.75	0.75	—	—	—	—	—	—
Egg albumin/g	—	—	0.75	0.75	—	—	—	—
Bovine albumin/g	—	—	—	—	0.75	0.75	—	—
Pectin/g	—	—	—	—	—	—	4	5
Xanthan gum/g	0.50	—	0.75	—	0.75	—	—	—
Guar gum/g	—	0.50	—	0.75	—	1.75	—	—
Menthol/g	0.25	0.25	0.25	0.25	0.25	0.25	0.25	0.25
σ-Tocopherol/g	0.50	0.50	0.50	0.50	0.50	0.50	0.50	0.50

续表

Materials	Formulation Code							
	F1	F2	F3	F4	F5	F6	F7	F8
Barbaloin/g	0.005	0.005	0.005	0.005	0.005	0.005	0.005	0.005
Glycerin/mL	—	—	—	—	—	—	5	—
Eugenol/mL	0.01	0.01	0.01	0.01	0.01	0.01	0.01	0.01
Methanol/mL	10	10	10	10	10	10	10	10
Distilled water to make/mL	50	50	50	50	50	50	50	50

The psoralen gel formulations were prepared by first mixing the polymer in water and stirring continuously at 37℃. This was followed by addition of gelling agent and continued mixing until a homogenous dispersion was attained. The required drug dissolved in methanol was then added followed by addition of antioxidant, preservatives and humectants. The mixture was then stirred until a homogenous mixture was obtained. All polymers used showed good compatibility with the drug. This is important as an interaction between the excipient and the drug formulation will likely affect the drug activity and could also pose some adverse health effects. This is not to say that all natural polymer excipient do not interact with the drug compound or psoralen in particular. The tendency of interaction between excipient and drug compound depends in the specific drug and specific polymer. While a polymer might show the desired biomechanical properties, release kinetics, bioactivity, etc. the applicability may be limited if there is interaction between the polymer being used as excipient and the active drug compound. For instance, nanofibrillar cellulose gels show good potential for drug delivery; however, nanofibrillar cellulose possesses various carboxyl and hydroxyl groups which may interact with drug compounds in different ways. This, therefore, must be investigated for every new formulation.

Interaction between excipient and drug compound is commonly evaluated using FTIR. Compatibility is indicated when the characteristic peaks of the pure drug are retained in the FTIR spectra of the drug formulation with the excipients present. Fig.4.3 shows the FTIR spectra of pure psoralen next to that of formulation of psoralen in albumin and xanthan gum as polymer excipients.

Of the polymers investigated, the formulation containing xanthan gum and egg albumin showed the best drug incorporation, release kinetics and in vitro antipsoriatic activity. Topical delivery system should possess sufficient pseudoplasticity and controllable release kinetics. Sodium alginate and derivatives of cellulose, sodium carboxymethyl cellulose, hydroxypropyl-methyl cellulose and hydroxypropyl cellulose when applied as excipients for topical delivery of fluconazole showed desirable pseudoplastic behaviour. This pseudoplastic behaviour is a shear thinning property that allows the topical formulation to be effectively spread with ease on the affected area while remaining in the required region for localised and sustained delivery. The release kinetics and viscosity also vary with the concentration of the polymer such that the release rate and viscosity can be varied as required by varying the concentration of the polymer as desired. Although a synthetic gelling agent, carbopol showed the best drug release profile and anti-fungi activity; the other polymers also had sufficient antifungal activity and drug release rate (Fig.4.4).

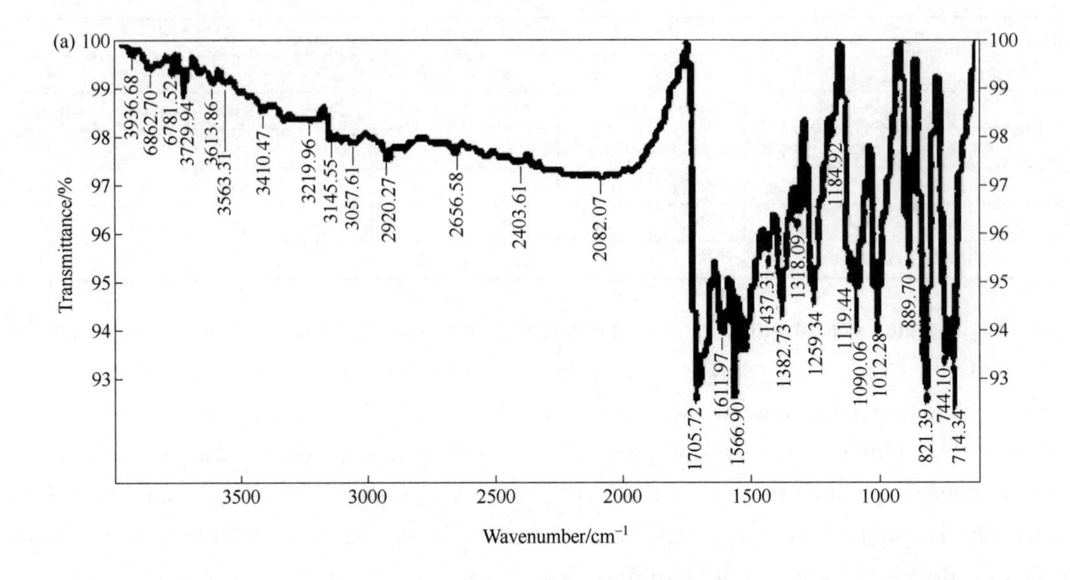

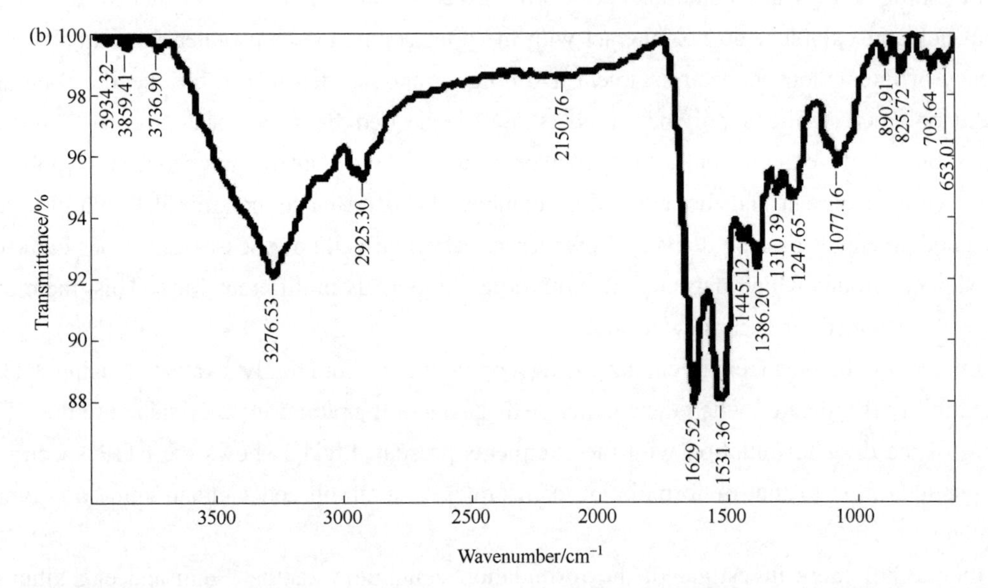

Fig.4.3 FTIR spectra of psoralen (a) and psoralen in a formulation of egg albumin and Xanthan gum (b).

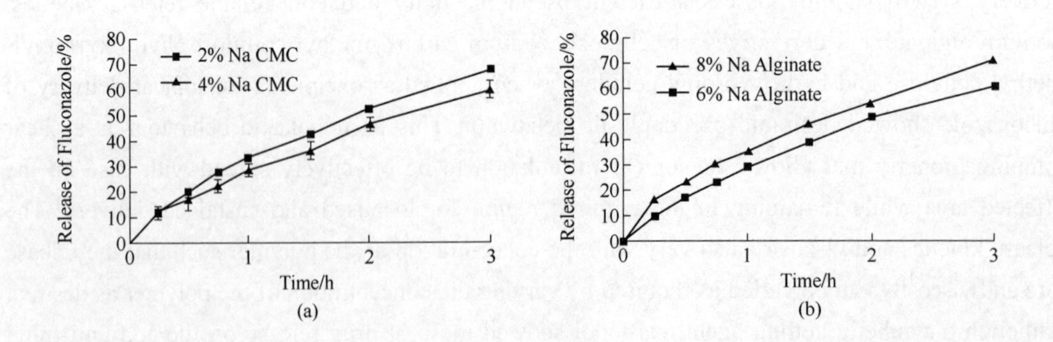

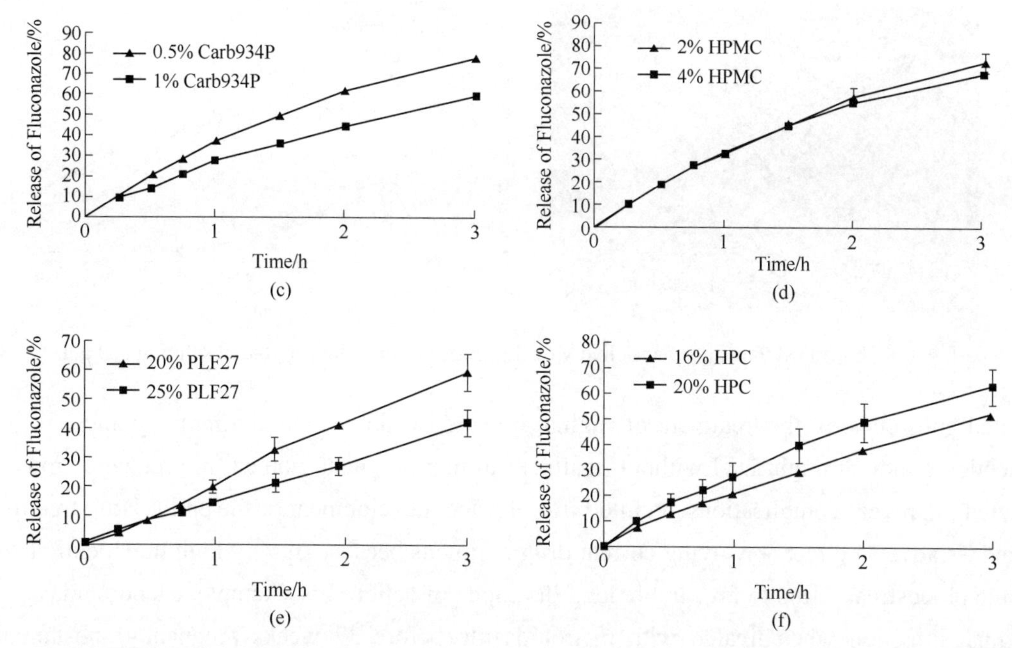

Fig.4.4 Effect of various polymer excipients on the release profile of fluconazole from prepared gel.

Over the 3 h observed, the release rate of the fluconazole increased as the concentration of the polymer id reduced. This can be attributed to increased porosity as polymer concentration reduces, allowing easier permeation of the drug compound through the polymer matrix of the gel.

The ability to control and predict release kinetics of drug formulation is important in the effective drug delivery. Here, we see gels from natural polymers showing controllable parameters comparable to that of synthetic polymers.

In the treatment of fungal keratitis, delivery and bioavailability of the antifungal agent can be enhanced by using chitosan-based formulations either in gel or solution. Topical formulations of fluconazole using chitosan solution and a gel system of chitosan with a thermoresponsive polymer poloxamer as vehicles showed improved bioavailability of fluconazole in the eye compared to aqueous solutions. The aqueous solutions used as eye drops have limited effectiveness due to the eye's inherent defence mechanism which prevents penetration of foreign substances (Fig.4.5). The chitosan-based formulation in solution and gel when tested on rabbit models in vivo and across porcine cornea ex vivo at a time of nearly 2 h retained the drug in the desired area allowing more of the drug to penetrate leading to increased bioavailability.

The mucoadhesive property of chitosan also makes it applicable for application in topical gels for localised and effective delivery topically. In the case of pregnant women where care must be taken to avoid systemic absorption of certain drugs such that the drug being administered to treat the mother does not get to the child as the drug, although beneficial to the mother might pose harm to the child. It is therefore desired that the drug be localised to the affected tissue as best as possible. An example is the delivery of clotrimazole using chitosan-

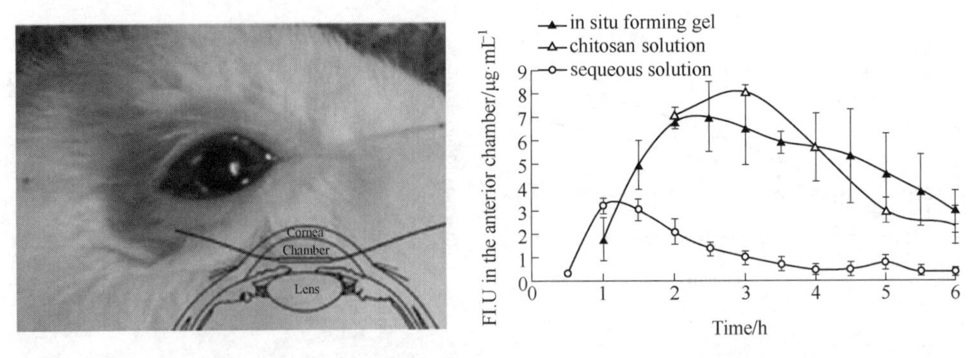

Fig.4.5 Enhanced topical ocular delivery of fluconazole using chitosan-based solution and gel.

coated liposomes for the treatment of vaginal infection which occurs during pregnancy. Vaginal infection although might heal without treatment in non-pregnant women, in pregnancy must be treated to prevent complications at child birth or affect development of the child. Drug treatment provides adverse effects involving current drug regimens because of very high therapeutic levels in the bloodstream despite favourable long durations of action. For example, trichomoniasis is a vaginal infection when treated with metronidazole before 37 weeks pregnancy substantially increases the adverse effects of preterm labour and low birth rate babies. The drug metronidazole is commonly prescribed to pregnant women in oral dosage form (500 mg or 250 mg). Further research in decreasing the adverse effects, maintaining an ideal therapeutic level and long sustainable duration of action for metronidazole is much sought after. Chitosan-coated liposomes containing 0.1%, 0.3% and 0.6% *W*/*V* concentration of chitosan and a drug concentration of 22 g/mg lipid of Clotrimazole, show good localised delivery of the drug. The retention of the drug in the vaginal tissue was also significantly increased by use of chitosan (Fig.4.6). Interestingly, it was also shown that the clotrimazole- containing liposome system with lower concentration of chitosan showed better mucoadhesive property than the higher concentration of chitosan. The topical formulation effectively adheres to the tissue preventing penetration into the systemic flow such that the drug does not cross the placenta to the child but remains in the tissue where it is needed to act.

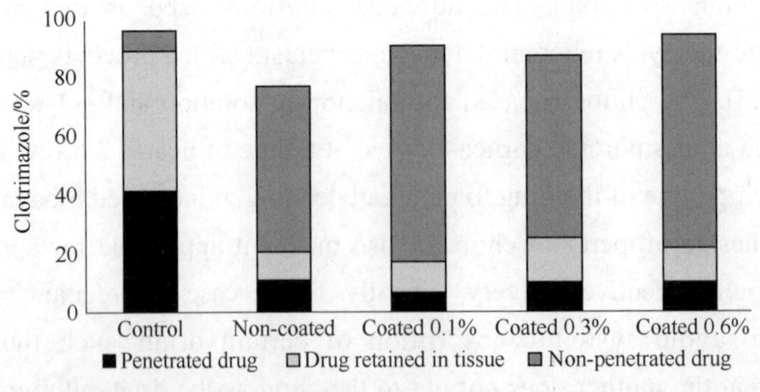

Fig.4.6 Comparing retention of the drug clomirazole at vaginal site using varying concentration of chitosan.

4.5 Oral Drug Delivery System

Drug delivery through the oral route is one of the most common forms of drug delivery into the body. Oral drug delivery refers to the intake of medicaments into the body through the mouth by swallowing, chewing or drinking. These 87 dosage forms could be in form of solid or liquids as tablets, capsules, powders, liquids. Oral dosage forms could be targeted at any tissue in the body and for a variety of purposes from general pain relief to regulation of insulin in diabetes patients.

4.5.1 Advantages and Disadvantages of the Oral Route

In certain cases, the oral route becomes more than just an alternative to other routes of drug delivery. For example, insulin delivery, where oral route provides an administration which is closer to the natural physiology of the body by delivering the drug into the liver which is the target tissue. The drug passes into the liver via the pancreatic β cells through the hepatic portal vein. This is unlike in the case of other delivery routes such as parenteral or nasal which aim to deliver the drug directly into the systemic circulation. While this route is favourable in avoiding the first pass metabolism, it is not in line with insulin's natural physiological pathway.

Protein drugs such as insulin pose a particular challenge in oral delivery as they are more likely to follow the paracellular route rather than the lipophilic membrane route which most other drugs follow. This makes them more susceptible to enzymatic degradation.

The main limitations in oral delivery route are the first pass metabolism and biodegradation in the gastrointestinal tract which adversely impacts on the bioavailability of the drug in vivo. Scientific focus in the area of oral drug delivery looks at developing oral formulations which can successfully pass through the gastrointestinal tract while still remaining potent, resist enzymatic degradation at the mucous membrane, can be effectively transported through the complex structure of the mucous membrane, get absorbed into the targeted tissue and have the desired pharmacological activity after passing through the mucous layer.

4.5.2 Current Challenges and Natural Polymer-Based Innovations in Oral Drug Delivery

Mucoadhesive polymers are applied with the aim of developing a delivery system which enables the drug to attach to the mucous membrane for enhanced permeation and sustained delivery. However, limitation of this procedure lies in the constant renewal of the mucous layer which inhibits mucoadhesive drug delivery systems.

Although rate of drug absorption using any kind of route e.g. oral, nasal, topical depends on factors such as age, diet and state of health of the patient, the drug properties also has a significant effect on the rate of absorption and effectiveness. The drug properties which affect oral delivery include molecular mass, particularly drugs with molecular mass greater than

500~700 Da like protein drugs such as insulin. Lower molecular mass drugs are generally easier to absorb. While their oral delivery would be of much significance pharmaceutically, delivery of proteins and peptides-based drugs orally have particularly proven challenging. This is mainly due to their generally large molecular mass and the tendency to be digested in the body without serving their purpose. There are about two known oral protein and peptide drugs in clinical development and these are Interferon-alpha and human growth hormone (HGH) while more are being studied for potential pharmaceutical application. Of much interest is the oral delivery of insulin. Research approaches in enhancing the oral delivery of proteins include reducing the particle size and using biodegradable nanoparticles .

4.6 Parenteral Drug Delivery Systems

Parenteral delivery concerns the delivery of drugs invasively through the skin, eye, vein, artery and spinal cavity. Substantial efforts in formulating a documented plan in the development of hypodermic needles was first initiated by Lafargue in 1836. Lafargue immersed the lancet in morphine and diluted the morphine by a once repeated immersion into water before self-injection. The hypodermic needle is a cylindrical tube with an elliptical shaped bevel end forming a sharp tip for the purpose of cutting into skin.

Hypodermic needles are conventionally fabricated from medical grade stainless steel. The luer lock is the plastic connector between the hypodermic needle and syringe body. Polyethylene and polypropylene are medical grade thermoplastics moulded into the luer lock. The syringe body is composed of medical grade plastic. Medical grade thermoplastics can be moulded into complex geometries in a process known as injection moulding. Medical grade plastics are regulated by USP with the aim of analysing if a grade of plastic reacts with mammalian cells cultures. There is no published material about syringes constructed from natural polymer materials. This is because most natural polymers may not be easily mouldable by injection moulding and end product assurance towards medical grade is less likely due to the risk of by-product toxicity if a reagent in a natural polymer blend is unstable despite high desirable yields.

Hypodermic needles and syringes are usually disposable and single use only. Complex blended natural polymers such as corn starch blended with clay, mineral montmorillonite and modified natural rubber latex were injection molded thus resulting in good tensile strength and elastic modulus properties. The constraints for complex blended polymers are of greater costs than conventional process, longer duration in process manufacturing and end product can be unaesthetically pleasing. In the past, syringes were constructed from borosilicate glass and autoclavable for reuse thus producing less of an ecological impact.

4.6.1 Advantages and Disadvantages of Parenteral Drug Delivery

Parenteral drug delivery is still a common and widely accepted route to drug administration. The main advantages are bypassing gastrointestinal tract metabolism, rapid drug delivery with target-based response and is an alternative route for patients with difficulty ingesting their medication or are completely sedated. The disadvantages are depth-related localised pain, likelihood of peripheral nerve injury and accidental piercing of a blood vessel at hypodermis level.

4.6.2 Properties of Parenteral Drug Molecules

Injection-based parenteral drug molecules are usually high molecular mass, more ring-based structural configuration, high counts for proton acceptors and lowest Log10 O/W. Fluid-based drug formulations are ideal for flow-based transfer along hollow hypodermic needles. Surface tension forces are the usual forces that allow fluid to travel along capillary tube. The volumetric flow rate of a fluid inside a microcapillary is defined by the Hagen-Poiseuille (Eq.4-1).

$$Q = \Delta P\left(\frac{\pi r^4}{8\mu L}\right) \tag{4-1}$$

where Q is the volumetric flow rate inside the hypodermic needle, ΔP is the pressure difference from. In Eq.4-2, F is the injection force, f is the frictional force from the tube and syringe walls, A is the interior cross-sectional area of the tube and r is the internal tube radius, μ is the fluid viscosity and L is the hypodermic needle length.

$$Q = \frac{F-f}{A}\left(\frac{\pi r^4}{8\mu L}\right) \tag{4-2}$$

The characteristic of fluid flow is expressed by the Reynolds number, Re (Eq.4-3).

$$\mathrm{Re} = \frac{\rho d V}{\mu} \tag{4-3}$$

where ρ is fluid density, d is the internal tube diameter, V is the fluid volume. A Reynolds number of 2100 or less indicates laminar flow and turbulent flow is above this value.

4.6.3 Current Proprietary Parenteral Devices

Parenteral devices are commonly injectables and examples of current devices available on the market are mentioned. Injectable devices for the delivery of soft implants subcutaneously are patented and commercially available from Rexam. Also pre-filled drug syringes is registered Safe ‘n’ Sound, patented and commercially available from Rexam. A self-injector trademarked SelfDose for the safe delivery of drugs is in the format of an adaptor for fitting syringe formats is commercially available from West Pharma. Another self-injector device has a window indicator regarding usage and is trademarked Project, patented and commercially available from Aptar.

4.6.4 Future Challenges of Parenteral Devices

Microneedles are used as minimal invasive parenteral devices because the needles are fabricated to penetrate a known depth in skin layers than a hypodermic needle. Natural polymers have the potential to support the sustained release of drugs in the skin and can prove advantageous for the biodegradable class of microneedles. However, the challenge arises to strengthen the microneedles with the result of all microneedles piercing the skin at a reproducible depth. Synthetic biodegradable polymer such as poly(di-lactic-co-glycolic acid) PLGA and poly(L-lactic acid) (PLLA) possess high mechanical strength. The possibility of enhancing the natural polymeric formulation with blended synthetic, polymeric fibers in providing improved mechanical strength properties is one direct solution.

4.7 Nasal Drug Delivery Systems

Nasal delivery is one of the oldest drug delivery systems originating from Ancient Indian Ayurveda called Nasya Karma. The mucosal epithelium inside the nasal cavity is an area for non-invasive drug delivery. This epithelial layer located in the inferior turbinate of the nasal cavity is highly vascularised with a significant absorption area (150 cm^2) and projections of microvilli in epithelial cells. The nasal cavity is covered with mucous membrane comprising of goblet cells, columnar cells and basal cells (Fig.4.7). Most cells of the nasal cavity have cilia apart from columnar cells in the anterior cavity (Fig.4.7). A collective group of microvilli are known as cilia.

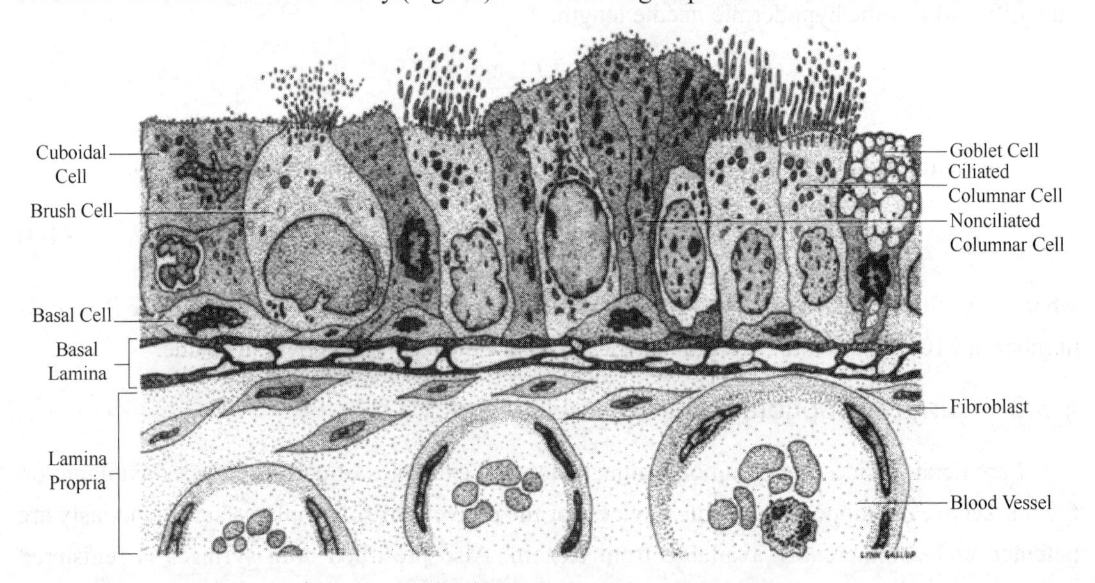

Fig.4.7 Histology of the nasal cavity morphology.

Cilia move rhythmically in waves with a function to clear mucus from the nasal cavity into the nasopharynx followed by the oesophagus before finally moving towards the gastrointestinal tract. The microvilli contribute to the large surface area thus highly desirable for effective drug absorption into the nasal mucosa. The nasal mucosa is neutral pH and permeable to numerous drug molecules. Mucosa is usually comprised of lipids, inorganic salts, mucin glycoproteins and water. The main functions of mucus are lubrication of surfaces and protection. The function of protection are goblet cells and mucus glands of nasal epithelium that prevent the absorption of foreign chemicals and decrease residence time for any applied drugs that are in surface contact with the epithelial lining. The purpose for nasal drug delivery is to target the drug systemically such as peptides or proteins in the bloodstream locally such as a nasal allergy, nasal congestion, sinus, and to target the central nervous system (CNS) such as bypassing the blood-brain barrier. The purpose of targeting the CNS is to develop drugs for rapid treatment of migraine, headaches, advanced neurodegenerative illnesses such as Alzheimer's and Parkinson's disease.

4.7.1 Advantages and Disadvantages of Nasal Drug Delivery

The advantages of nasal drug delivery are avoidance of potential gastrointestinal and hepatic first pass metabolism, low molecular mass drugs have a good bioavailability via the nasal route, straight forward self-administration and protein-based drugs are able to absorb through the nasal mucosa as an alternative to parenteral drug delivery. The disadvantages of nasal drug delivery are possibility of irreversible cilia damage on the nasal mucosa caused by the drug formulation, high molecular mass molecules and polar molecules may not permeate or result in low permeation thorough nasal membranes (molecular mass threshold: 1 kDa), clearance of mucosa frequently by cilia has the potential to decrease or prevent full drug absorption, nasal mucosa could denature and change the structure of some drugs through enzymes and possible incompatibility observed between drug and nasal mucosa interaction.

4.7.2 Natural Polymers in Nasal Drug Delivery

Mucoadhesive microspheres, liposomes, solutions, gels and Mucoadhesive hydrogels are vehicles commonly adopted in the nasal delivery of drugs. Starch, Chitosan, alginate, dextran, hyaluronic acid and gelatin are natural polymers adopted for nasal drug delivery. Mucoadhesion is defined as the contact between the drug formulation and the mucin surface. The concept of mucoadhesion is to allow sustained drug delivery in nasal membranes by prolonging the contact time between the drug formulation and nasal mucosa layers in the cavity. Mucoadhesion promotes drug absorption and lowers the chances of complete mucociliary clearance.

Starch is a biodegradable polysaccharide which can be readily processed into microspheres. Starch microspheres commercially available under the name Spherex are used in nasal drug delivery. Drugs such as Insulin, morphine, inactivated influenza and Salbutamol are examples of starch loaded intranasal drugs at developmental stage.

Chitosan is a natural polysaccharide with mucoadhesive properties thus it has very good binding properties to nasal epithelial cells and the covering mucus layer. The cationic nature of chitosan readily permits the electrostatic attraction with the negatively charged mucosal surface. There has been a wealth of research published on chitosan for intranasal delivery according to variable salt forms, degrees of acetylation, variation in derivatives, variation in molecular masses and variation in physical form such as gel, microspheres. Drugs such as Loratadine, Zolmitriptan and Insulin are examples of loaded chitosan-based intranasal drugs at developmental stage. However, there is yet to be a marketed nasal drug product containing chitosan as the drug absorption enhancer. A morphine intranasal formulation containing chitosan (Rylomine) has already published phase 2 clinical trials and has already pursued phase 3 clinical trails.

Alginate is a divalent cation-induced rapid gelation, natural polysaccharide with greater mucoadhesion strength as compared with chitosan, PLA and carboxymethyl cellulose. Usually alginate gel is blended with one or more mucoadhesive polymers in order to improve the strength and drug loading efficiency of the vehicle. Drugs and macromolecules such as bovine serum albumin in representing a water soluble antigen Carvedilol and Terbutaline Sulphate are examples of alginate loaded intranasal drugs at developmental stage. There appears to be no significant proprietary drugs containing alginate as the vehicle for intranasal drug delivery.

4.8 Hydrogel-Based Drug Delivery Systems

The need in optimised semi-solid, biocompatible, polymeric formulations in drug loading and routes of entry in the human body is still a growing area in pharmaceutics. Gel- and ointment-based drug formulations are normally oily and thick in appearance. A common purpose of such semisolid, polymeric gel/ointment formulations are to enhance the viscoelasticity and improve target-based pharmacokinetics such as enhanced permeability of luteinizing-hormone, releasing hormone (LH-RH) from polycarbophil hydrogels inside the vagina compared with solution. In terms of viscoelasticity, an example in enhancing pseudoplastic properties for oral Ibuprofen is Carbopol-based hydrogels. Limitations for semi-solid formulations such as topical applications concerning transdermal drug routes of delivery are one common area. Hydrogels can be considered as a semi-solid matrix for the purpose of controlled drug release. Hydrogels can change structural configuration during certain temperature or pH-induced environments in bodily systems. Usually, hydrogels are known to release trapped drug molecules by swelling in watery plasma solvent. Distinct variability from conventional swelling mechanism of active molecule release are thermoresponsive hydrolysis of block copolymers hydrogels and full dissociation of polycationic poly(allylamine) hydrochloride and polyanionic polystyrene sulfonate complex microgel during increase pH. A growing demand for hydrogel-based drug delivery since 1980 onwards shows increasing treads (Fig.4.8).

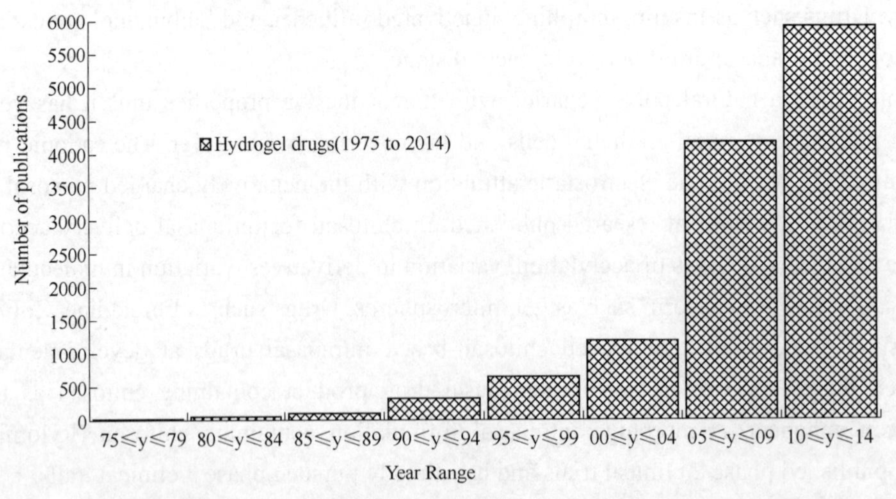

Fig.4.8 The number of hydrogel drug publications according to year range.

This section focuses on hydrogels obtained from natural polymers. It outlines the structure and function of natural polymeric hydrogels in the area of Pharmaceutics-based drug delivery. The distinct sub-classification of a less common form of hydrogels, known as microgels, explains this difference. Also, another area of this review focuses on the physico-chemical properties of hydrogels as a drug delivery system with ideal pharmacokinetic targeting areas.

A hydrogel is a solid or semi-solid hydrophilic matrix comprising of polymeric macromolecules crosslinked by varying combinations of hydrogen bonding, van der Waals, ionic electrostatic-based and covalent-based intermolecular interactions. Hydrogels possess matrix swelling or shrinkage properties in physico-chemical solvent media such as pH, temperature and ionic strength of electrolytes in solution. Usually, solvent ion concentrations at medium ionic strengths allow for ion exchange between polyelectrolyte gel and solvent ions resulting in osmotic pressure increases inside hydrogel and thus causing swelling.

Polymeric hydrogels such as *N*-isopropylacrylamide (NIPAAm) are influenced by higher ionic strength of electrolytes and temperature in solution and they can swell above their critical solution thresholds. NIPAAm has swelling properties as the nitrogen groups' hydrogen bond with water at NIPAAm lower critical solution temperature of 34℃. Also the deprotonation of carboxylic acid groups in hydrogels such as polyacrylic acid (pAA) and interpenetrating network of chitosan combined poly(sodiumacrylate-co-hydroxyethyl methacrylate) (SCPSC) in high pH results in ionic repulsion and induces swelling (Fig.4.9).

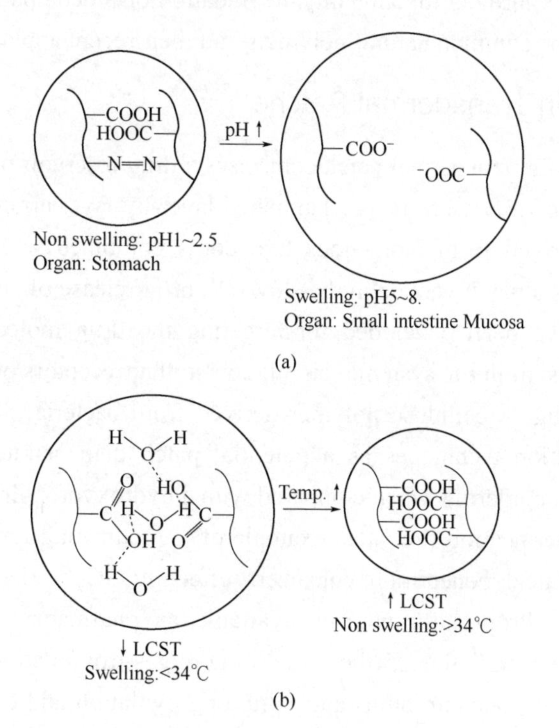

Fig.4.9 (a) A schematic representation of swelling according to a pH with a hydrogels such as crosslinked azo polyacrylic acid (pAA). (b) Temperature with hydrogels such as PNIPAm with PNIPAm/IA.

The equilibrium swelling ratio of SCPSC was significantly 1.6 folds greater in pH 7 buffer medium when compared with pH 3.9. The polymeric macromolecules in hydrogels can be cationic, anionic or entirely neutral with regard to interacting with another macromolecule or drug molecules. The crosslinking of hydrogels combines highly desirable characteristics such as mechanical strength, pseudoplasticity, drug and macromolecular intermolecular interactions and plasma swelling. The porosity of the crosslinked hydrogel matrix determines aqueous solvent adsorption and rate of drug release. However, the insolubility of hydrogels to water is attributed to the networking arrangement of crosslinks between polymer chains thus maintaining physical structure. Nevertheless control in the polymeric swelling release rates of drugs with possible subsequent degradation of hydrogels is very much a sought after challenge in matching the duration of drug release across a therapeutic range and target specificity in the body. There are constraints and possible major limitations in stabilizing porous combination of polymers in defined mass ratios in attaining desirable controlled release of drug molecules.

However, the complex chemical structures of hydrogels can pose challenging in synthesis coupled with mass reproducibility and end product purification. Although synthetic polymers seem to have largely dominated over natural polymers in the past decade due to their relatively long service life, high water absorption capacity and gel strength and the possibility of tailored degradation and functionality, natural polymers are highly sort after for their biocompatibility, availability and low cost. Hydrogels from natural source usually require inclusion of synthetic components as, for example crosslinkers or the hydrogel could be a blend of both natural and synthetic polymers for improved functionality, degradation or biocompatibility. In the following sections we look at some common natural polymers and their recent applications as hydrogels.

4.8.1 Hydrogels in Transdermal Patche

The architecture of a transdermal patch comprises a drug reservoir or a polymerdrug matrix trapped between two polymeric layers as a laminated layer-by-layer arrangement. A study of the effect of mucilage derived from indigenous taro corns combined with hydroxypropylmethyl cellullose (HPMC) was a patch vehicle in the slow Ⅳ drug release of an antihypertensive drug, Diltiazem. Patches have been developed for targeting the drug molecules through full skin thickness passive diffusion in the systemic circulation so that receptors or pathogens in the body are affected by the drug. A cellulose polymer derived from bacteria, plasticised with glycerol using solvent evaporation techniques as a potential patch demonstrated a reduced lidocaine permeation flux in skin epidermis when compared with a hydroxypropylmethylcellulose gel. The observation of a low permeation flux is an example of implementing further optimisation-based studies by using chemical penetration enhancers effecting SC barrier properties at possible higher concentrations. Proprietary patches available as pharmaceuticals are Nicorette® in nicotine delivery to wean addiction, Ortho Evra® in norelges-tromin/ethinyl estradiol delivery to decrease blood levels of gonadotrophins and inhibiting ovulation and chances of pregnancy an Exelon® Patch in rivastigmine delivery to inhibit cholinesterase by reversible inhibition in delaying the progression of Alzheimers disease. Those proprietary patches mentioned are examples outlining three completely different therapeutic before current growing trends emerged since the 1980s. The major benefits of patch-based delivery are reduction in adverse effects such as gastrointestinal disturbances caused by high dose oral rivastigmine as compared with a rivastigmine patch, reduction in peak plasma concentrations and interventions of prior dose adjustments periodically from oral and fast intraveneous delivery.

The sensitivity of patient's skin to transdermal patches is a major concern because of the likelihood of allergic reactions if the patched skin area is left covered for a long duration.

4.8.2 Nanoparticles for Controlled Delivery

The controlled release of active drug molecules sustained at therapeutic thresholds in specific targets of the body according to the length of treatment is a major focus in pharmaceutics research. A significant gap for nanoparticle mediated drugs to enter the pharmaceutical drugs market exists because of the sophisticated pathological targeting mechanisms and therefore traditional pharmacology cannot distinctly characterise nanoparticle drugs. Themoresponsive Poly(NIPAAm-co-AAm) hydrogels were shown to have a z-diameter of 156.0 nm after encapsulating gold-silica nanoparticles and forming nanoshells by collapsing to absorb the gold-silica at 40~45℃ at 780 nm. A chemotherapeutic agent, doxorubicin was loaded into the Poly (*N*-isopropyl acrylamide-co-Acrylamide) [Poly(NIPAAm-co-AAm)] nanoshells by 1.12 folds greater than without nanoshells arrangement. The crosslinkers in NIPAAm-co-AAm hydrogels can reversibly collapse into a dehydrated globular conformation above their lower critical solution temperature, normally above physiological body temperatures, to release the drug. Poly(NIPAAm-co-AAm) is a synthetic polymer. Nevertheless Poly(NIPAAm) has been commonly crosslinked with natural chitosan because of pH sensitive properties of the amino groups. The cytotoxicity of Poly(NiPAAm-co-chitosan) containing 5 mg•mL^{-1} NIPAAm nanoparticles encapsulated with paclitaxel resulted in 60 % viability of human lung cancer cells thus proving favourable toxicity.

Complementing the 60 % cell viability, the cumulative release of Paclitaxel was increased by 1.86 fold in extracellular tumour conditions of pH 6.8 compared with pH 7.4 at the same physiological temperatures. Nanoparticle drugs are usually between 10nm and 200 nm in size with generally high efficacies. Liposomes are mainly natural phospholipids nanoparticles as highly advantageous drug delivery vehicles because of the potential to deliver hydrophobic drugs and biocompatible properties. Liposomal synthesised PEG nanoparticles loaded anti-cancer carfilzomib allowed the inhibition in tumour growth and subsequently proved to be up to fourfolds more cytotoxic to tumours compared with unloaded carfilzomib.

Liposomes synthesised with PEG prevents any aggregation of nanoparticles and adsorption of plasma-based serum proteins that promote immediate clearance. The advantage of drug nanoparticles in drug therapy is the reduction in systemic toxicity and greater drug loading in nanospheres. A huge vacuole still remains for research into drug hydrogel nanoparticles containing higher concentrations of ideal naturally sourced polymers.

4.8.3 Hydrogels for Wound Dressing

Wound dressing is an immediate first aid response in superficial and chronic skin wounding injuries. The general treatment of skin wounds is to minimise scarring, microbial infection, pain, protection from further trauma and absorption of excess exudates from open lacerations. Conventional gauzes and pads based on cotton and synthetic rayon polyester bandages need regular changing and tend to be more expensive than modern dressings. Also conventional bandages are known to keep the wound bed dry and slow down the natural skin healing process due to restricted new cell migration and healthy tissue removal when bandage requires changing. Hydrogels are an ideal dressing material for absorbing excess exudates, allowing enough moisture of the wound bed and filling irregular-shaped wound cavities. A synthesized gelatine-hydroxyphenylpropionic acid hydrogel was studied because of well-known biocompatible and tissue adhesive properties. A gelatine-hydroxy phenyl propionic acid hydrogel loaded with human dermal fibroblast resulted in a 1.9 folds wound closure in mice compared with phosphate buffer solution control after four days. The focus on hydrogels for wound dressing may seem irrelevant in the area of traditional pharmaceutics as defined in the section Portals of drug administration in the human body. The importance of a new area of study relating to emergency trauma shows the need for the application hydrogels compounds.

4.8.4 Polymeric Crosslinking in Hydrogels

An important characteristic of a hydrogel is the polymeric strand crosslinking. Crosslinking of hydrogels with morphologically cross-hatched or entangled macromolecular architecture allows a 3D structure and avoids immediate dissolution of separate macromolecular strands in hydrophilic solvent.

Physical crosslinking of polypeptides are attributed to ionic bonding, hydrogen bonding and hydrophobic interactions in aid of bipolymeric crosslinking. Physically crosslinked hydrogels are inhomogeneous due to more than one type of intermolecular-based interaction.

Chemically crosslinked hydrogels involve covalent linkages in bridging two different polymeric strands and the use of crosslinking agents that can react with specific functional groups in polymeric macromolecules. Chemically crosslinked hydrogels permit bigger volume increases during sol-gel transition than physically crosslinked hydrogels. The use of chemical crosslinking agents to bind-specific functional groups for crosslinking polymers is shown in Table 4.2. The process and target application of hydrogel and microgel polymers is outlined in Table 4.3.

Table 4.2 Chemical agents for the chemical crosslinking of functional groups in hydrogel

Crosslinker	Functional Groups	Reaction or Functional Group Interactions	Chemical Reaction Conditions
Glutaraldehyde	Di-aldehydes	Imine group formed by Schiff base formation Acetal group formation from hydroxyl groups	No heat required and slow addition is usual

续表

Crosslinker	Functional Groups	Reaction or Functional Group Interactions	Chemical Reaction Conditions
Poly(ethylene glycol)-propion dialdehyde (PEG-diald)	Amine	Azide addition	Unimolecular addition of PEG-diald and polymer in ambient temperature conditions
Methylene bis-acrylamide	Acrylamide, ethylene	Variable	
Genipin	Amino acid groups and secondary amino group in acidic and neutral pH	Amino acid groups Condensation reactions in acidic or neutral conditions and aldol condensation in basic conditions	Set pH conditions

Table 4.3 Recent examples of hydrogels developed in drug delivery

Polymer	Composition	Process	Target/delivery
Casein	100%	Temperature-based gelation	BSA molecule into buffered solution
Poly(*N*-isopropyl acrylamideco-acrylamide)	poly(NIPAAm-co-AAm) NIPAAm and AAm, 83.3 : 16.7 (% mol ratio)	Poly(NIPAAm-co-AAm) synthesis: free radical copolymerization with AIBN initiator. Microsphere process: W/O emulsification and copolymer solubilised by acidic DI water and crosslinking using glutaraldehyde	Propranolol and lidocaine
guaraeyAlginate (Monomer unit: 1,4-linked β-D-mannuronicacid and a-L-guluronic acid)	Methacrylated alginate (5.7%~45.3 %)	Photocrosslinking of methacrylated alginate at 365 nm and 0.05 % *W*/*V* Irgacure D-2959 photo initiator	Bovine chondrocytes for cytocompatibility for cell culture
Sodium carboxymethylcellulose	(NaCMC)∶cellulose NaCMC: cell (5∶5~9∶1 by wt).	A hydrogel film Solubilisation of cell and NaCMC and crosslinking with epichlorohydrin (ECH)	In vitro release of Bovine Serum Albumin (BSA)
Poly(ethylene oxide)-Poly (propylene oxide)-Poly (ethylene oxide) (PF127) and Poly (methyl vinyl ether-co-maleic anhydride) (GZ)	GZm/PF127: molar ratio from 1 to 20 (GZm is the monomer, methyl vinyl ether-co-maleic anhydride)	Esterification between carboxyl groups of maleic anhydride and hydroxyl groups of PF127. Subsequent solvent evaporation of tetrahydrofuran followed by precipitate copolymer filtration and collection	BSA, glucoprotein rKPM-11 and dextran in PBS (pH 7.4)

Note: The process are the main experimental conditions, reagents or crosslinking reagents in preparing hydrogels. The target/delivery is the active molecule or drug studied for encapsulation or controlled release from hydrogel

4.8.5 Natural Polymers in Hydrogels

Polysaccharides such as hyaluronic acid, chondroitin sulphate, chitosan, carboxymethylcellulose, hydroxypropylmethylcellulose, methylcellulose, bacterial cellulose and sodium alginate are common examples of carbohydrate derived polymers in hydrogels. Examples of proteins used in hydrogels include gelatine, collagen, elastin, ovalbumin, β-lactoglobulin and silk fibroin from both plant and animal sources. Polymer strands from natural, synthetic and partially synthetic sources are acquired as drug delivery vehicles. Polypeptides have straight chained or helical assemblies in their gross macromolecular arrangement such as β-pleated sheets and α-helix respectively. Amino acids in polypeptides, containing Ala, Glu, Lys and Gln occur more in α-helices compared with Thr and Val in β-pleated sheets, in-conjunction to Gly and Pro usually located in the turn area of molecule. Two hydrophobic regions in the macromolecular structure of anti-parallel conformation assemble to form the β-pleated sheet (Fig.4.10).

Polypeptide structure hydrogels overall are the most suitable in mimicking natural extracellular crosslinking matrix .

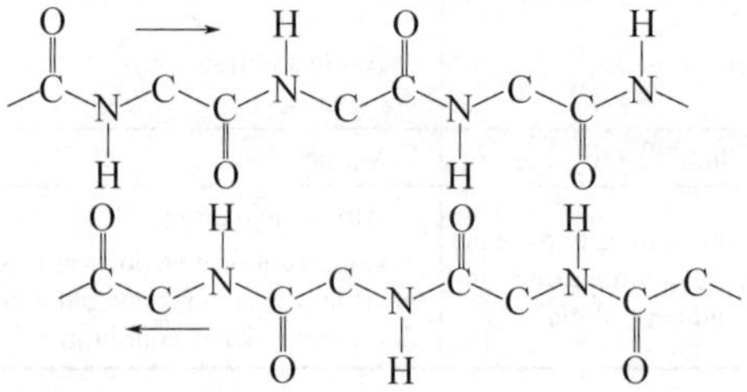

Fig.4.10 An anti-parallel orientation for a β-pleated sheet.

Hyaluronic acid (HA)-based hydrogel particles have been investigated for drug delivery using trimethoprim (TMP) and naproxen as model drugs. Hyaluronic acid was modified with an aqueous solution of sodium bis (2-ethythexyl) sulfosuccinate (AOT)-Isoctane microemulsion system. This formed hyaluronic acid particle which were further modified by oxidizing to aldehyde (HA-O) using treatment with $NaIO_4$ followed by reacting with cysteamine thus forming thiol ligands onto the surface of the HA particles. The final HA-based hydrogel particles were formed by radical polymerization of the HA particles with anionic and cationic monomers 2-acrylamido-2-acrylamido-2-methyl-propanosulfonic acid and 3-acrylamidopropyl-trimethyl-ammonium chloride, respectively. The HA-based hydrogel particles derived demonstrated good pH dependent size variation an swelling properties. This is important for applications such as controlling and tuning the rate of drug delivery in different parts of the body. This takes advantage of the remarkable ability of HA to demonstrate variety of swelling kinetics in different pH environment. Other natural polymers which tend to form hydrogels with pH dependent swelling kinetics include alginate. Arginine grafted alginate hydrogels are also potential carriers for protein drugs enabling oral delivery. This can be used to orally deliver proteins while limiting the effect of metabolism in the gastrointestinal tract prior to reaching the target area.

Nanocellulose has had increasing application in the pharmaceutical area in recent times. Current interests in exploring the industrial application of nanocellulose extend to their use as hydrogels for drug delivery. Nanofibrillar cellulose derived from wood pulp was developed into injectable hydrogel for localised and controlled release of large and small compounds in vivo. Although further studies are required to establish the nature and possibility of interaction between the hydrogel material and the active drug, studies carried out so far show that nanofibrillar cellulose has good potential as an injectable hydrogel drug delivery system.

This application exploits the shear thinning property of nanofibrillar cellulose hydrogel which makes it possible to inject with ease using a syringe while still maintaining its viscosity . This allows for localised and targeted delivery to easily assessable regions using injections. Nanofibrillar cellulose hydrogel also has the advantage of ease of preparation without need for an external source of gel activation unlike most other hydrogels being explored for the same application. The external activators could be chemicals or irradiation methods which could invoke toxicity or complication of the delivery process. Nanofibrillar cellulose-based hydrogels, however possess intrinsic pseudoplasticity which makes them suitable an injectable hydrogels.

Chitosan and its various derivatives have also been expired as hydrogels for drug delivery. Due to the robust chemical property, chitosan can be crosslinked using a crosslinker such as genipin and glutaraldehyde with a variety of other natural polymers to obtain desired functionality. For example, chitosan is crosslinked with gelatin for improved rigidity and with starch for improved flexibility and cohesion.

Cellulose is a highly abundant natural polymer in plants, bacteria, algae and fungi phylum. The unbranched chains consist of 1,4 glycosidic linkage of monomer units, D-glucopyranose (DGP) and presence of three hydroxyl groups per DGP monomer. Cellulose polymers consist of amorphous and crystalline arrangements in which the hydrolysis properties of cellulose are found to be more unfavourable in higher crystalline arrangements.

Sodium carboxymethylcellulose (NaCMC) is a cellulose derived water soluble polymer. NaCMC is grossly anionic because of the negative electron density with respect to the carboxymethyl substitution region. Hence, polyanionic NaCMC has the potential to electrostatically interact with gelatine below its isoelectric point. NaCMC and gelatine are biocompatible as NaCMC is biologically excreted and gelatine is degraded by natural enzymes. NaCMC is able to hydrogen bond with water molecules hence hydrogel NaCMC crosslinked gelatine possesses swelling properties. Individual polymers of NaCMC and gelatine have the tendency to swell in ambient temperature water. As far as we know there is no published literature comparing swelling rates of individual NaCMC and gelatine with post bipolymeric NaCMC：gelatine microgel. Ionic interactions are dominant intermolecular forces in crosslinking polyanionic NaCMC with polycationic polymers such as polyvinylamine (PVAm). The degree of substitution (DS) defines this structure when hydroxyl groups in the glucopyranose monomer are replaced with carboxymethyl groups in which the number of substituted hydroxyls accounts to the degree of substitution. The higher the DS and quite significantly the lower the M_w of NaCMC allows for increased in ionic conductivity.

The discharge capacity of NaCMC (0.9 DS and 250 kDa) up to 0.5 current density (C-rate) was 165 mAh•g^{-1} compared with NaCMC (0.9 DS and 700 kDa) at 155 mAh•g^{-1}. Potentiometric titration with hydrochloric acid as a carboxylate proton donor coupled with Infrared spectroscopy in knowing the relative amount of carboxyl groups is implemented in calculating DS.

Gelatin is another natural polymer which finds wide application as hydrogels for drug delivery. Hydrogel made from gelatin and polyvinyl alcohol (PVA) has been developed for application in delivery of anti-cancer drug Cisplatin. The anti-cancer drug encapsulated within the macrocycle cucurbit(7)uril was incorporated in hydrogel formulations containing between 0 and 4 % PVA. The hydrogel formed demonstrated a controllable swelling and degradation rate which was PVA concentration dependent. As the concentration of PVA in the hydrogel formulation increases, the release rate of encapsulated drug decreases such that the release rate of the drug can be controlled by varying the concentration of the PVA in the hydrogel formulation. Hydrogel containing gelatin only inhibited cancer cell growth by 80% while hydrogel containing 2% PVA inhibited cell growth by 4%. At 4% cell growth inhibition was 20%. When compared to intraperitoneal injection of free cisplatin at high dose of 150 μg, subcutaneous implantation of the gelatin PVA hydrogels at just 30 μg of cisplatin achieved the same effectiveness such that the use of the gelatin/PVA hydrogel improved the effectiveness of the anti-cancer drug .

A globular whey protein of high abundance from cow's milk is β-lactoglobulin which has the potential in binding hydrophobic molecules via hydrogen bonding and van der Waals interactions. Chitosan forms a complex coacervates with β-lactoglobulin at pH 6.5 .

Pectin which is an anionic polysaccharide coacervates with β-lactoglobulin and has an apparent mean particle diameter below 1000 nm and zeta potential reaching −40 mV above pH 6 for formulations containing pectin 0.5% *W*/*W*. Here, very low zeta potential values outline particle repulsion and minimal particle aggregation. As far as we know there is currently studies performed in the encapsulation and release of drugs using β-lactoglobulin as a co-polymer in a hydrogel.

4.8.6 The Preparation Techniques of Hydrogels

There are numerous valid engineering techniques in the preparation of natural hydrogels. Natural polymers such as gelatin, κ-carrageenan, agarose and gellan gum in hot solutions undergo random coil to helix transitions with the support of ionic salts such as Na^+ which lowers the repulsive forces between same electrostatic charges, allowing ionic interaction and the polymeric crosslinking to occur (Fig.4.11). The polymer κ-carrageenan can further form a superhelical network when a number of helices aggregate in the presence of ions and a gel is formed.

Polymers possessing charged functional groups such as chitosan, carboxymethylcellulose, gellan, gelatin, alginates and pectin can be crosslinked with multivalent ions of opposite charges, which is known as ionotropic gelation. Polyanionic molecules such as alginic acid and L-carrageenan can be reversibly crosslinked by cations such as Ca^{2+}, Zn^{2+} and Fe^{3+}. An ionotropic crosslinking interaction between a divalent cation and polyanionic groups between two chains of sodium alginate is by chelate complex with glucuronic acid groups (Fig.4.11) .

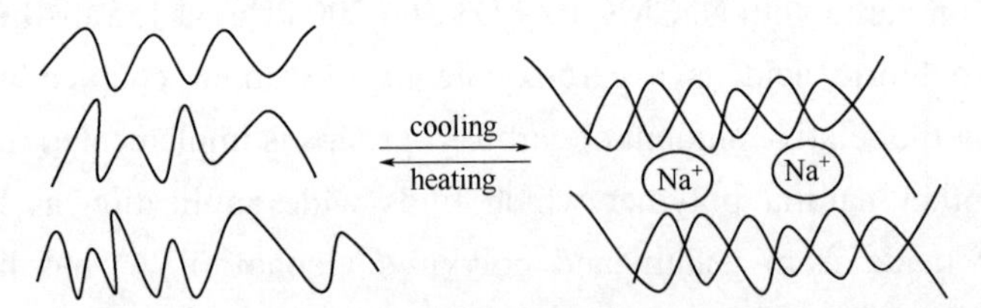

Fig.4.11 Illustration of random coil to helical transition of anionic, natural polymers, e.g. gellan gum during the cooling of a hot polymeric solution.

A study combined polyanionic sodium carboxymethylcellulose with polycationic albumin via Al^{3+} ions to induce electrostatic interactions by ionotropic gelation prior to chemical crosslinking using glutaraldehyde. The entrapment efficiency of a drug, simvastatin, was between 74% and 82% in a bipolymeric sodium carboxymethylcellulose and albumin hydrogel network. Slightly different to ionotropic gelation, a process known as complex coacervation involves the electrostatic attraction of oppositely charged polyelectrolytes such as precipitate or gel in solution because of change of factors such as pH, ionic strength and polymeric mass ratios. An alginate/β-lactoglobulin lipid droplets contained in hydrogel matrices were complex coacervated with alginate ($-NH_3^+$) and cationic chitosan ($-COO^-$) at acidic pH ranges of 3.5~6.5 in the formation of beads for gastrointestinal active molecule delivery. An example of a complex coacervate bipolymer is sodium carboxymethylcellulose and gelatin in the formation of a complex coacervate.

Chemical crosslinking of two non-ionic polymers can be enzyme catalysed in the addition of a crosslinking agent forming covalent bonds on specific functional groups in forming a hydrogel. An example is a crosslinker, 1,2,3,4-butanetetracarboxylic dianhydride (BTCA) forming ester linkages with the hydroxyl groups on β-mannose or α-galactose monomers present in guar gum with enzyme, 4-dimethylaminopyridine (DMAP) (Fig.4.12) .

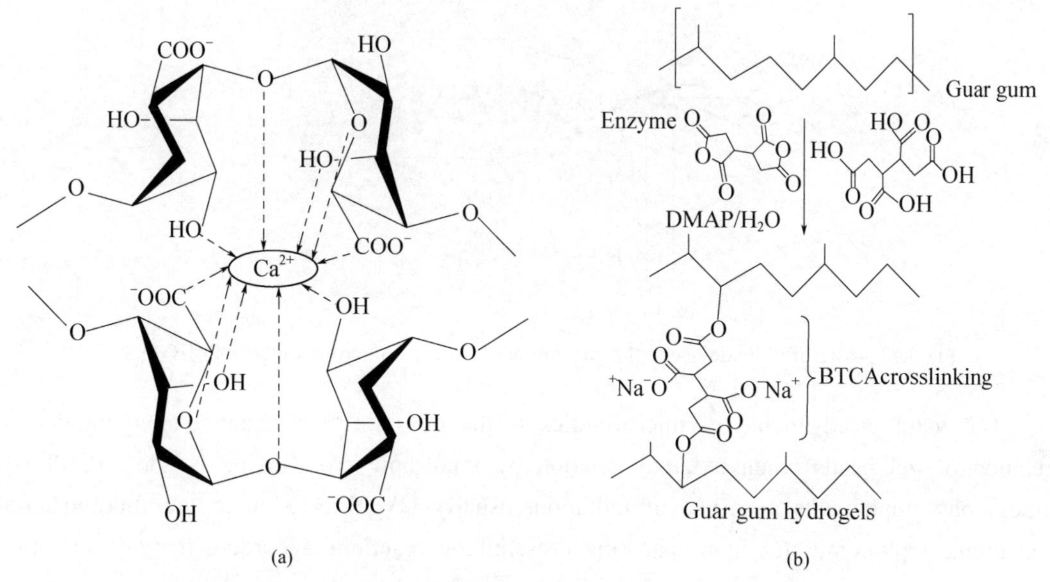

Fig.4.12 Schematic outlines in the preparation of hydrogels: (a). An ionotropic interaction formed by chelation between Ca^{2+} and alginic acid, and, (b). Chemical crosslinking of guar gum with BTCA crosslinking agent and enzyme.

Monomers of low molecular mass can undergo radical polymerization using photoinitiators forming photopolymerized hydrogels such as dextran and glycidyl acrylate. The advantages of photopolymerization are rapid curing rates during processing, lower production of heat and spatial and temporal control of process polymerization reactions. Hyaluronic acid is radically polymerized with methacrylic anhydride under basic conditions in producing methacrylated hyaluronic acid.

The manufacturing considerations in bulk production of hydrogels of bead morphology use microengineering processes in attempting to optimise control and batch wise consistency requirements. Micromolding is an engineering process recently employed in the production of hydrogel microneedles composed of NIPAAm particles suspended in 50/50 polylactic-co-glycolic acid (PLGA). The micromould implemented in the fabrication of NIPAAm hydrogel microneedles were poly-di-methyl siloxane and molten PLGA was added to pre-filled NIPAAm particles in the mould followed by curing at 150℃ and −100 kPa pressure in a vacuum oven. Microfluidics is a special area concerned with the fluid dynamics and engineering of micron scale confinement of flowing fluids. A phospholipid polymer, poly(2-methacryloyloxyethyl phosphorylcholine (MPC)-co-*n*-butyl methacrylate (BMA)-co-4-vinylphenyl boronic acid (PMBV) and poly (vinyl alcohol) (PVA) were crosslinked with the aid of a microfluidic device (Fig.4.13) . The PMBV and PVA were separately injected and droplets were pinched off after gelation induced by contact, the flow rate ratio between paraffin oil and polymer was high in order to decrease the diameter of hydrogel droplets .

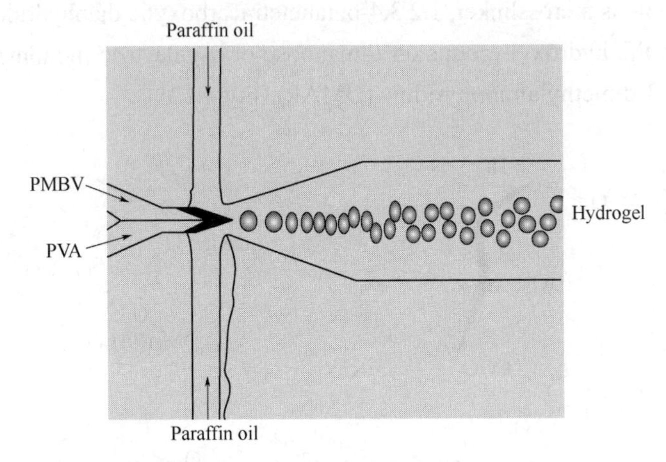

Fig.4.13 Microfluidic device in the generation of hydrogel microparticles of PMBV/PVA.

The main disadvantage of microfluidics is the possibility of channel clogging due to gelation of gel beads when external gelation by ionotropic crosslinking is adopted. Photolithography implements a source of radiation, usually UV, directed onto the fluid material containing a photoinitiator in propagating crosslinking reactions according to polymerization kinetics via transparent areas of the photomask that outline the pattern (Fig.4.14).

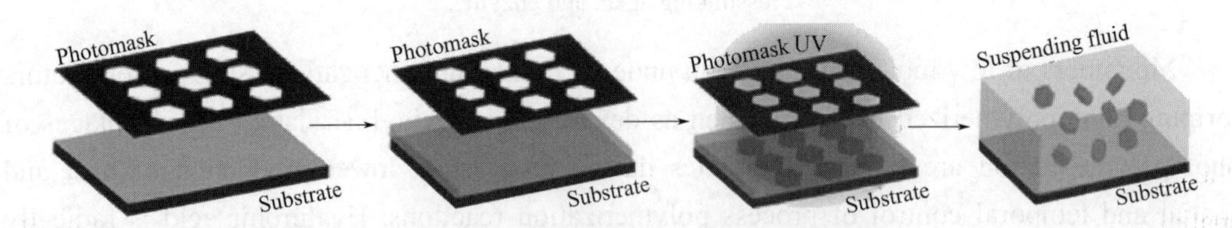

Fig.4.14 Outline of photolithography for a polymeric hydrogel.

Hydrogels made photoresponsive can evoke changes in degree of swelling, shape, viscosity or elasticity properties. They are functionalised as photoresponsive when a polymer is modified with supramolecular interacting groups, formation of photoresponsive low molecular mass gelators into a supramolecular hydrogel and addition of photoresponsive groups in hydrogel modification. A study fabricated methacrylated gelatine and silk fibroin interpenetrating polymer network hydrogels using 2-hydroxy-1-[4-(hydroxyethoxy)-phenyl]-2-methyl-l-propanone as the photoinitiator under UV radiation. The mass ratios of crystallised silk fibroin crosslinked with methacrylated gelatine defined the mechanical stiffness and the rate of degradation. Photolithography and micromoulding require a lot of capital investment relating to the precision fabrication of photomasks, photocrosslinking reagents and moulds.

Membrane emulsification involves injection of the dispersed phase through a microporous membrane into the continuous phase of an immiscible liquid under pressure. The purpose of membrane emulsification is to obtain monodisperse particles from controlled membrane pore size and pore size distributions for average emulsion diameters. Chitosan-coated calcium alginate particles with a diameter of 4.4 μm were produced from a W/O emulsion using Shirasu porous glass (SPG) membranes.

4.8.7 Microgels

Microgels are hydrogel microparticles that are colloidally stable in aqueous solutions. Temperature-responsive microgels undergo a rapid change in hydrodynamic particle diameters in temperature-based hydrating or dehydrating polymers in aqueous solution at the lower critical solution temperature. Techniques for the preparation of hydrogels can be copied or adopted for microgels as long as there is no non-particulate morphology such as film or deviation towards a pure polymeric formulation. There are three important factors in using microgels in drug delivery. The first factor concerns the stability of microgels as a stable dispersion in physiological conditions mimicking blood plasma because the microgel drug has to circulate systemically before significant controlled release of the drug. The second factor is the degradation kinetics in allowing sustainable release leading to clearance after complete degradation of the microgel. The third factor is controlling the microgel particle diameter to less than 200 nm in diameter to pass blood vessels or enter cells membranes.

4.8.8 Microgels from Natural Polymers in Drug Delivery

Current research in formulating and pharmacokinetic-based testing of microgel drugs is still being pursued. Most recently, plasmid DNA macromolecules were loaded in microgels by an inversion microemulsion polymerization technique with ethylene glycol diglycidyl ether (EGDE) crosslinking reagent for cancer research therapy. A novel pH sensitive microgel was prepared using a salt bridge interaction between polyanionic carboxymethylcellulose (CMC) and tertiary amide of cationic (2-hydroxyethyl) trimethylammonium chloride benzoate (TMACB) linked with β-Cyclodextrin (β-CD) at pH 8.0. β-CD was crosslinked with CMC using TMACB and a model drug, calcein was loaded successfully .

References

[1] P. Agulhon, V. Markova, M. Robitzer, F. Quignard, T. Mineva. (2012) Structure of alginate gels: interaction of diuronate units with divalent cations from density functional calculations. Biomacromolecules 13:1899~1907.

[2] S. P. Ahirrao, P. S.Gide, B. Shrivastav, P. Sharma (2014) Ionotropic gelation: a promising cross linking technique for hydrogels. Res Rev J Pharm Nanotechnol 2:1~6.
[3] E. M. Ahmed. (2015) Hydrogel: preparation, characterization, and applications. J Adv Res 6:105~121.
[4] S. Ahmed, M. Baig. (2014) Biotic elicitor enhanced production of psolaren in suspension cultures of Psoralea corylifolia L. Saudi J Biol Sci 21:499~504.
[5] T. Aikawa, T. Konno, M. Takai, K. Ishihara. (2012) Spherical phospholipid polymer hydrogels for cell encapsulation prepared with a flow-focusing microfluidic channel device. Langmuir 28: 2145~2150.
[6] K. Akamatsu, W. Chen, Y. Suzuki, T. Ito, A. Nakao, T. Sugawara, R. Kikuchi, S. Nakao. (2010) Preparation of monodisperse chitosan microcapsules with hollow structures using the SPG membrane emulsification technique. Langmuir 26:14854~14860.
[7] A. Alhalaweh, S. Andersson, S. P. Velaga. (2009) Preparation of zolmitriptan-chitosan microparticles by spray drying for nasal delivery. Eur J Pharm Sci 38:206~214.
[8] A. Allahham, P. Stewart, J. Marriott, D. E. Mainwaring. (2004) Flow and injection characteristics of pharmaceutical parenteral formulations using a micro-capillary rheometer. Int J Pharm 270:139~148.
[9] Aptar pharma Inc. (2015) Parenteral drug delivery device.
[10] S. Arkvanshi, N. Akhtar, S. S. Bhattacharya. (2014) Transdermal delivery a preclinical and clinical perspective of drugs delivered via patches. Int J Pharm Pharm Sci 6:26~38.

Chapter 5

Hydrophobic Polymers of Pharmaceutical Significance

Over the recent years, the use of hydrophobic polymers for drug delivery applications has dramatically increased. These materials offer particular promise for controlled/sustained-drug release, thereby enhancing the pharmacological effects of the drug. These controlled/sustained release drug delivery systems can result in considerable clinical and economic advantages. The physicochemical properties of the hydrophobic polymers and the design of the drug delivery system both affect the mechanism by which a drug diffuses from the polymeric system. This chapter provides an overview of the different types of pharmaceutical hydrophobic polymers, and drug delivery applications of these polymers.

5.1 Introduction

The use of hydrophobic polymers in the pharmaceutical industry has received great attention within the last few decades, due principally to the clinical and economic benefits that can be achieved from the use of these polymers. Hydrophobic polymers can be used to produce drug delivery systems that can sustain or control drug release for longer periods in the gastrointestinal tract (GI tract) and hence result in more reproducible and uniform drug blood levels. As a result of extending the pharmacological effects of the drug, both drug dose and frequency of administration can be reduced, thereby resulting in a lower incidence of side effects. These effects can enhance the cost effectiveness and safety profiles of drugs and thus improve patient compliance.

Sustained release dosage forms based on hydrophobic polymers can be fabricated either as matrix systems, in which the drug is dispersed homogeneously in the polymeric matrix, or as reservoir systems, in which the polymer surrounds the drug core, release being controlled by the nature of membrane. The hydrophobic properties of these polymers reduce/minimize water penetration into and through the system, resulting in retarded drug diffusion and hence lowered drug release rate.

Moreover, hydrophobic polymers can be used efficiently in drug colon targeting applications with combination to the pH-dependent hydrophilic polymers to provide more controlled drug release behavior in colonic fluid medium. Additionally, hydrophobic polymers have been used efficiently to target the highly water soluble polysaccharides such as pectin and chitosan to colon, where enzymatic degradation can take place.

Hydrophobic polymers may be broadly categorized as biodegradable or nonbiodegradable. Biodegradable polymers have advantages over the nonbiodegradable polymers when employed as controlled release implants as there is no need to remove the implants physically after completion of therapy.

Within the pharmaceutical literature, different manufacturing technologies have been utilized to prepare hydrophobic polymer based drug delivery systems. These include wet granulation, direct compression, hot melt extrusion (HME), injection molding, wet mass extrusion–spheronization, and film coating. Under certain circumstances, drug release from hydrophobic polymeric platforms may be insufficient to ensure that the required mass of drug is released over the normal residence time of the drug delivery system within the GI tract. Therefore, the inclusion of hydrophilic polymer(s) within such formulations may be used to enhance the drug release rate, thereby ensuring complete drug release within 12~24 h. Conversely, the inclusion of hydrophobic polymers within hydrophilic polymer platforms containing highly water soluble drugs can be a useful strategy to ensure successful controlled drug delivery.

Examples of pharmaceutically approved hydrophobic polymers that are currently being investigated for drug delivery applications include: ammoniomethacrylate copolymers (Eudragit® RL/RS), poly(ethylacrylate–methylmethacrylate) (Eudragit® NE30D), ethylcellulose, polyvinylacetate, polyethylenevinylacetate, and poly(ecaprolactone).

5.2 Examples of Hydrophobic Polymers and Their Drug Delivery Applications

There are several examples of hydrophobic polymers that have been/are currently being used for pharmaceutical applications. This section seeks to describe the key properties of these polymers and to provide examples of their pharmaceutical uses. In the interest of brevity, only the most common polymer and uses have been described.

5.2.1 Ammoniomethacrylate Copolymers

Eudragit® RL100/RLPO and RS100/RSPO are referred to as ammoniomethacrylate copolymers (Types A and B, respectively) in the USP/NF 23 monograph. RL100 and RS100 are colorless clear to cloudy granules, whereas RLPO and RSPO are white fine powders. RL100/RLPO and RS100/RSPO have a slight amine-like odor and are freely soluble in acetone and alcohols. They are copolymers of ethyl acrylate, methyl methacrylate, and a low content of a methacrylic acid ester with quaternary ammonium groups (trimethyl ammonioethyl methacrylate chloride) (Fig.5.1) of average molecular weight about 150,000 $g \cdot mol^{-1}$. Eudragit® RL100/RLPO and RS100/RSPO are chemically defined as poly (ethyl acrylate, methyl methacrylate, trimethyl ammonio ethyl methacrylate chloride) in the ratio of monomers, viz., 1∶2∶0.2 and 1∶2∶0.1, respectively. Eudragit® RL100/RLPO contains a higher percentage of ammoniomethacrylate units (8.85%~11.96% on dry substance) compared to Eudragit® RS100/RSPO (4.48%~6.77% on dry substance) (USP/NF).

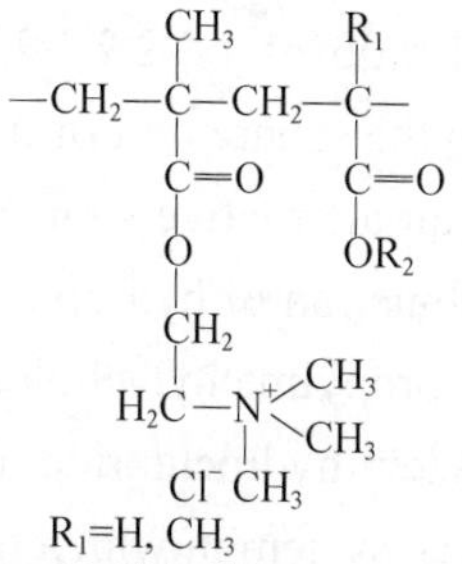

R_1=H, CH_3
R_2=CH_3, C_2H_5

Fig.5.1 Chemical structure of ammoniomethacrylate copolymers.

The ammonium groups are present as salts, making these polymers water permeable and physiologically pH-independent. As a result of the difference in the content of ammonium groups, RL100/RLPO and RS100/RSPO differ in their water permeation rate, which is considerably higher in case of RL100/RLPO. It was reported that films prepared from RL are freely permeable to water, whereas films prepared from RS are only slightly permeable to water.

5.2.1.1 Hydrophobic Matrices Based on Ammoniomethacrylate Copolymers

The hydrophobic and the pH-independent properties of RL/RLPO and RS100/RSPO polymers make them suitable for producing oral controlled/sustained release drug delivery systems. Different approaches can be used for producing controlled/ sustained release dosage forms. Among these different approaches, matrix tablets are still considered to be one of the most efficient and interesting from both the economic and the process development aspects. It was reported that RLPO has better compaction properties than RSPO, and hence matrix tablets of excellent flow and physical properties can be produced using RLPO by direct compression even in absence of other excipients.

Due to the differences in their permeability, RLPO and RSPO can be used in combination to produce matrix tablets of desired sustained release drug profiles. RLPO and RSPO have been combined to modulate the release kinetics of theophylline from matrix tablets prepared by direct compression. The release rate of theophylline from RLPO matrix tablets was observed to be too high; however, this was effectively modified by the addition of RSPO to the matrix. The use of an RLPO-RSPO ratio of 1∶9 resulted in formulations that offered a similar release profile of theophylline to that observed with commercial formulations.

As previously described, the release of a water soluble drug from a water soluble polymeric platform is often rapid, and therefore, hydrophobic polymers may be included within the matrix formulation to offer greater control of drug release. In this scenario, hydrophobic polymers reduce the rapid diffusion of the dissolved drug through the hydrophilic gel network. For this purpose, Eudragit® RSPO was combined with a hydrophilic polymer, Methocel K100M, to reduce water penetration and retard the drug release of venlafaxine from matrix tablets prepared by direct compression. The desired 16h sustained release profile was achieved using drug/polymer/ratio of 1∶2.2∶2.2 and followed Higuchi kinetics, with cumulative release proportional to the square root of time.

In an alternative strategy, matrix tablets have been developed by direct compression based on combination of hydrophobic polymers (RSPO and RLPO) and a gelling hydrophilic polymer, hydroxypropylmethylcellulose (HPMC 60SH) to achieve a (20 h) sustained release formulation of diltiazem hydrochloride. Film coating of the matrix tablets using Eudargit® NE30D produced a delivery system in which the release of diltiazem hydrochloride was pH-independent (from 1.2 to 7.4). Eudragit® RSPO was employed to delay the penetration of dissolution medium into the matrix, thereby decreasing the drug release rate. Inclusion of RLPO/RSPO as hydrophobic polymers in matrix tablets containing gelling hydrophilic polymers such as HPMC can be used efficiently to further sustain the release of water soluble drugs. The hydrophobicity of these polymers reduces water penetration into the matrix tablets, which results in further retardation of drug release rate.

5.2.1.2 Solid Dispersions Based on Ammoniomethacrylate Copolymers

It has been demonstrated that the use of hydrophobic matrices containing solid drug dispersions, i.e., systems in which a drug is molecularly dispersed within a polymeric platform, is a valuable strategy in the production of sustained release products. For example, solid dispersions of verapamil hydrochloride with Eudragit® RLPO or Kollidon® SR, a directly compressible polymeric blend composed primarily of polyvinylacetate (PVAc) and povidone (PVP), were prepared using solvent evaporation method and then compressed into tablets. Drug release from RLPO tablets at drug-polymer ratio (1∶3) were able to sustain the drug release up to 12h, whereas Kollidon® SR tablets at similar drug-polymer ratio only sustained drug release up to 8h. In a similar fashion, solid dispersions of promethazine hydrochloride and Eudragit® RLPO and RS100 at different drug-polymer ratios (1∶1 and 1∶5) by solvent evaporation were prepared. It has been reported that drug release from the solid dispersions was highly dependent on the type and amount of the polymer used.

Dissolution of the drug from RLPO matrices was greater than from systems containing RS100, due principally to the higher swelling and permeation characteristics of RLPO over a range of pH values (1.2~7.4). Tablets prepared by direct compression from RLPO solid dispersions (ratio of 1∶5) displayed extended release of drug for 12 h.

In a similar fashion, solid dispersions of metoprolol tartrate (7%, 15%, or 25% *W*/*W*) were prepared by melting and solvent methods using different ratios of RLPO-RSPO (0∶10, 3∶7,

5 ∶ 5, 7 ∶ 3, and 10 ∶ 0). At high drug contents (15% and 25% *W*/*W*), a significant initial burst effect was observed as a result of a nonhomogeneous dispersion of the drug in the polymer and the unincorporated metoprolol around the matrix, which immediately dissolves in the medium. Solid dispersions containing higher ratios of RSPO showed slower release rates than those containing higher ratios of RLPO due to its lower permeability. Drug release from dispersions was slower than from the physical mixtures (Fig.5.2). Furthermore, at similar drug-polymer ratios, drug release from systems prepared using the fusion method was significantly slower than that from systems prepared using the solvent method.

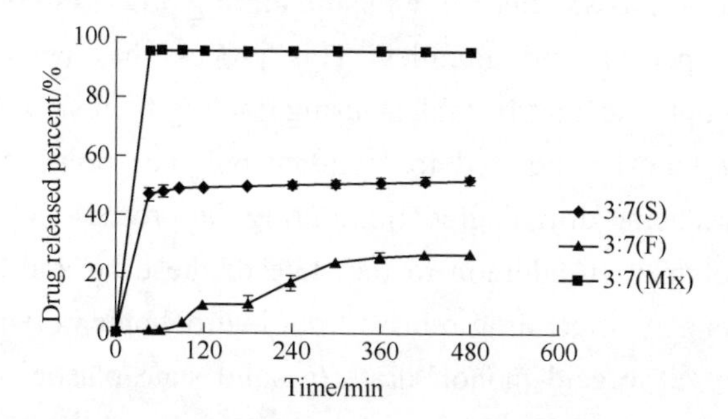

Fig.5.2 Drug release profiles from solid dispersions (3 ∶ 7 Eudragit RLPO-RSPO containing 7% metoprol tartrate, particle size of 200 mm) prepared with fusion method (F) or solvent method (S) compared with physical mixture of drug with the same ratios of polymers in phosphate buffer solution (pH 6.8) ($n = 3$).

Figure 5.3(a) and (b) show the effect of particle size on the drug release profiles from platforms containing 7% drug loading and composed of two different ratios of RLPO and RSPO (3 ∶ 7 prepared using the fusion method and 5 ∶ 5 prepared using the solvent method). Solid dispersions with a particle size of 100 mm containing (7% *W*/*W*) of metoprolol and 5 ∶ 5 ratio of RLPO-RSPO prepared by solvent method or 3 ∶ 7 ratio of RLPO-RSPO with fusion method had similar release pattern to Lopressor® sustained-release tablets up to 8 h.

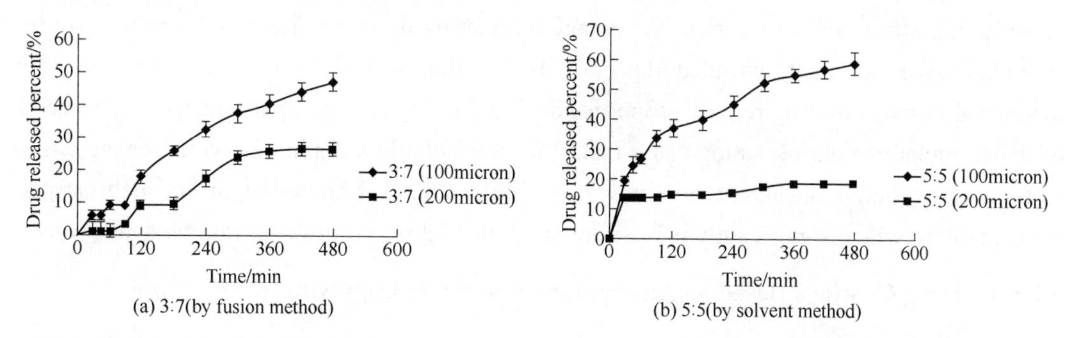

Fig.5.3 Drug release profiles from solid dispersions of (a) 3 ∶ 7 (by fusion method), (b) 5 ∶ 5 (by solvent method) Eudragit RLPO-RSPO with different particle sizes containing 7% metoprolol tartrate in phosphate buffer solution (pH 6.8) ($n = 3$).

RLPO and RSPO can be used alone or in combination as efficient carriers in solid dispersions to produce sustained drug release profiles. Matrix tablets prepared from RLPO/RSPO solid dispersions usually have slower drug release profiles than matrix tablets prepared from the corresponding physical mixtures. The intimate contact between the drug and the polymers within the solid dispersions plays an important role in such further retardation effects. Preparation method, drug loading, RLPO-RSPO ratio, and particle size of the solid dispersions are important factors that can affect the drug release profile, which need to be considered in optimizing the final formulation to achieve the desired drug release profile.

5.2.1.3 Hot Melt Extrusion Based on Ammoniomethacrylate Copolymers

Hot melt extrusion (HME) technology has received great interest in the pharmaceutical industry over the last decade as an alternative manufacturing process to produce different dosage forms such as tablets, pellets, and granules. This process has been employed to prepare controlled release hydrophobic matrix tablets using Eudragit® RSPO. For efficient hot melt extrusion, a plasticizer may be required to facilitate polymer flow and to lower processing temperatures, thereby reducing drug degradation during the process and increasing the stability of the drug and the polymer. In addition to the state of the drug and the composition of the formulation, plasticizers may affect drug release from hydrophobic polymeric matrices prepared by HME. For example, citric acid monohydrate (a solid state plasticizer) has been shown to enhance the release of diltiazem hydrochloride from melt extruded RSPO tablets, due principally to enhanced pore formation, improved drug dispersion in the plasticized polymer and the increase in the polymer aqueous permeability. The influence of the level of triethylcitrate (TEC), a water soluble plasticizer, on drug release from hot melt extrudates containing RSPO and highly water soluble drugs, chlorpheniramine maleate and diltiazem hydrochloride has been described. In these studies, chlorpheniramine maleate, but not diltiazem hydrochloride, was observed to offer solid state plasticization of Eudragit® RSPO on RSPO. The release of chlorpheniramine maleate from the melt extrudate was increased with increasing TEC concentration. The authors attributed these observations to the leaching out of triethyl citrate from the melt extrudates, thereby creating channels within the melt extruded matrix tablet. Therefore, it is important to consider the effects of the plasticizer type and amount used on the drug release profiles during formulation of hot melt extruded dosage forms. Additionally, these formulations must be monitored in terms of drug release properties during stability as liquid plasticizers may leach out of the formulations during storage as a result of their volatility. Additionally, the change in the amount of plasticizer during storage may affect the mechanical properties of hot-melt extruded matrix tablets, which significantly may affect the drug release profiles and drug bioavailability.

5.2.1.4 Drug Coatings Based on Ammoniomethacrylate Copolymers

Eudragit® RL and RS are commonly used in film coating of solid dosage forms to produce controlled release reservoir systems. Typically, the coating is applied as an organic polymer solution or as an aqueous colloidal polymer dispersion. Aqueous colloidal dispersions of Eudragit® RL30D and RS30D (30% *W*/*W* dry substance) form water-insoluble pH- independent swellable films across which drug release may be effectively controlled. Film coatings that have been deposited from aqueous polymeric dispersions are frequently associated with physical aging during storage. The mechanism by which films are formed from aqueous polymeric dispersions is more complex than those encountered with organic polymer solutions. In aqueous systems, the coalescence of individual colloidal particles and the inter diffusion of polymeric particles must occur to form a continuous film. Since the coalescence of the colloidal polymer particles into a homogeneous film is often incomplete after coating with aqueous polymeric dispersions, further coalescence of the colloidal particles occurs during storage as the polymer relaxes toward an equilibrium state (termed physical aging). These effects result in a decreased void volume (porosity), an increase in polymer tortuosity, and a resultant decrease in the rate of drug release. Therefore, with aqueous colloidal polymer dispersions, a thermal after-treatment (curing) at elevated temperatures (above the T_g of the polymer) is often recommended to complete film formation and to avoid changes in the release profiles during storage. Various strategies have been proposed to overcome aging of hydrophobic coatings of pharmaceutical dosage forms. For example:

① Storage of the dosage form at high humidity may increase the physical aging of polymeric films and therefore this factor must be controlled.

② The degree of coalescence of the colloidal polymeric particles increases as the mass of plasticizer in the coating increases, producing less pronounced aging effects.

③ Thermal treatment along with high concentrations of micronized talc stabilizes drug release from pellets coated by Eudragit® RS/RL30D plasticized with TEC.

④ Curing of Eudragit® RS30D film coated pellets at an elevated temperature (above the glass transition temperature), T_g of the polymeric film, after coating is efficient in reducing the aging effects and hence in stabilizing drug release rate.

⑤ Inclusion of the enteric polymer of high T_g, e.g., Eudragit® L100-55, to RS30D results in a more stabilized drug release rate from theophylline coated pellets during storage compared with the pellets coated with RS30D alone without Eudragit® L100-55. These results are mostly attributed to the network change in the RS polymeric films as a result of the formation of miscible system between Eudragit® L100-55 and RS30D, which has a single higher T_g than the T_g of RS polymeric films without L100-55. This increase in the T_g of RS30D polymeric films with the inclusion of L100-55 results in a greater restriction of the mobility of RS30D polymeric films. Hence, the aging effects during storage under accelerated conditions are minimized.

⑥ Addition of a hydrophilic polymer, hydroxyethylcellulose (HEC) to Eudragit® RS30D dispersions stabilizes the drug release rates during storage. This stabilization effect of HEC is related to the formation of an immiscible secondary phase surrounding the colloidal particles, which interferes with further coalescence of the colloidal RS particles.

⑦ The ionic electrostatic interactions between lactic acid (LA) and aqueous dispersions of RL and RS minimize the physical aging effects, resulting in stabilized paracetamol release from the coated tablets during storage at different accelerated storage conditions.

To reduce the physical aging problems associated with the incomplete coalescence of polymer particles after coating from aqueous colloidal dispersions, it is important to consider the coating formulations in terms of adding optimum type and amount of plasticizer that can aid in coalescence of polymer particles during coating process. Consequently, more reproducible drug release properties can be achieved following storage. Additionally, as described in the aforementioned examples, additives can play important roles in such stabilization based on their miscibility and interactions with the coating polymer. Regarding the process, it is better to leave the batch after the coating process for a certain period of time for drying at a specific temperature above the glass transition temperature of the coating polymer to improve the coalescence of polymer particles.

5.2.2 Poly(ethylacrylate-methylmethacrylate)

Eudragit® NE 30D and Eudragit® NE 40D (30% and 40% *W*/*W* dry substance, respectively) are aqueous colloidal dispersions based on neutral (nonionic) poly (ethylacrylate-methyl methacrylate) (2∶1) copolymers (Fig.5.4) that are prepared by emulsion polymerization. Highly flexible film coatings may be produced by poly(ethylacrylate-methylmethacrylate) aqueous dispersions without using any plasticizers, as the polymer has low minimum film-forming temperature (5℃). This film flexibility is related to strong interchain interactions. The nonionic hydrophobic properties of poly(ethylacrylate-methylmethacrylate) make this polymer suitable to be used as a sustained release coating material for solid dosage forms forming pH-independent drug release profiles over entire pH range of the GI tract. There have been several examples of the use of this polymer, as a coating, for the development of controlled drug release formulations. For example, It has been reported the formulation of a controlled release pellet formulation containing venlafaxine hydrochloride that was coated with Eudragit® NE30D. Using this approach, a once daily sustained release formulation was possible. Similarly, improved release profiles of ofloxacin were obtained from pellets that had been coated with Eudragit® NE30D and Eudragit® L30D55, at a ratio of 1∶8 (*W*/*W*) and containing the plasticizer diethyl phthalate (DEP). Finally, using a combination of Eudragit NE 30D and Eudragit L30D-55 as coatings, the delayed release of verapamil hydrochloride was successfully reported.

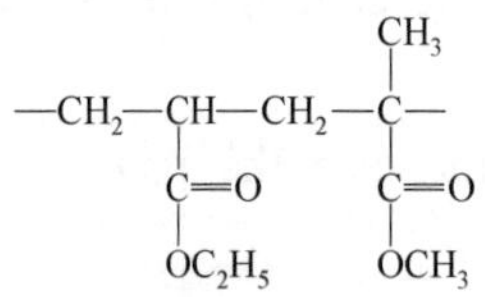

CH_3
—CH_2—CH—CH_2—C—
C=O   C=O
OC_2H_5 OCH_3

Fig.5.4 Chemical structure of ethylacrylate and methylacrylate copolymer.

Poly(ethylacrylate-methylmethacrylate) has additionally been successfully been employed for colonic delivery of therapeutic agents. For example, in one study theophylline pellets were coated with Eudragit® NE30D aqueous dispersions, containing various pectin HM-Eudragit® RL30D ionic complexes. It was shown that without pectinolytic enzymes, the release of theophylline from the coated pellets, after aninitiallatency phase, occurslinearly as a function oftime and was dependent on the pectin HM content. The lowest theophylline release from the coated pellets was obtained whenever the pectin HM content was 20.0% *W*/*W* (related to Eudragit® RL), i.e., when the complexation between pectin HM and Eudragit® RL is optimal. Consequently, the coating permeability and the release of theophylline from the coated pellets was also minimal resulting in the slowest drug release rate. The degradation and leaching of pectin from the coatings in presence of the pectinolytic enzymes resulted in the solvation, swelling of Eudragit® RL. This, in turn, induced stresses in the film coatings, thereby facilitating (increasing) the release of theophylline.

Film coatings based on poly(ethylacrylate-methylmethacrylate) are efficient in sustaining release profiles of different drugs. The composition of the cores can determine the permeability of the surrounding film-coatings and consequently the drug release profiles through these films.

5.2.3 Ethylcellulose

Ethylcellulose (EC) is a semisynthetic cellulose derivative that widely used in the pharmaceutical industry as a pharmaceutical coating due to its excellent film forming properties, good mechanical strength, and relatively low cost. This polymer has been widely used to control drug release from solid dosage forms.

5.2.3.1 Hydrophobic Matrices Based on Ethylcellulose

Owing to its hydrophobic properties, ethylcellulose reduces the penetration of water into the solid polymeric matrix, hence reducing drug release. Combination of this polymer with hydroxypropylmethylcellulose (HPMC K100M) in matrix tablets prepared by direct compression was reported to sustain the release of venlafaxine hydrochloride up to 16h at drug-ethylcellulose-HPMC ratio of 1∶2.2∶2.7 with the release following Higuchi kinetics. Similarly, inclusion of ethylcellulose, at a concentration of 14% *W*/*W* of the matrix of tablets composed of RLPO and RSPO prepared by wet granulation resulted in an extended release profile of zidovudine for up to 12 h, whereas only a maximum of 6 h sustained release profile was achieved from the matrix tablets formulated without ethylcellulose. These observations were accredited to the effects of ethylcellulose on the hydrophobicity of the tablet matrix and the concomitant reduction in the rate of penetration of dissolution fluid into the matrix. Moreover, the addition of ethylcellulose to the RLPO and RSPO formulations reduced the initial burst release of zidovudine in the acidic medium, hence decreasing the chances of dose dumping.

The drug release kinetics the indicated a combined effect of diffusion and erosion release mechanisms from the matrix tablets. In a further example, ibuprofen mini tablets containing HPMC or ethylcellulose were produced by direct compression. At similar drug/polymer ratio, tablets based on ethylcellulose provided a greater sustained release of drug than the HPMC counterpart. For example, the use of ethylcellulose involves the formulation of sustained-release bilayer matrix tablets of propranolol hydrochloride. In these, one layer was formulated (using superdisintegrants) to offer rapid release to provide the loading dose of the drug. The second layer was composed of Eudragit® RLPO, Eudragit® RSPO and ethylcellulose, and which was designed to provide controlled release. The hydrophobic polymers were added at three different drug/polymer ratios (1∶0.5, 1∶1, and 1∶1.5). The increase in the proportion of ethylcellulose resulted in a decrease in drug release. Using drug/ethylcellulose ratios of 1∶1 and 1∶1.5 resulted in the desired release profile over the test period of 12 h, whereas only the drug/polymer ratio of 1∶1.5 for RLPO and RSPO polymers resulted in the 12-h sustained release profiles.

The previously mentioned examples suggest a high efficiency of ethylcellulose to act as a drug release retardant from matrix tablets when it is used alone or in combination with other hydrophobic or hydrophilic retardant polymers to further sustain the drug release profiles. This sustained drug release effect by ethylcellulos is highly dependent on the concentration of polymer used and is mostly related to its hydrophobicity.

Sustained-release hot-melt extruded matrices of ibuprofen and metoprolol tartrate have been produced using ethylcellulose. No plasticizer was required to process ibuprofen containing formulations due to the solid state plasticization effect of ibuprofen. Conversely, dibutyl sebacate was added as a plasticizer at 50% (*W*/*W*) of the ethylcellulose concentration to metoprolol tartrate formulations.

Controlled drug release was achieved from these formulations. Furthermore, the authors illustrated that the release of each drug may be enhanced by the addition of xanthan gum due to the higher liquid uptake into these formulations.

The effect of the concentration and molecular weight of a hydrophilic polymer (polyethylene glycol or polyethylene oxide) on metoprolol tartrate release kinetics from ethylcellulose minimatrices were described that increasing the concentration of either hydrophilic polymer increased the rate of drug release, whereas the influence of molecular weight of the hydrophilic polymers was dependent on its concentration. The mechanisms of drug release from these formulations were strongly dependent on the composition of the formulations. At high concentrations of polyethylene glycol or polyethylene oxide (20%) and/or intermediate concentrations (2.5%~10%) of low molecular weight variants of these hydrophilic polymers (~100,000 Da), drug diffusion was the predominant release mechanism. Conversely, at low concentrations of these hydrophilic polymers contents and intermediate concentrations of high molecular weight (1,000,000 and 7,000,000 Da.) variants of these polymers, changes in matrix porosities significantly affected the diffusion of the drug from the ethylcellulosebased minimatrices.

A hot melt extruded ethylcellulose pipe surrounding a drug-containing hydroxypropyl methylcellulose (HPMC)-Gelucire® 44/14 core was developed. The developed drug delivery system was successful in producing a sustained zero-order, erosion controlled, release profile that was independent of drug solubility. Shortening the length of the ethylcellulose cylinder accelerated drug release, while modifying the diameter did not affect the drug release rate. These unexpected results were justified that based on calculating the relative surface area of the matrix core of the matrix-in-cylinder system, where the drug release rate was independent of the diameter of the matrix-in-cylinder and it was only a function of the length of ethylcellulose cylinder. A randomized cross over in vivo study in dogs revealed that the matrix-in-cylinder system containing propranolol hydrochloride has an ideal sustained release profile with constant plasma levels maintained over 24 h. Moreover, administration of the matrix-in-cylinder system resulted in a fourfold increase in propranolol bioavailability when compared with a commercial sustained release formulation (Inderal®).

Sustained release injection molded matrix tablets of metoprolol tartrate were developed based on ethylcellulose. Formulations containing ethylcellulose, plasticized with dibutyl sebacate, and low substituted hydroxypropylcellulose (L-HPC) were first melt extruded and subsequently injection molded into tablets at different temperatures. Incomplete drug release (<50%) within 24 h was observed from tablets containing only ethylcellulose (4 cps). Increasing the mass of L-HPC in the formulation resulted in significantly higher drug release rates due to its hydrophilic swelling properties, achieving complete drug release after 16 h.

Based on the Ritger-Peppas classification, the release mechanism shifted from diffusion controlled, in formulations without L-HPC, toward anomalous transport at higher L-HPC concentration. In the later case, the drug release occurred via drug diffusion through the micro capillary network formed after dissolution of metoprolol clusters in the tablet and by disruption of the matrix structure due to the swelling properties of L-HPC. Similarly, the effect of hydroxypropylmethylcellulose (HPMC) on the release of metoprolol tartrate from ethylcellulose matrix tablets prepared by combination of HME and injection molding was studied.

Tablets containing 30% metoprolol and 70% ethylcellulose showed an incomplete drug release within 24h (<50%). Substituting part of the ethylcellulose fraction with HPMC (HPMC-ethylcellulose ratios of 2∶5 and 1∶1) resulted in faster and constant drug release rates. Formulations containing 50% HPMC had a complete and first order drug release profile with drug release controlled via the combination of diffusion and swelling/erosion.

These examples suggest the efficiency of ethylcellulose to act as a controlled release platform in hot melt extrusion using either liquid plasticizers or the solid state plasticizing effects of the drug. HME based on ethylcellulose was efficient in producing sustained release dosage forms such as tablet matrices, minimatrices, matrix-in-cylinder, injection molded tablets. Incorporation of hydrophilic additives such as xanthan gum, polyethylene glycol/polyethylene oxide, L-HPC, and HPMC can enhance the diffusion of drug from ethylcellulose matrix due to the hydrophilicity of these additives that result in increasing water penetration through the matrix and hence increasing drug diffusion.

5.2.3.2 Coatings Based on Ethylcellulose

Ethylcellulose is widely used to coat solid dosage forms, being applied either as organic polymeric solutions or aqueous colloidal polymeric dispersions. Two aqueous colloidal polymeric dispersions are available for use in controlled release applications (Aquacoat® ECD and Surelease®). These are highly concentrated (30% and 25% *W*/*W* ethylcellulose, respectively) with relatively low viscosities.

It has been reported that theophylline release rates from pellets coated with aqueous ethylcellulose dispersions (Aquacoat®) were too slow due to the poor permeability of ethylcellulose films to drug diffusion. This problem was overcome by the inclusion of a low percentage of a water soluble poly (vinyl alcohol)-poly (ethylene glycol) (PVA-PEG) graft copolymer. Even at low loadings, this hydrophilic copolymer significantly increased the rate and extent of water uptake and hence the permeability of the films to drug diffusion. It was demonstrated that a broad spectrum of pH-independent drug release rates were obtained from drug loaded pellets by simply varying the PVA-PEG graft copolymer content. In addition to its contribution in enhancing theophylline release rate from EC coated pellets, it has been shown that the addition of small amounts of PVA-PEG graft-copolymer to aqueous EC dispersions provides long term stable drug release patterns even upon open storage under stress conditions. The presence of this hydrophilic compound can be expected to trap water within the film coatings during coating and curing, thus facilitating polymer particle coalescence.

Drug release from ethylcellulose-coated pellets (coated using Surelease®) has been reported to be highly dependent on drug type and coating level. As the coating load of Surelease® increased the rate of drug release decreases. It was shown that at similar coating load metoclopramide hydrochloride (a water-soluble cationic drug) release from the ethylcellulose coated pellets was slower than from diclofenac sodium (a sparingly soluble anionic drug) coated pellets. The slower release of metoclopramide hydrochloride may be due to an in situ formation of a poorly soluble complex of the cationic drug and the anionic ammonium oleate present in Surelease® as a surfactant.

This complex, because of its large molecular size, may diffuse more slowly through the film, causing a reduction in the release rate of metoclopramide hydrochloride. A dry powder coating technology using micronized ethylcellulose has been developed for application in a fluidized bed coater (Glatt® GPCG-1, Wurster insert), and this technology has been applied as a pellet coating to produce a sustained-release propranolol hydrochloride dosage form. The dry coating process (batch size = 1.2 kg) was performed using a powder feed rate of 10~14 g•min^{-1} at a product temperature 45~47℃. It was shown that despite its high T_g (133.4℃), micronized ethylcellulose powder can be used for dry powder coating (with the aid of a plasticizer). Although the coated pellets had an uneven surface, an extended drug release profile was produced using a coating level of 15% *W*/*W* after curing the pellets at 80℃ for 24h. This curing step was required for complete coalescence of the polymer particles. Both the cured and uncured ethylcellulose coated pellets showed unchanged drug release profiles upon storage at room temperature for 3 years mostly because of its high T_g (Fig.5.5). In a subsequent study, it was shown that EC coated

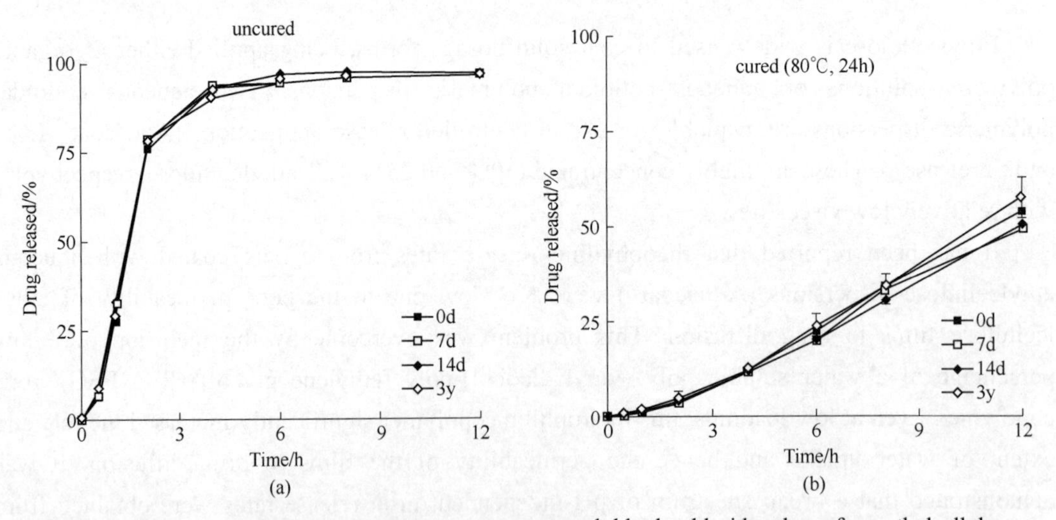

Fig.5.5 Effect of storage at room temperature on propranolol hydrochloride release from ethylcellulose powder coated pellets: (a) uncured pellets; and (b) cured pellets at 80℃ for 24 h (coating level, 30.3%; 40% acetylated monoglyceride).

pellets prepared using the dry powder coating required a higher coating level than pellets coated with the aqueous dispersions, Aquacoat® ECD, to achieve a similar extended release profile. In addition, a higher amount of plasticizer was used with the dry powder coating. Pellets coated with the ethanolic ethylcellulose solution had the slowest release.

The use of ethylcellulose as a coating polymer is highly efficient in producing sustained drug release profiles. Drug release from ethylcellulose based filmcoatings depends on the coating level, drug solubility, and the form in which the polymer is applied in the coating process, i.e. as powder, aqueous colloidal dispersion, or organic solution. Addition of hydrophilic additives such as PVA-PEG graft copolymer may increase the drug release rate and the stability of ethylcellulose-based coated dosage forms.

5.2.4 Polyvinylacetate

Polyvinylacetate (PVAc) is a thermoplastic synthetic amorphous homopolymer (Fig.5.6) with a relatively low glass transition temperature (T_g = 32.7℃ and 35.9℃ for PVAc of molecular weight 12,000 and 45,000, respectively). It is a predominantly water insoluble polymer that is used to produce controlled release drug delivery systems.

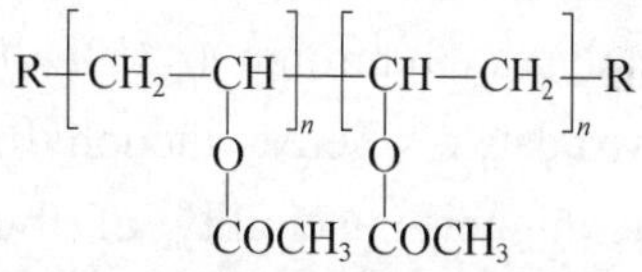

$$R{-}\left[CH_2{-}\underset{\underset{COCH_3}{|}}{\underset{O}{|}}{CH}\right]_n\left[\underset{\underset{COCH_3}{|}}{\underset{O}{|}}{CH}{-}CH_2\right]_n{-}R$$

Fig.5.6 Chemical structure of polyvinylacetate (PVAc).

5.2.4.1 Hydrophobic Matrices Based on Polyvinylacetate

The thermoplastic properties of PVAc and its relatively low glass transition temperature make this polymer particularly suitable for HME. Sustained-release theophylline matrix tablets based on PVAc have been prepared by HME at a temperature of 70℃. The cylindrical extrudates were either cut into tablets or ground into granules and compressed with other excipients into tablets. Theophylline was present in the extrudate in its crystalline form and was released from the tablets by diffusion. Increasing the granule size resulted in a significant decrease in drug release rate as a result of the longer diffusion pathway. Higher drug loading levels resulted in faster drug release due to presence of drug clusters on the surface of the matrix. Inclusion of water soluble polymers such as hydroxypropylcellulose and polyethylene oxide in the matrix resulted in significant increase in drug release rates due to the formation of highly porous structure as a result of the dissolution of these hydrophilic polymers, and hence more drug diffusion through the matrix system.

Kollidon® SR is an extended release excipient based on polyvinylacetate and polyvinylpyrrolidone (8∶2) and is used in preparation of sustained-release matrix tablets. Extended-release matrix tablets containing ZK 811 752, a weakly basic drug, were successfully prepared based on Kollidon® SR. Addition of the (highly swellable) maize starch and (the water-soluble) lactose accelerated the drug release in a more pronounced manner compared to the water insoluble calcium phosphate. Drug release rate from the matrix tablets prepared by wet granulation was faster than the tablets prepared by direct compression. Stability studies conducted at 25℃/60% RH, 30℃/70% RH and 40℃/75% RH showed no drug degradation upon storage at 25℃/60% RH, 30℃/70% RH and 40℃/75% RH for up to 6 months. Reproducible drug release patterns were obtained for matrix tablets stored at 25℃/60% RH, 30℃/70% RH for up to 6 months that remained almost unchanged compared to the initial release profiles. On the contrary, drug release from matrix tablets stored at 40℃/75% RH decreased slightly. After 6h initially 53% drug has been released versus only 47% and 43% drug release after storage for 3 and 6 months, respectively. These changes in the in vitro drug release for tablets stored at 40℃/75% RH may be related to the increase in the hardness of the tablets as a result of storing the matrix tablets at 40℃ above the glass transition (T_g) of the polymer (PVA/PVP) (T_g = 35℃) and then storing them at room temperature before dissolution testing.

In another study, hot-melt extruded extended-release minimatrices were developed based on Kollidon® SR using ibuprofen and theophylline as model drugs. It was shown that ibuprofen had solid-state plasticization effects on Kollidon® SR, whereas no such effects were observed on the polymer by theophylline. Increasing ibuprofen concentrations resulted in a significant decrease in the T_g of Kollidon® SR and the torque generated during the HME process. According to the differential scanning calorimetry (DSC) and X-ray diffraction (XRD) analyses, ibuprofen (<35% *W*/*W*) remained in an amorphous or dissolved state within the extrudates, whereas theophylline was dispersed in the polymer matrix. This can be related to the higher miscibility of ibuprofen with Kollidon® SR than theophylline especially that ibuprofen showed solid state plasticization effects on Kollidon® SR, whereas theophylline did not show such effects. The drug release rates were increased with increasing amounts of ibuprofen or theophylline in the hot-melt extrudates.

A higher processing temperature resulted in decreasing theophylline release rate, which was most probably due to the formation of extrudates that have more dense structure and lower porosity as a result of decreasing the melt viscosity during HME with increasing the processing temperature. Conversely, ibuprofen release rate increased with increasing extrusion temperature, which was mostly due to the increase in the water uptake by the extrudates as a result of the plasticizing effects of ibuprofen on Kollidon® SR. Inclusion of a plasticizer (Triethyl citrate) at 5% *W*/*W* in theophylline/Kollidone® SR formulations was sufficient to improve their processibility. Theophylline release rate from hot melt extrudates decreased with increasing TEC level due to formation of a denser matrix. Inclusion of Klucel® LF as a hydrophilic additive to the melt extrudates resulted in increasing ibuprofen and theophylline release rates as a result of its leaching out from the extrudates, creating a more porous matrix.

The low glass transition temperature of PVAc makes Kollidon® SR suitable for HME. Sustained-release hot melt extruded matrix tablets and minimatrices can be produced efficiently based on Kollidon® SR. The drug release from Kollidon® SR formulations depends on the drug loading, type and amount of hydrophilic additives added to the formulations. Attention must be taken upon storage of PVAc based matrices at temperatures above its glass transition temperature. The effect of extrusion temperature on the drug release profiles from Kollidon® SR matrices depends on the type of drug used and its miscibility and interactions with Kollidon® SR, which as well may affect the solid state properties of the drug within the matrix tablets.

5.2.4.2 Coatings Based on Polyvinylacetate

In addition to its use as a tablet matrix, polyvinylacetate and its copolymers have been used as a coating for solid dosage forms. For example, the properties of Kollicoat® SR, an aqueous colloidal dispersions based on polyvinylacetate (27%, *W*/*V*) and polyvinylpyrrolidone (2.5%, *W*/*V*) were investigated, they can be used for sustained release coatings. The release of propranolol HCl from Kollicoat® SR 30D coated pellets was reported to decrease with increasing coating level. Using this approach, a 12-h controlled release profile was obtained for propranolol hydrochloride, in which the mass of the coating was from 10% to 15% *W*/*W*. Based on their mechanical and dissolution properties, coatings based on chitosan/Kollicoat® SR 30D films have been proposed for use as colon targeted systems. The extent of film digestion in simulated colonic fluid (SCF) by rat cecal bacterial enzymes or β-glucosidase was directly proportional to the amount of chitosan present within the film. The release rate of theophylline from tablets coated by different chitosan/Kollicoat® RS 30D blends was influenced by the amount of chitosan present in the film and the coating levels. Increasing the coating levels resulted in a decrease in the release rate due to the increased diffusion pathway. The drug release was faster in simulated gastric fluid (SCF) than that in simulated intestinal fluid (SIF) since chitosan dissolved and leached from the coating in acidic medium as a result of the protonation of its amine groups. Faster drug release in SCF than in simulated gastric fluid (SGF) or SIF demonstrated the susceptibility of chitosan to degradation by the bacterial enzymes in SCF. Drug release was controlled by polymer relaxation. The in vivo pharmacokinetic studies of the coated tablets in rats showed delayed $T_{\max}$, decreased $C_{\max}$, and prolonged mean residence time (MRT), indicating the efficiency of Chitosan/Kollicoat® SR30D coating system to target drug delivery to the colon. SAG/ZK, a potent drug candidate for the oral treatment of inflammatory diseases, has a short biological half life and it exhibits a pH-dependent solubility because of its weakly basic nature. Drug release from conventional pellet formulations decreased with increasing pH values of the dissolution medium.

Extended drug release pellets were prepared by extrusion/spheronization followed by film coating with Kollidon® SR. To achieve pH-independent drug release, different organic acids were incorporated into the core pellets. The addition of fumaric acid was found to lower the pH values within the core pellets during the release of SAG/ZK in phosphate buffer pH 6.8. Therefore, increased release rates at higher pH values were observed leading to pH-independent drug release. Conversely, drug release remained pH-dependent for pellets containing tartaric and adipic acid as a result of their lower acidic strength and higher aqueous solubility of these acids.

These aforementioned examples indicate that Kollicoat® SR can produce efficient sustained release film-coatings. Drug release rate from these film-coatings depends on the coating level and the composition of the cores and the films.

5.2.5 Polyethylenevinylacetate

Polyethylenevinylacetate (EVA) copolymer is composed of long chains of ethylene and

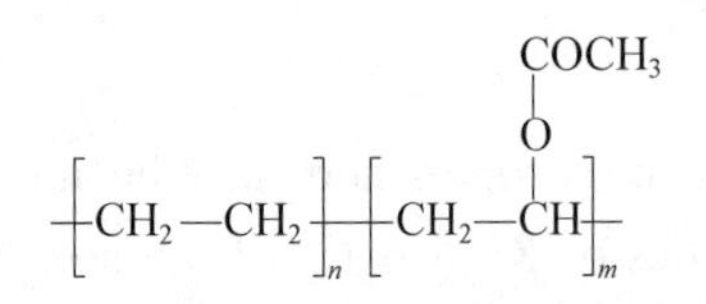

Fig.5.7 Chemical structure of polyethylenevinylacetate copolymer (EVA).

vinyl acetate groups randomly distributed throughout the chains (Fig.5.7). The weight percent of vinyl acetate usually varies from 10% to 40%, with the remainder being ethylene. EVA is inexpensive, nonbiodegradable, biocompatible polymer that has been approved for human use by Food and Drug Administration.

EVA copolymer has been widely used as a polymeric matrix to produce controlled release polymeric implant devices of high (macromolecules) or small molecular weight compounds. HME is a highly viable method of producing implants of EVA as it is thermoplastic, heat processable and flexible. The vast majority of papers that have employed EVA have been designed as biomedical implants, some of which are described in this section. Implanon® is a controlled-release implants based on EVA copolymer developed by Organon using HME technology. Implanon® is designed to release progestagen for a period of 3 years. The properties of the polymer can be adapted by varying the amount of vinyl acetate. By increasing the amount of vinyl acetate, the crystallization process of the polyethylene segments is disturbed. As a consequence, the copolymer becomes less crystalline and thus more permeable.

Recently the very small uniformly sized implants were developed based on microextrusion technology. Using microextrusion it was possible to incorporate small masses of drugs into EVA polymer and consequently this technology was successfully used to produce controlled release implant devices (based on EVA polymer) of several compounds such as alpha-methyl-*p*-tyrosine, diazepam, quinolinic acid, and phencyclidine. Each substance was slowly released from the polymer for up to 120 days at daily rates varying from 18.4 mg for phencyclidine to 97.6 mg/day for diazepam. Release was dependent on the hydrophilic properties of the drug. Drug release resulted in the formation of pores formation within the EVA matrix thereby facilitating further release.

An implantable stent containing 5-fluorouracil (5-FU) was fabricated by coating a film, composed of one 5-FU-containing EVA copolymer layer and one drug free EVA protective layer, around a commercial self-expandable nitinol stent with the drug free EVA layer facing the lumen of the stent. The stents with various drug loadings were implanted into rabbit esophagus. Quantitative analysis of 5-FU showed that the 5-FU concentration adjacent to the esophageal tissue was overwhelmingly higher than that in serum or liver at all the investigation time points until 45 days. So, the 5-FU-loaded esophageal stent offered long term, effective local drug delivery.

A biocompatible sustained-release subretinal implant was developed successfully based on coating nitinol, poly(methyl methacrylate), or chromic gut core filament with a drug eluting polymer matrix composed of a mixture of poly(butyl methacrylate) and poly(ethylene-co-vinyl acetate). Triamcinolone acetonide and sirolimus were used as model drugs. The implant had the ability to elute triamcinolone acetonide for a period of at least 4 weeks without eliciting an inflammatory response or complications, suggesting good biocompatibility and efficacy.

Controlled DNA delivery systems based on EVA implantable polymer matrices were developed and characterized. Herring sperm DNA and bacteria phage lambda DNA were encapsulated as a model system. Released DNA concentration was determined by fluoroassays. Agarose electrophoresis was used to determine the dependence of release rate on DNA size. Both small and large DNA molecules (herring sperm DNA, 0.1~0.6 kb; GFP, 1.9 kb; lambda DNA, 48.5 kb) were successfully encapsulated and released from EVAc matrices. Release from the DNA-EVAc systems was diffusion controlled. When coencapsulated in the same matrix, the larger lambda DNA was released more slowly than herring sperm; the rate of release scaled with the DNA diffusion coefficient in water. The chemical and biological integrity of released DNA was not changed. These low cost and adjustable controlled DNA delivery systems, using the FDA approved implantable/injectable material, EVAc could be useful for in vivo gene delivery, such as DNA vaccination and gene therapy.

The sublingual formulation of buprenorphine has been associated with variable drug blood levels and requires frequent dosing that limits patient compliance. Sustained release buprenorphine implants based on EVA (26 mm in length and 2.4 mm in diameter) were developed to improve patient compliance. In vitro studies showed a steady-state drug release (0.5 mg/implant/day). In vivo pharmacokinetic studies conducted in beagle dogs showed that peak buprenorphine concentrations were generally reached within 24 h after implantation. Steady state plasma levels were attained between 3 and 8 weeks, and were maintained for study duration (52 weeks), with a calculated mean release rate of 0.14±0.04 mg/implant/day. There were no serious adverse effects reported. These results suggest that this delivery system can provide long term stable systemic buprenorphine levels that may improve patient compliance, thereby improving outcome for opioid dependent patients.

Pharmacotherapy treatment for alcoholism is limited by poor compliance, adverse effects, and fluctuating drug levels after bolus administration. To overcome compliance issues, a sustained release implant based on EVA and containing nalmefene, an opioid antagonist used for treatment of alcoholism was developed by melt extrusion and characterized. The extrudates of 2.8 mm×27 mm rods were further coated with EVA to optimize release. In vitro release after subcutaneous implantation into rats was high from the uncoated rods, and they were depleted of drug fairly quickly; however, EVA coatings maintained release over longer periods. The 25 wt% coated rods provided in vitro release of 0.36 mg/day/rod and in vivo release of 0.29 mg/day/rod over 6 months, and showed dose dependent nalmefene plasma concentrations. After explanation, nalmefene plasma concentrations were undetectable by 6 h. A sustained release nalmefene rod provided 6 months of drug with no adverse effects.

A nonbiodegradable polymeric episcleral implant based on EVA was developed as a controlled intraocular delivery system of betamethasone (BM) to the posterior pole of the eye. For in vivo studies, the implants were placed on the sclera of the eyes of rabbits so that the drug releasing surface could attach to the sclera at the posterior pole. The implant released BM in a zero-order fashion for 4 weeks, thereby maintaining drug concentrations in the retina-choroid above the concentrations effective for suppressing inflammatory reactions, suggesting that the episcleral implant can be a useful drug carrier for the intraocular delivery of BM to the posterior part of the eye. These examples indicate the efficiency of EVA copolymer as a sustained release carrier in implantable devices. Its ability to produce long term drug release profiles make EVA based implants applicable in wide range of drug delivery systems such as ocular, subcutaneous, sublingual, gene delivery, esophageal for local or systemic effects.

5.2.6 Poly(ε-caprolactone)

Poly(ε-caprolactone) (PCL) is an aliphatic, biodegradable, semicrystalline polyester (Fig.5.8) that has found several applications as a biodegradable drug delivery system. PCL is highly permeable to small drug molecules of molecular weight less than 400 Da. This high permeability of PCL coupled with the long controllable induction period prior to polymer weight loss enables the development of delivery devices that offer diffusion-controlled drug delivery. Increasing PCL crystallinity reduces the permeability of the polymer, thereby lowering the drug release rate. In particular, the slow rate of degradation of PCL renders it suitable for long-term delivery of therapeutic agents, for periods greater than 1 year. Conversely, for many drug delivery applications, the degradation rate of PCL is too slow to directly influence drug release. Therefore, to increase the rate of degradation, PCL has been copolymerized with other more hydrophilic cyclic esters, including L-LA and g-butyrolactone. Additionally, it has been demonstrated that copolymerization of ε-caprolactone with hydrophilic segments such as poly (ethylene glycol) (PEG), methoxy poly (ethylene glycol) (MPEG) or poly(ethylene oxide)-poly(propylene oxide)-poly(ethylene oxide) (PEOPPO-PEO) can effectively change the hydrophobicity and improve the biodegradability of PCL. The properties of these copolymers can be controlled to exhibit various degradation rates and permeability behaviors. Moreover, the degradation rate of PCL may also be manipulated by addition of acidic and basic additives, which can facilitate polymer degradation. For example, the incorporation of oleic acid caused an increase in the rate of degradation that was proportional to the amount of acid added. An increased rate of degradation was also obtained with the addition of decylamine, as the amine functional group can launch a nucleophilic attack on the ester bond. Primary, secondary and tertiary alkylamines were incorporated into PCL. Primary alkylamines caused the most rapid degradation, whereas the effect of the tertiary alkylamines was not significant. Inclusion of a polar phospholipid, phosphorylcholine (PC), into PCL (PC-PCL) effectively improved the hydrophilicity and biodegradability of PCL. It was shown that ibuprofen release rate was faster from PC-PCL matrix systems than that from PCL systems with drug release mechanism governed mainly by diffusion kinetics. PCL has been employed for drug delivery to the eye. For example, a drug delivery system based on combinations of different molecular weights of PCL was developed. Increasing the proportion of high-M_w PCL to low-M_w PCL decreased the rate of degradation of PCL and consequently resulted in slowing down 5-FU release. These devices showed promising in vivo results for the treatment of proliferative retinopathy. Similarly, implants composed of PLGA-co-PCL were investigated in vivo and were found to inhibit inflammation in experimental uveitis by controlling cyclosporine A release, with no systemic or local toxicity.

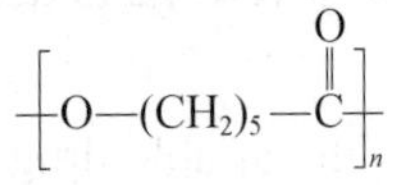

$$\left[O - (CH_2)_5 - \overset{O}{\overset{\|}{C}} \right]_n$$

Fig.5.8 Chemical structure of poly(ε-caprolactone) (PCL).

A sustained release monolithic implant based on gabapentin (GBP)-loaded PCL matrices that produces constant plasma levels over a 1-week period was designed by melt-molding/compression procedure to overcome the clinical problems associated with the chronic treatment with gabapentin (GBP). In vitro release studies showed that the uncoated implants displayed release profiles according to a pseudo-first order model. In order to further regulate the release, two sided coated implants where drug free layers would perform as membranes controlling the delivery rate were prepared. A more moderated burst effect and a relatively linear (zero-order) release between days 1 and 7 were apparent. Implants were investigated in vivo by inserting them subcutaneously in mice and the plasma levels monitored during 10 days. Findings indicated that after a more pronounced release during day 1 and the achievement of the levels in blood comparable to a twice a da intraperitoneal management, relatively constant levels were attained until day 7.

Overall results support the usefulness of this manufacturing method for the production of implants to attain more prolonged GBP release profiles. Combination of twin-screw extruder and injection molding technologies were efficient in producing a long term sustained release PCL implants containing praziquantel (PZQ), a broad-spectrum antiparasite drug. Uniform dispersion of the drug within the polymeric matrices were achieved whatever the drug loadings (6.25%~50%). X-ray diffraction analyses showed that PZQ exists primarily in its crystalline state in the fabricated implants. In vitro release studies showed that all implants, regardless of drug loading, exhibit similar release patterns and about 70% of PZQ was released after 365 days from PCL implants. It has been demonstrated that the release of PZQ was based on a gradual diffusion of the drug from the exterior to interior of the implants.

Poly(ε-caprolactone) can be used to fabricate implantable devices that can sustain drug release profiles for moderate and long terms. The drug release from poly(ε-caprolactone) based implants is diffusion controlled and depends on the molecular weight, crystallinity, and chemical modification of the polymer. Additionally, inclusion of additives to poly(ε-caprolactone) based implantable devices can modify the degradation rate of the polymer making these devices more flexible to be used in many drug delivery applications.

5.3 Conclusions

This chapter provides an overview of examples of hydrophobic polymers that are used in pharmaceutical industry and focuses on the applications of these polymers in the drug delivery systems. The diversity in the physicochemical properties of these hydrophobic polymers provides potential opportunities to develop drug delivery systems with tailored drug release profiles that can meet different clinical conditions and can improve the patient compliance, the cost effectiveness, and the safety profile of the drug. Based on the valuable clinical and economic benefits that the hydrophobic polymers can provide, it is expected that the significance of hydrophobic polymers as drug delivery systems will continue to evolve in the near future.

References

[1] M. Abbaspour, R. Sadeghi, H. A. Garekani. (2005) Preparation and characterization of ibuprofen pellets based on Eudragit RSPO and RLPO or their combination. Int J Pharm 303:88~94.

[2] M. Abbaspour, R. Sadeghi, H. A. Garekani. (2007) Thermal treating as a tool to produce plastic pellets based on Eudragit RSPO and RLPO aimed for tableting. Eur J Pharm Biopharm 67:260~267.

[3] A. Akhgari, F. Sadeghi, H. A. Garekani. (2006) Combination of time-dependent and pH-dependent polymethacrylates as a single coating formulation for colonic delivery of indomethacin pellets. Int J Pharm 320:137~142.

[4] K. Amighi, A. J. Moes (1996) Influence of plasticizer concentration and storage conditions of the drug release rate from Eudragit® RS 30 D film-coated sustained-release theophylline pellets. Eur J Pharm Biopharm 42(1):29~35.

[5] N. R. Beeley, J. M. Stewart, R. Tano, L. R. Lawin, R. A. Chappa, G. Qiu, A. B. Anderson, E. de Juan, S. E. Varner. (2005) Development, implantation, in vivo elution, and retrieval of a biocompatible, sustained release subretinal drug delivery system. J Biomed Mater Res 76A(4):690~698.

[6] R. Bodmeier, O. Paeratakul. (1994) Mechanical properties of dry and wet cellulosic and acrylic films prepared from aqueous colloidal polymer dispersions used in the coating of solid dosage forms. Pharm Res 11(6):882~888.

[7] H. Borhani, M. H. Rahimy, G. A. Peymann. (1993) Sustained release of 5-FU from biodegradablematrix delivery for intraocular application: in vitro and in vivo evaluation. Invest Ophthalmol Vis Sci 34:1488 (Suppl.).

[8] H. Boyapally, R. K. Nukala, D. Douroumis. (2009) Development and release mechanism of diltiazem HCl prolonged release matrix tablets. Drug Deliv 16(2):67~74.

[9] J. Breitenbach. (2002) Melt extrusion: from process to drug delivery technology. Eur J Pharm Biopharm 54:107~117.

[10] G. Buckton, D. Ganderton, R. Shah. (1988) In vitro dissolution of some commercially available sustained-release theophylline preparations. Int J Pharm 42:35~39.

[11] L. D. Bruce, N. H. Shah, A. W. Malick, M. H. Infeld, J. W. McGinity. (2005) Properties of hot-melt extruded tablet formulations for the colonic delivery of 5-aminosalicylic acid. Eur J Pharm Biopharm 59:85~97.

Chapter 6

Polymers in Drug Delivery

Drug delivery has experienced an outstanding advance in the last few decades. Two key elements have contributed in a large extent to such a progress: the better knowledge of the physio/pathological environments through which the drugs have to pass through to reach their targets, and the development of novel excipients that actively participate in the accomplishment of the aimed delivery. In this context polymers occupy an outstanding position due to the versatility of the synthesis routes and the possibility of tuning their features and performances to fulfill the needs of every particular application. Polymers can finely regulate the site and the rate at which the drug is released from the formulation, improve drug solubility, contribute to the stability in the physiological environment, and help the drug to overcome cellular barriers, facilitating diagnostics with the therapeutic contact. This chapter reviews the role of polymers on the evolution of drug delivery systems and the current performances they are expected to play in improving the efficiency and safety of the treatments with both old and novel active pharmaceutical ingredients (APIs). An analysis of how polymers themselves are contributing to optimize classical methods of preparing drugs dosage forms and to envision advanced drug nanocarriers is also included. Whenever possible, the information was organized trying to offer structure-property-functionality relationships, with examples of commercially available materials.

6.1 Evolution of Drug Dosage Forms

Since ancient times humans have dedicated strong efforts to identify remedies that could ameliorate wounds and diseases and to implement suitable ways to administer those remedies in an efficient way. Dosage forms appeared as a tool to facilitate the administration of drugs that could not be directly applied or ingested, for example due to the small amount required for the effect or its unpleasant taste. Ebers papyrus (16th century BC) and other antique documents describe hundreds of recipes based in natural remedies that required, for example, the previous boiling in water to extract the active substances and to enable the intake, or the mixing with components (mostly fats) for a prolonged permanence on the application site (cataplasms or ointments) or an easier swallow (solid preparations). Thus, together with the remedy containing the active components, some other materials named *excipients* (derived from the Latin verb *excipere*, which literally means to mix) had to be incorporated to the medicines in order to make easier their administration and to maintain their stability for a while. The initial role of excipients was that of acting as vehicles, and as such the most used ones were milk, honey and wine. Despite the efforts spent for many centuries in having adequate medicines, relevant advances in composition and preparation did not occur until the scientific method was introduced in the age of Enlightenment.

Evolution of knowledge in chemistry, physiology and pharmacology together with the industrial revolution in the 19th century made it possible the design and manufacture of more complex dosage forms, which are still used nowadays such as tablets or capsules, for isolated or newly synthesized active pharmaceutical ingredients (APIs). Industrial revolution was a milestone for the wide access of the population to medicines, and prompted the search of excipients able to face up to the growing challenges of the new manufacturing procedures. Advances in pharmacokinetics and born of biopharmacy in the mid 20th century, made pharmacists and clinicians to realize about the importance of preparing dosage forms that not only contain the adequate amount of active substance, but that also can release it at the adequate rate when the medicine enters into contact with the physiological fluids. For systemic treatments, the amount and the rate at which unaltered drugs reaches blood stream (*bioavailability*) was set as a main criterion of quality of a medicine. As a consequence, the requirement of drug dissolution test for solid dosage forms was established for first time in 1970.

Current pharmacological treatments aim to adapt drug administration to the therapeutic needs of the patient so that, using the lowest possible dose, the disease process can be cured or its symptoms alleviated. Local, systemic and targeted release dosage forms are the main approaches to afford this aim (Fig.6.1). Acute processes require that once the medicine is administered either locally or systemically, the drug is immediately transferred from the dosage form to the body. Conversely, the treatment of a chronic condition demands a prolonged supply of drug in order to maintain effective therapeutic drug levels for a sufficiently long period of time. Targeting is particularly useful when the disease requires the use of drugs that are very unstable in the biological medium and thus should be protected from degradation before reaching the site of action, drugs that are highly toxic for the non-target tissues and may cause untoward effects if systemically distributed (e.g. antitumor agents), or drugs that should reach cellular structures that are not easily accessible from the general circulation (e.g. cell nucleus in gene therapy). Thus, depending on the particular therapeutic aim and the administration route, drug dosage forms have to fulfill a wide variety of highly demanding performances.

When the medicine is administered through a non-parenteral route, the drug has to be transferred from the dosage form to the neighbor biological environment. Dissolved drug molecules can either accumulate in the tissue in contact with the dosage form [Fig.6.1(a); e.g., skin, ophthalmic or bronchial application] for local treatment, or can pass through the membranes and enter into the bloodstream [Fig.6.1(b)], from where the drug will be distributed throughout the body, reaching the site of action (systemic effect) and the elimination organs.

Intravenous solutions allow direct incorporation of the drug to the blood stream [Fig.6.1(c)]. Parenteral route offers also the possibility of administering colloidal (nanometric) carriers with the drug encapsulated for specific delivery to specific tissues or cells (targeting) [Fig.6.1(d)].

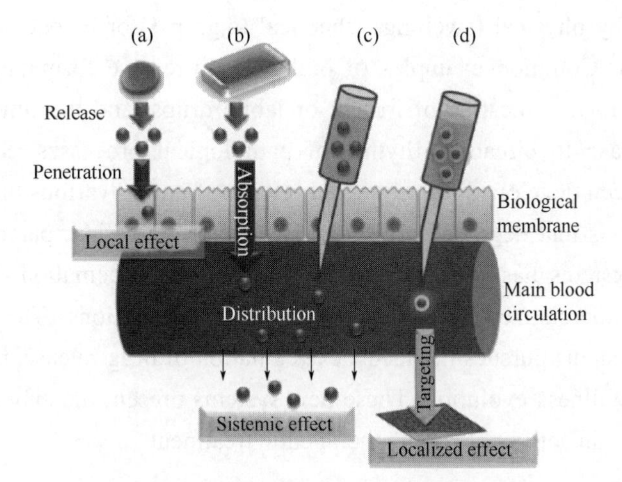

Fig.6.1 Medicines can be designed: (i) to exert a local effect on the application site by means of drug release and penetration in surrounding cells (a), (ii) to exert a systemic effect by releasing the drug in a place suitable for drug absorption (b) and/or biodistribution (c), (iii) or to target the drug to profound specific tissues or cells via direct transport of the drug (inside nanocarriers) through the blood stream towards the target (d). In either case, the drug has to be released from the dosage form to get access to the pharmacological target and exert the therapeutic effect, but the release occurs in different sites and at different rates.

Except for target delivery systems, drug release and passage through biological membranes are steps that occur sequentially. As a consequence, the slower process determines the overall rate of transfer of drug from the dosage form to the blood circulation; that is, the slower process is the limiting step for drug absorption. If both drug absorption and clearance are rapid, repeated administrations at short time intervals may be required to maintain effective plasma levels. The possibility of modulating the rate of absorption of a drug by applying technological means able to adjust the release from the dosage form is the basis of the *modified release dosage forms*. This group comprises prolonged (sustained)-release, delayed-release and pulsate-release forms (Table 6.1).

Prolonged release dosage forms, also known as rate-programmed release systems, pursue simplification of dosing regimens, by spacing the intervals between successive administrations of doses, and attenuation of peaks in drug levels. The second generation of controlled release

Table 6.1 Classification of drug dosage forms according to drug release profile as stated in the USP 37/NF 32 (2014)

Dosage Form	Drug Release Pattern	Performance
Conventional	Immediate release	Release depends on the physico-chemical properties of the drug
Modified release	Prolonged	Release occurs at slower rate than from conventional dosage form
	Delayed	Release does not occur until a certain lag time or a certain stimulus triggers it
	Pulsate	Sequential release pulses of one or more therapeutic substances

systems, called activation-modulated ones, appeared to regulate not only the rate, but also the site at which drug release should occur. The first devices were developed for site-specific release of orally administered medicines, but rapidly adapted to other routes. In these systems, drug release is activated by physical (swelling), chemical (e.g., pH) or biochemical (e.g., enzymatic degradation) stimuli. Common examples of oral delayed-release formulations are the enteric systems for release in the intestine of irritant or labile drugs, and the time-retard systems for fitting of drug release to circadian rhythm of pathological processes. Some pulsate-release formulations have been developed for sequential release of one or various therapeutic substances by combining particles that degrade at different rates after activation; parenteral administration of pulsate-release systems has been successfully applied for vaccination. Most modified-drug release devices commercialized belong to the two first generations. The third generation of controlled release systems pursues the feedback-regulation of drug release, by testing the patient condition, especially illness evolution. These new systems present the singularity of integrating components that play an active role in the therapeutic treatment.

6.2 Polymer Excipients

As introduced above, the roles that a given dosage form is expected to play are very varied, but can be gathered into three large groups: (i) facile drug dosing and patient acceptability; (ii) enhanced safety of the treatment and reproducibility of the therapeutic response; and (iii) improved drug efficacy (Table 6.2). The dosage form has to provide the adequate physical properties to facilitate the administration, but also have to correct/modulate physicochemical and biopharmaceutical properties of the drug, such as solubility or permeability, which are critical for the effectiveness of treatment. Poor aqueous solubility is a common problem among new chemical entities, because the current drug discovery methods favor the selection of highly hydrophobic molecules. In other cases, hydrophilicity and large molecular weight hinder the pass through biological membranes. Current diversity of requirements regarding spatio-temporal regulation of drug release also make design and development of drug dosage forms a highly challenging task. To ensure the quality of medicines, the dosage form has to contribute to the chemical and physical stability of the drug, both in the course of the manufacturing process and during storage.

Table 6.2 **Main functions of the drug dosage forms**

Category	Performance
Facile drug dosing and patient acceptability	Adequate physical format for administration Simple and rational dosing Correction of unpleasant organoleptic properties
Enhanced safety of the treatment and reproducibility of the therapeutic response	Improved physical and chemical stability of the drug Protection against microbial contamination Minimization of toxicity and untoward effects
Improved drug efficacy	Modulation of drug solubility, release rate and permeability Enhanced bioavailability Modulation of drug biodistribution Easy adherence to the treatment

The diversity of roles that dosage forms are called to play is only possible thanks to strong efforts in the finding/synthesis of suitable excipients able to face up to the increasing complexity of production processes and increasingly stringent quality requirements. The United States Pharmacopeia (USP37/NF32, 2014) classifies the excipients in more than 40 functional categories according to their most common application in the formulation of pharmaceuticals. Such large list of functions can be reorganized in main four categories:

① Processability, i.e., they facilitate the manufacture of the dosage form;

② Stabilization, to overcome problems related to the stability of the drug or the dosage form;

③ Correction of organoleptic properties;

④ Modulation of drug bioavailability (altering solubility, permeability or drug release rate) or the access (targeting) to specific tissues or cell structures.

As material science progresses, novel and everyday more sophisticated materials are being evaluated as excipients for drug dosage forms. It should be noticed that, among other materials, polymers occupy a relevant position as versatile excipients able to perform one or more functions simultaneously. Multifunctional excipients are particularly attractive because they simplify development and manufacturing processes and reduce costs. Currently, natural, semi-synthetic and synthetic polymers are common excipients (Fig.6.2). Natural polymers, mainly polysaccharides and protein derivatives, have been largely used for many years to cover the demands of classical dosage forms (tablets, capsules, gels) and are still widely used today because the good biocompatibility of most of them, biodegradability in the body or in the environment, and obtaining from renewable sources.

Improvements in the extraction and characterization methods make natural polymers reliable materials for quality standards. Moreover, some natural polymers perform as multifunctional or can be easily derivatized to modulate their solubility and biodegradability patterns or to endow them with responsiveness to certain stimuli. Synthetic polymers have the advantage of being prepared on demand to fit to specific requirements, setting a priori their functional groups and molecular weight. Outstanding examples of biodegradable and non-biodegradable polymers are poly(*dl*-lactic-co-glycolic acid) (PLGA) and polyethylene glycol (PEG), respectively. In the next sections, examples of polymer excipients suitable for specific functionalities are analyzed.

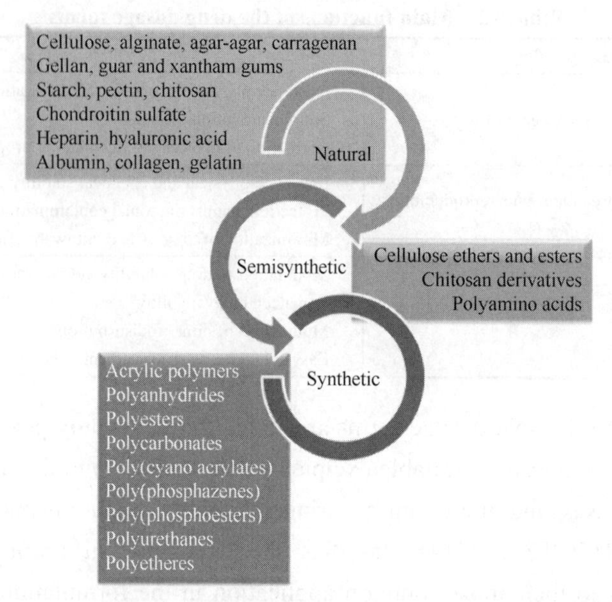

Fig.6.2 Examples of natural, semi-synthetic and synthetic polymers used to prepare drug dosage forms.

6.3 Polymers as Manufacture Aids and Sustained Release

6.3.1 Formulations Components

Processing requirements are notably different depending on the physical state of the dosage form. Manufacture of dosage forms requires excipients that are able to overcome deficiencies of the drug regarding processing and contribute to rapid and reproducible production of manageable medicines. Hydrophilic polymers may also contribute to slow down drug release by increasing the viscosity of liquid formulations or by forming gel matrix networks when solid formulations become wet. In contrast, hydrophobic polymers can help regulate drug release by means of hydrolytic or enzymatic degradation mechanisms.

6.3.1.1 Cellulose and Cellulose Derivatives

(1) Cellulose

Cellulose is an organic compound with the formula $(C_6H_{10}O_5)_n$, a polysaccharide consisting of a linear chain of several hundred to many thousands of β-(1→4) linked D-glucose units (Fig.6.3). Cellulose is an important structural component of the primary cell wall of green plants, many forms of algae and the oomycetes. Some species of bacteria secrete it to form biofilms. Cellulose is the most abundant organic polymer on Earth. The cellulose content of cotton fiber is 90%, that of wood is 40%~50%, and that of dried hemp is approximately 57%.

Cellulose is mainly used to produce paperboard and paper. Smaller quantities are converted into a wide variety of derivative products such as cellophane and rayon. Conversion of cellulose

Fig.6.3 Cellulose, a linear polymer of D-glucose units (two are shown) linked by β-(1→4)-glycosidic bonds.

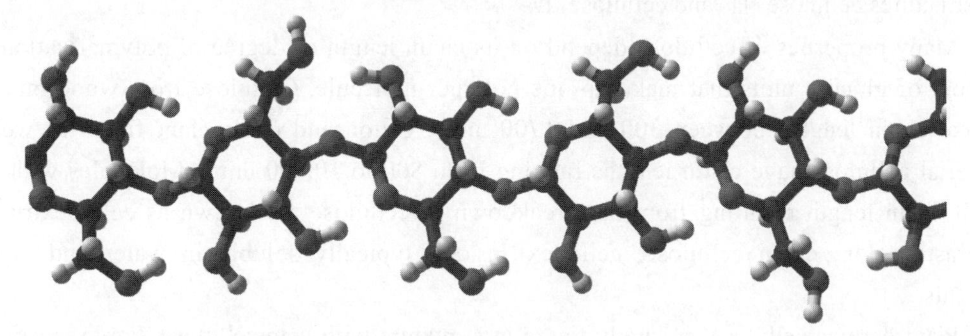

Fig.6.4 Three-dimensional structure of cellulose.

from energy crops into biofuels such as cellulosic ethanol is under investigation as an alternative fuel source. Cellulose for industrial use is mainly obtained from wood pulp and cotton.

Some animals, particularly ruminants and termites, can digest cellulose with the help of symbiotic micro-organisms that live in their guts, such as Trichonympha. In human nutrition, cellulose is a non-digestible constituent of insoluble dietary fiber, acting as a hydrophilic bulking agent for feces and potentially aiding in defecation.

① Structure and properties

Cellulose has no taste, is odorless, is hydrophilic with the contact angle of 20~30 degrees (Fig.6.4), is insoluble in water and most organic solvents, is chiral and is biodegradable. It was shown to melt at 467℃. It can be broken down chemically into its glucose units by treating it with concentrated mineral acids at high temperature.

Cellulose is derived from D-glucose units, which condense through β-(1→4)-glycosidic bonds. This linkage motif contrasts with that for α-(1→4)-glycosidic bonds present in starch and glycogen. Cellulose is a straight chain polymer: unlike starch, no coiling or branching occurs, and the molecule adopts an extended and rather stiff rod-like conformation, aided by the equatorial conformation of the glucose residues. The multiple hydroxyl groups on the glucose from one chain form hydrogen bonds with oxygen atoms on the same or on a neighbor chain, holding the chains firmly together side-by-side and forming microfibrils with high tensile strength. This confers tensile strength in cell walls, where cellulose microfibrils are meshed into a polysaccharide matrix. Cotton fibres represent the purest natural form of cellulose, containing more than 90% of this polysaccharide.

Compared to starch, cellulose is also much more crystalline. Whereas starch undergoes a crystalline to amorphous transition when heated beyond 60~70℃ in water (as in cooking), cellulose requires a temperature of 320℃ and pressure of 25 MPa to become amorphous in water. Several different crystalline structures of cellulose are known, corresponding to the location of hydrogen bonds between and within strands. Natural cellulose is cellulose Ⅰ, with structures $Ⅰ_{\alpha}$ and $Ⅰ_{\beta}$. Cellulose produced by bacteria and algae is enriched in $Ⅰ_{\alpha}$ while cellulose of higher plants consists mainly of $Ⅰ_{\beta}$. Cellulose in regenerated cellulose fibers is cellulose Ⅱ. The conversion of cellulose Ⅰ to cellulose Ⅱ is irreversible, suggesting that cellulose Ⅰ is metastable and cellulose Ⅱ is stable. With various chemical treatments it is possible to produce the structures cellulose Ⅲ and cellulose Ⅳ.

Many properties of cellulose depend on its chain length or degree of polymerization, the number of glucose units that make up one polymer molecule. Cellulose from wood pulp has typical chain lengths between 300 and 1700 units; cotton and other plant fibers as well as bacterial cellulose have chain lengths ranging from 800 to 10,000 units. Molecules with very small chain length resulting from the breakdown of cellulose are known as cellodextrins; in contrast to long-chain cellulose, cellodextrins are typically soluble in water and organic solvents.

Plant-derived cellulose is usually found in a mixture with hemicellulose, lignin, pectin and other substances, while bacterial cellulose is quite pure, has a much higher water content and higher tensile strength due to higher chain lengths.

Cellulose is soluble in Schweizer's reagent, cupriethylenediamine (CED), cadmium ethylenediamine (Cadoxen), *N*-methylmorpholine *N*-oxide, and lithium chloride/dimethyl acetamide. This is used in the production of regenerated celluloses (such as viscose and cellophane) from dissolving pulp. Cellulose is also soluble in many kinds of ionic liquids.

Cellulose consists of crystalline and amorphous regions. By treating it with strong acid, the amorphous regions can be broken up, thereby producing nanocrystalline cellulose, a novel material with many desirable properties. Recently, nanocrystalline cellulose was used as the filler phase in bio-based polymer matrices to produce nanocomposites with superior thermal and mechanical properties.

② Processing

Given a cellulose-containing material, the carbohydrate portion that does not dissolve in a 17.5% solution of sodium hydroxide at 20℃ is α cellulose, which is true cellulose. Acidification of the extract precipitates β cellulose. The portion that dissolves in base but does not precipitate with acid is γ cellulose.

Cellulose can be assayed using a method described by Updegraff in 1969, where the fiber is dissolved in acetic and nitric acid to remove lignin, hemicellulose, and xylosans. The resulting cellulose is allowed to react with anthrone in sulfuric acid. The resulting coloured compound is assayed spectrophotometrically at a wavelength of approximately 635 nm. In addition, cellulose is represented by the difference between acid detergent fiber (ADF) and acid detergent lignin (ADL). Luminescent conjugated oligothiophenes can also be used to detect cellulose using fluorescence microscopy or spectrofluorometric methods.

(2) Cellulose Derivatives

The hydroxyl groups (—OH) of cellulose can be partially or fully reacted with various reagents to afford derivatives with useful properties like mainly cellulose esters (Table 6.3) and cellulose ethers (—OR) (Table 6.4). In principle, though not always in current industrial practice, cellulosic polymers are renewable resources.

Table 6.3 Ester derivatives

Cellulose ester	Reagent	Example	Reagent	Group R
Organic esters	Organic acids	Cellulose acetate	Acetic acid and acetic anhydride	H or —(C=O)CH_3
		Cellulose triacetate	Acetic acid and acetic anhydride	—(C=O)CH_3
		Cellulose propionate	Propanoic acid	H or —(C=O)CH_2CH_3
		Cellulose acetate propionate (CAP)	Acetic acid and propanoic acid	H or —(C=O)CH_3 or —(C=O)CH_2CH_3
		Cellulose acetate butyrate (CAB)	Acetic acid and butyric acid	H or —(C=O)CH_3 or —(C=O)$CH_2CH_2CH_3$
Inorganic esters	Inorganic acids	Nitrocellulose (cellulose nitrate)	Nitric acid or another powerful nitrating agent	H or —NO_2
		Cellulose sulfate	Sulfuric acid or another powerful sulfuring agent	H or —SO_3H

Table 6.4 Ether derivatives

Cellulose ethers	Reagent	Example	Reagent	Group R = H or	Water solubility	Application	E number
Alkyl	Halogeno alkanes	Methyl cellulose	Chloro methane	—CH_3	Cold water-soluble		E461
		Ethyl cellulose	Chloro ethane	—CH_2CH_3	Water-insoluble	A commercial thermoplastic used in coatings, inks, binders, and controlled-release drug tablets	E462
		Ethyl methyl cellulose	Chloro methane and chloro ethane	—CH_3 or —CH_2CH_3			E465
Hydroxyalkyl	Epoxides	Hydroxyethyl cellulose	Ethylene oxide	—CH_2CH_2OH	Cold/hot water-soluble	Gelling and thickening agent	
		Hydroxypropyl cellulose (HPC)	Propylene oxide	—$CH_2CH(OH)CH_3$	Cold water-soluble		E463
		Hydroxyethyl methyl cellulose	Chloro methane and ethylene oxide	—CH_3 or —CH_2CH_2OH	Cold water-soluble	Production of cellulose films	
		Hydroxypropyl methyl cellulose (HPMC)	Chloromethane and propylene oxide	—CH_3 or —$CH_2CH(OH)CH_3$	Cold water-soluble	Viscosity modifier, gelling, foaming and binding agent	E464
		Ethyl hydroxy ethyl cellulose	Chloro ethane and ethylene oxide	—CH_2CH_3 or —CH_2CH_2OH			E467
Carboxyalkyl	Halogenated carboxylic acids	Carboxymethyl cellulose (CMC)	Chloroacetic acid	—CH_2COOH	Cold/Hot water-soluble	Often used as its sodium salt, sodium carboxy methyl cellulose (NaCMC)	E466

The cellulose acetate and cellulose triacetate are film- and fiber-forming materials that find a variety of uses. The nitrocellulose was initially used as an explosive and was an early film forming material. With camphor, nitrocellulose gives celluloid.

The sodium carboxymethyl cellulose can be cross-linked to give the croscarmellose sodium (E468) for use as a disintegrant in pharmaceutical formulations.

(3) Applications in Pharmaceutical Field

Cellulose is the main structural component of cell walls in higher plants and consists in linear, unbranched polysaccharide of β-1,4-linked D-glucose units (Fig.6.5). Cellulose chains form long crystalline microfibrils aligned with each other's to provide structural support to the cell wall. The strong structure of cellulose provides a high resistance against enzymatic reactions that degrade the chains. It is also insoluble in water and indigestible by the human gastrointestinal system. Modifications of cellulose structure leads to materials widely used in the pharmaceutical field to facilitate the manufacture of drugs.

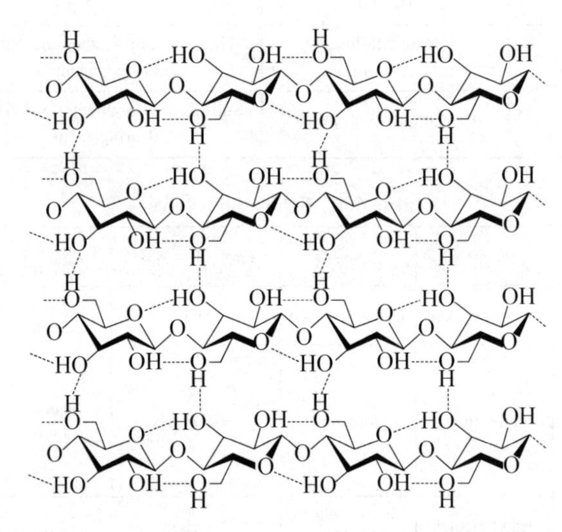

Fig.6.5 A strand of cellulose (conformation I_{α}), showing the hydrogen bonds (dashed) within and between cellulose molecules.

① Microcrystalline cellulose (MCC)

It is obtained by treatment of cellulose with hydrochloric acid to partially depolymerise the structure. After purification by filtration and spray-drying, a free-flowing powder of small aggregates of cellulose fibers is obtained. Polymer compactability can be modified by varying the degrees of polymerization and of crystallinity. It is commonly used as diluent in wet granulation and as filler in capsules. The most known commercial brands are Avicel® and Emcocel®.

② Silicified MCC

Commercialized as Prosolv SMCC®, it is produced by co-processing MCC (98 %) with colloidal silicon dioxide (2%) and commonly used as filler binder and glidant. This excipient has better properties for direct compaction, such as flowability and packaging properties. In addition, silicified MCC has higher bulk density than does regular MCC and maintains compactation properties after wetting and drying.

③ Cellulose ethers

Derivatives more hydrophilic than cellulose and thus that can easily disperse in water are obtained via substitution of one or more hydroxyl groups of the D-glucose units with alkyl, hydroxyalkyl and carboxyalkyl molecules.

Applications of cellulose derivatives depend on the degree of substitution because it determines the strength of the interaction with water. Non-ionic cellulose ethers, such as hydroxypropylcellulose (HPC) and hydroxypropyl methylcellulose (HPMC) are commonly used in tableting as binders and filmcoating components. Gelling features of medium-high substituted cellulose ethers make them suitable to create gel barriers when tablets enter into contact with physiological fluids, which can sustain drug release. They are also suitable as thickener agents for the formulation of solutions and suspensions. Low substituted non-ionic cellulose ethers behave as disintegrants of tablets and pellets. Co-processed excipients containing MCC, HPMC and crospovidone show excellent flowability, compressibility and mixing ability, and therefore notably improve manufacturability of solid dosage forms.

6.3.1.2 Starch and Modified Starches

(1) Starch

Starch is a storage carbohydrate consisting of glucose monomers organized as amylose (Fig.6.6) and amylopectin (Fig.6.7) chains. Starches are used in solid oral formulations as diluents, binders and disintegrants. Deficient flow properties of natural starches can be overcome by substitution with acetate groups, prepared by partial reaction of the hydroxyl groups of starch with acetic acid anhydride in an acetyl esterification reaction. Alternatively, flow properties can be improved via chemical and/or mechanical processing to break all or part of the starch granules (namely, the bonds between amylose and amylopectin molecules) in a process named pregelatinization. Partial pregelatinization renders the starch flowable, directly compressible, and suitable for preparing oral capsules. Both partial and fully pregelatinized can be used in wet granulation processes. Combination of pregelatinized starch with amylose and amylopectine leads to multifunctional excipients that are effective as binders, disintegrants, flow aids and lubricants. Most known commercial brands are Starch 1500®, Sepistab ST200® and C*PharmGel®.

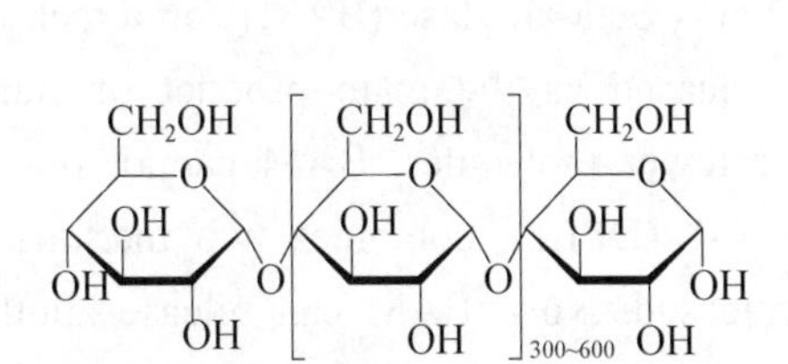

Fig.6.6 Structure of the amylose molecule.

Fig.6.7 Structure of the amylopectin molecule.

Pure starch is a white, tasteless and odorless powder that is insoluble in cold water or alcohol. Depending on the plant, starch generally contains 20% to 25% amylose and 75% to 80% amylopectin by weight. Glycogen, the glucose store of animals, is a more highly branched version of amylopectin.

① Biosynthesis

Plants produce starch by first converting glucose 1-phosphate to ADP-glucose using the enzyme glucose-1-phosphate adenylyltransferase. This step requires energy in the form of ATP. The enzyme starch synthase then adds the ADP-glucose via a 1,4-alpha glycosidic bond to a growing chain of glucose residues, liberating ADP and creating amylose. Starch branching enzyme introduces 1,6-alpha glycosidic bonds between these chains, creating the branched amylopectin. The starch debranching enzyme isoamylase removes some of these branches. Several isoforms of these enzymes exist, leading to a highly complex synthesis process.

Glycogen and amylopectin have similar structure, but the former has about one branch point per ten 1,4-alpha bonds, compared to about one branch point per thirty 1,4-alpha bonds in amylopectin. Amylopectin is synthesized from ADP-glucose while mammals and fungi synthesize glycogen from UDP-glucose; for most cases, bacteria synthesize glycogen from ADP-glucose (analogous to starch).

In addition to starch synthesis in plants, starch can be synthesized from non-food starch mediated by an enzyme cocktail. In this cell-free biosystem, beta-1,4-glycosidic bond-linked cellulose is partially hydrolyzed to cellobiose. Cellobiose phosphorylase cleaves to glucose 1-phosphate and glucose; the other enzyme——potato alpha-glucan phosphorylase can add a glucose unit from glucose 1-phosphorylase to the non-reducing ends of starch. In it, phosphate is internally recycled. The other product, glucose, can be assimilated by a yeast. This cell-free bioprocessing does not need any costly chemical and energy input, can be conducted in aqueous solution, and does not have sugar losses.

② Degradation

Starch is synthesized in plant leaves during the day and stored as granules; it serves as an energy source at night. The insoluble, highly branched starch chains have to be phosphorylated in order to be accessible for degrading enzymes. The enzyme glucan, water dikinase (GWD) phosphorylates at the C-6 position of a glucose molecule, close to the chains 1,6-alpha branching bonds. A second enzyme, phosphoglucan, water dikinase (PWD) phosphorylates the glucose molecule at the C-3 position. A loss of these enzymes, for example a loss of the GWD, leads to a starch excess (sex) phenotype, and because starch cannot be phosphorylated, it accumulates in the plastids.

After the phosphorylation, the first degrading enzyme, beta-amylase (BAM) can attack the glucose chain at its non-reducing end. Maltose is released as the main product of starch degradation. If the glucose chain consists of three or fewer molecules, BAM cannot release maltose. A second enzyme, disproportionating enzyme-1 (DPE1), combines two maltotriose molecules. From this chain, a glucose molecule is released. Now, BAM can release another maltose molecule from the remaining chain. This cycle repeats until starch is degraded completely. If BAM comes close to the phosphorylated branching point of the glucose chain, it can no longer release maltose. In order for the phosphorylated chain to be degraded, the enzyme isoamylase (ISA) is required.

The products of starch degradation are predominantly maltose and smaller amounts of glucose. These molecules are exported from the plastid to the cytosol, maltose via the maltose transporter, which if mutated (MEX1-mutant) results in maltose accumulation in the plastid. Glucose is exported via the plastidic glucose translocator (pGlcT). These two sugars act as a precursor for sucrose synthesis. Sucrose can then be used in the oxidative pentose phosphate pathway in the mitochondria, to generate ATP at night.

③ Structure and Properties

a. Structure

Rice starch seen on light microscope(Fig.6.8). Characteristic for the rice starch is that starch granules have an angular outline and some of them are attached to each other and form larger granules. While amylose was thought to be completely unbranched, it is now known that some of its molecules contain a few branch points. Amylose is a much smaller molecule than amylopectin. About one quarter of the mass of starch granules in plants consist of amylose, although there are about 150 times more amylose than amylopectin molecules.

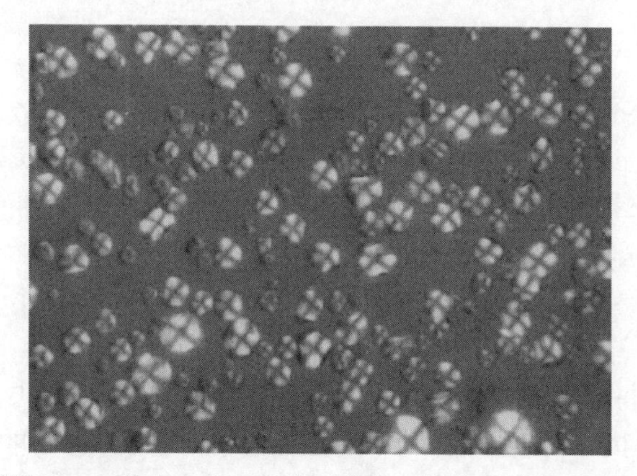

Fig.6.8 Starch, 800x magnified, under polarized light, showing characteristic extinction cross.

Starch molecules arrange themselves in the plant in semi-crystalline granules. Each plant species has a unique starch granular size: rice starch is relatively small (about 2 μm) while potato starches have larger granules (up to 100 μm).

Starch becomes soluble in water when heated. The granules swell and burst, the semi-crystalline structure is lost and the smaller amylose molecules start leaching out of the granule, forming a network that holds water and increasing the mixture's viscosity. This process is called starch gelatinization. During cooking, the starch becomes a paste and increases further in viscosity. During cooling or prolonged storage of the paste, the semi-crystalline structure partially recovers and the starch paste thickens, expelling water. This is mainly caused by retrogradation of the amylose. This process is responsible for the hardening of bread or staling, and for the water layer on top of a starch gel (syneresis).

Some cultivated plant varieties have pure amylopectin starch without amylose, known as waxy starches. The most used is waxy maize, others are glutinous rice and waxy potato starch. Waxy starches have less retrogradation, resulting in a more stable paste. High amylose starch, amylomaize, is cultivated for the use of its gel strength and for use as a resistant starch (a starch that resists digestion) in food products.

Synthetic amylose made from cellulose has a well-controlled degree of polymerization. Therefore, it can be used as a potential drug deliver carrier.

Certain starches, when mixed with water, will produce a non-newtonian fluid sometimes nicknamed “oobleck”.

b. Hydrolysis

The enzymes that break down or hydrolyze starch into the constituent sugars are known as amylases.

Alpha-amylases are found in plants and in animals. Human saliva is rich in amylase, and the pancreas also secretes the enzyme. Individuals from populations with a high-starch diet tend to have more amylase genes than those with low-starch diets; Beta-amylase cuts starch into maltose units. This process is important in the digestion of starch and is also used in brewing, where amylase from the skin of seed grains is responsible for converting starch to maltose (Malting, Mashing).

c. Dextrinization

If starch is subjected to dry heat, it breaks down to form dextrins, also called “pyrodextrins” in this context. This break down process is known as dextrinization. (Pyro)dextrins are mainly yellow to brown in color and dextrinization is partially responsible for the browning of toasted bread.

d. Chemical tests

Granules of wheat starch, are stained with iodine, photographed through a light microscope (Fig.6.8).

A triiodide (I_3^-) solution formed by mixing iodine and iodide (usually from potassium iodide) is used to test for starch; a dark blue color indicates the presence of starch. The details of this reaction are not yet fully known, but it is thought that the iodine (I_3^- and I_5^- ions) fit inside the coils of amylose, the charge transfers between the iodine and the starch, and the energy level spacings in the resulting complex correspond to the absorption spectrum in the visible light region. The strength of the resulting blue color depends on the amount of amylose present. Waxy starches with little or no amylose present will color red. Benedict’s test and Fehling’s test is also done to indicate the presence of starch.

Starch indicator solution consisting of water, starch and iodide is often used in redox titrations. In the presence of an oxidizing agent the solution turns blue, in the presence of reducing agent the blue color disappears because triiodide (I_3^-) ions break up into three iodide ions, disassembling the starch-iodine complex. A 0.3% *W/W* solution is the standard concentration for a starch indicator. It is made by adding 3 grams of soluble starch to 1 liter of heated water; the solution is cooled before use (starch-iodine complex becomes unstable at temperatures above 35℃).

Each species of plant has a unique type of starch granules in granular size, shape and crystallization pattern. Under the microscope, starch grains stained with iodine illuminated from behind with polarized light show a distinctive Maltese cross effect (also known as extinction cross and birefringence).

(2) Modified starches

A modified starch is a starch that has been chemically modified to allow the starch to function properly under conditions frequently encountered during processing or storage, such as high heat, high shear, low pH, freeze/thaw and cooling.

The modified food starches are E coded according to the International Numbering System for Food Additives (INS):

1400 Dextrin
1401 Acid-treated starch
1402 Alkaline-treated starch
1403 Bleached starch
1404 Oxidized starch
1405 Starches, enzyme-treated
1410 Monostarch phosphate
1412 Distarch phosphate
1413 Phosphated distarch phosphate
1414 Acetylated distarch phosphate
1420 Starch acetate
1422 Acetylated distarch adipate
1440 Hydroxypropyl starch
1442 Hydroxypropyl distarch phosphate
1443 Hydroxypropyl distarch glycerol
1450 Starch sodium octenyl succinate
1451 Acetylated oxidized starch

INS 1400, 1401, 1402, 1403 and 1405 are in the EU food ingredients without an E-number. Typical modified starches for technical applications are cationic starches, hydroxyethyl starch and carboxymethylated starches.

(3) Applications

① Use as food additive

As an additive for food processing, food starches are typically used as thickeners and stabilizers in foods such as puddings, custards, soups, sauces, gravies, pie fillings, and salad dressings, and to make noodles and pastas. Function as thickeners, extenders, emulsion stabilizers and are exceptional binders in processed meats.

Gummed sweets such as jelly beans and wine gums are not manufactured using a mold in the conventional sense. A tray is filled with native starch and leveled. A positive mold is then pressed into the starch leaving an impression of 1,000 or so jelly beans. The jelly mix is then poured into the impressions and put onto a stove to set. This method greatly reduces the number of molds that must be manufactured.

② Use in pharmaceutical industry

In the pharmaceutical industry, starch is also used as an excipient, as tablet disintegrant, and as binder.

In the pharmaceutical industry, starch is also used as an excipient, as tablet disintegrant, and as binder. IP/BP GRADE STARCH is produced from higher quality Maize and packed in extremely hygienic condition under very strict quality condition process to make it suitable for I.P. & B.P grade. Pharma Grade Starch is a mixture of Polysaccharides containing about 75% to 80% Amylopectin & 20%/25% Amylose. The Pharmaceutical industry is an important user of Maize Satrch (Corn Starch). IP/BP (pharmaceutical) grade maize starch is useful as a tablet binder and disintegrating agent in dispersible tablets. It is used as abase in the preparation of vitamin-C. It is also used as filler in Pills (Tables) & Capsules. It is used in coating the capsules.

6.3.1.3 Dextran

(1) Dextran

Dextran is a complex branched glucan (polysaccharide made of many glucose molecules) composed of chains of varying lengths (from 3 to 2000 kilodaltons).

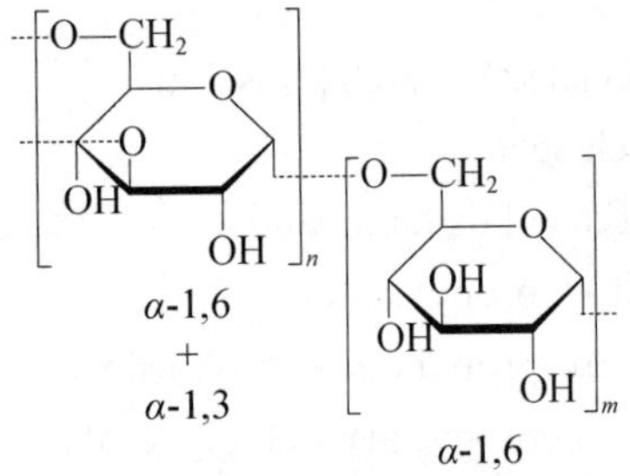

The straight chain consists of α-1,6 glycosidic linkages between glucose molecules, while branches begin from α-1,3 linkages (Fig.6.9). This characteristic branching distinguishes a dextran from a dextrin, which is a straight chain glucose polymer tethered by α-1,4 or α-1,6 linkages.

Fig.6.9 Dextran.

Dextran was first discovered by Louis Pasteur as a microbial product in wine, but mass production was only possible after the development by Allene Jeanes of a process using bacteria. Dextran is now synthesized from sucrose by certain lactic acid bacteria, the best-known being Leuconostoc mesenteroides and Streptococcus mutans. Dental plaque is rich in dextrans.

Medicinally it is used as an antithrombotic (antiplatelet), to reduce blood viscosity, and as a volume expander in hypovolaemia. Dextran 70 is on the WHO Model List of Essential Medicines, the most important medications needed in a health system.

(2) Applications of Dextran

① Microsurgery

These agents are used commonly by microsurgeons to decrease vascular thrombosis. The antithrombotic effect of dextran is mediated through its binding of erythrocytes, platelets, and vascular endothelium, increasing their electronegativity and thus reducing erythrocyte aggregation and platelet adhesiveness. Dextrans also reduce factor Ⅷ-Ag Von Willebrand factor, thereby decreasing platelet function. Clots formed after administration of dextrans are more easily lysed due to an altered thrombus structure (more evenly distributed platelets with coarser fibrin). By inhibiting α-2 antiplasmin, dextran serves as a plasminogen activator, so possesses thrombolytic features.

Outside from these features, larger dextrans, which do not pass out of the vessels, are potent osmotic agents, thus have been used urgently to treat hypovolemia. The hemodilution caused by volume expansion with dextran use improves blood flow, thus further improving patency of microanastomoses and reducing thrombosis. Still, no difference has been detected in antithrombotic effectiveness in comparison of intra-arterial and intravenous administration of dextran.

Dextrans are available in multiple molecular weights ranging from 3,000 to 2,000,000 Da. The larger dextrans (>60,000 Da) are excreted poorly from the kidney, so remain in the blood for as long as weeks until they are metabolized. Consequently, they have prolonged antithrombotic and colloidal effects. In this family, dextran-40 (M_w = 40,000 Da), has been the most popular member for anticoagulation therapy. Close to 70% of dextran-40 is excreted in urine within the first 24 hours after intravenous infusion, while the remaining 30% are retained for several more days.

② Other medical uses

It is used in some eye drops as a lubricant and in certain intravenous fluids to solubilize other factors, such as iron (in a solution known as Iron Dextran).

Intravenous solutions with dextran function both as volume expanders and means of parenteral nutrition. Such a solution provides an osmotically neutral fluid that once in the body is digested by cells into glucose and free water. It is occasionally used to replace lost blood in emergency situations, when replacement blood is not available, but must be used with caution as it does not provide necessary electrolytes and can cause hyponatremia or other electrolyte disturbances. It also increases blood sugar levels.

③ Laboratory uses

Dextran is used in the osmotic stress technique for applying osmotic pressure to biological molecules. It is also used in some size-exclusion chromatography matrices; an example is Sephadex. Dextran has also been used in bead form to aid in bioreactor applications.

Dextran has been used in immobilization in biosensors.

Dextran preferentially binds to early endosomes; fluorescent-labelled dextran can be used to visualize these endosomes under a fluorescent microscope.

Dextran can be used as a stabilizing coating to protect metal nanoparticles from oxidation and improve biocompatibility.

Dextran coupled with a fluorescent molecule such as fluorescein isothiocyanate can be used to create concentration gradients of diffusible molecules for imaging and allow subsequent characterization of gradient slope.

Solutions of fluorescently-labelled dextran can be perfused through engineered vessels to analyze vascular permeability. Dextran is used to make microcarriers for industrial cell culture.

(3) Side effects

Although relatively few side effects are associated with dextran use, these side effects can be very serious. These include anaphylaxis, volume overload, pulmonary edema, cerebral edema, or platelet dysfunction.

An uncommon but significant complication of dextran osmotic effect is acute renal failure. The pathogenesis of this renal failure is the subject of many debates with direct toxic effect on tubules and glomerulus versus intraluminal hyperviscosity being some of the proposed mechanisms. Patients with history of diabetes mellitus, renal insufficiency, or vascular disorders are most at risk. So, It is recommend the avoidance of dextran therapy in patients with chronic renal insufficiency.

6.3.1.4 Dextrin

(1) Dextrin

Dextrins are a group of low-molecular-weight carbohydrates produced by the hydrolysis of starch or glycogen. Dextrins are mixtures of polymers of D-glucose units linked by α-(1→4) or α-(1→6) glycosidic bonds (Fig.6.10).

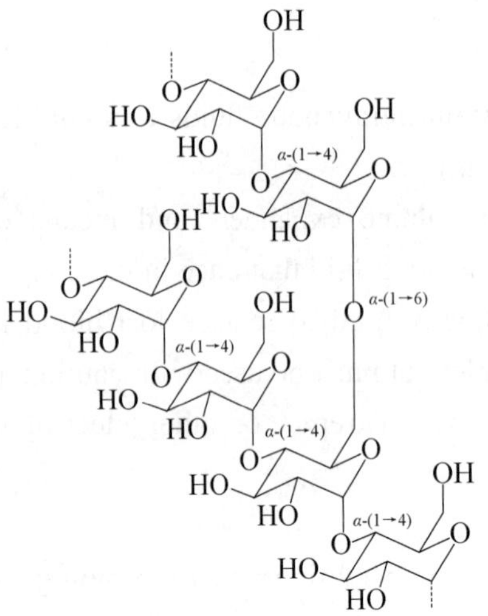

Dextrins can be produced from starch using enzymes like amylases, as during digestion in the human body and during malting and mashing, or by applying dry heat under acidic conditions (pyrolysis or roasting). The latter process is used industrially, and also occurs on the surface of bread during the baking process, contributing to flavor, color and crispness. Dextrins produced by heat are also known as pyrodextrins. The starch hydrolyses during roasting under acidic conditions, and short-chained starch parts partially rebranch with α-(1,6) bonds to the degraded starch molecule.

Fig.6.10 A dextrin with α-(1→4) and α-(1→6) glycosidic bonds.

Dextrins are white, yellow, or brown powders that are partially or fully water-soluble, yielding optically active solutions of low viscosity. Most of them can be detected with iodine solution, giving a red coloration; one distinguishes erythrodextrin (dextrin that colors red) and achrodextrin (giving no color). White and yellow dextrins from starch roasted with little or no acid are called British gum.

(2) Uses

Yellow dextrins are used as water-soluble glues in remoistable envelope adhesives and paper tubes, in the mining industry as additives in froth flotation, in the foundry industry as green strength additives in sand casting, as printing thickener for batik resist dyeing, and as binders in gouache paint and also in the leather industry.

White dextrins are used as:

a crispness enhancer for food processing, in food batters, coatings, and glazes, (INS number 1400);

a textile finishing and coating agent to increase weight and stiffness of textile fabrics;

a thickening and binding agent in pharmaceuticals and paper coatings;

a pyrotechnic binder and fuel; this is added to fireworks and sparklers, allowing them to solidify as pellets or "stars";

a stabilizing agent for certain explosive metal azides, particularly Lead(Ⅱ) azide.

Owing to the rebranching, dextrins are less digestible; indigestible dextrins are developed as soluble stand-alone fiber supplements and for adding to processed food products.

6.3.1.5 Other Types

(1) Maltodextrin

Maltodextrin is a polysaccharide that is used as a food additive. It is produced from starch by partial hydrolysis and is usually found as a white hygroscopic spray-dried powder. Maltodextrin is easily digestible, being absorbed as rapidly as glucose and might be either moderately sweet or almost flavorless. It is commonly used for the production of soft drinks and candy. It can also be found as an ingredient in a variety of other processed foods.

① Structure

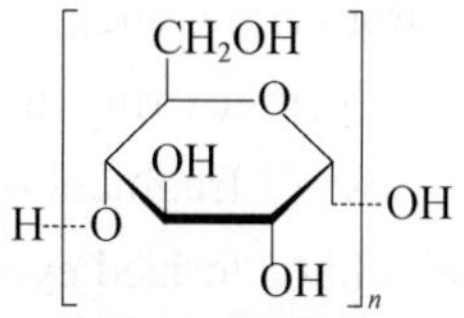

Maltodextrin consists of D-glucose units connected in chains of variable length (Fig.6.11). The glucose units are primarily linked with *α*-(1→4) glycosidic bonds. Maltodextrin is typically composed of a mixture of chains that vary from three to 17 glucose units long.

α-1,4

$2<n<20$

Fig.6.11 Structure of Maltodextrin.

Maltodextrins are classified by DE (dextrose equivalent) and have a DE between 2 and 20. The higher the DE value, the shorter the glucose chains, the higher the sweetness, the higher the solubility, and the lower heat resistance. Above DE 20, the European Union's CN code calls it glucose syrup; at DE 10 or lower the customs CN code nomenclature classifies maltodextrins as dextrins.

② Production

Maltodextrin can be enzymatically derived from any starch. In the US, this starch is usually corn; in Europe, it is commonly to use wheat. Some individuals suffering from gluten-related disorders may be concerned by the presence of wheat derived maltodextrin but it is highly unlikely to contain significant (20 $mg \cdot kg^{-1}$ or 20ppm) amounts of gluten. Maltodextrin derived from wheat is exempt from labeling, as set out in Annex Ⅱ of EC Directive No 1169/2011. However, wheat-derived maltodextrin is not exempt from allergen declaration in the United States per FALCPA, and its effect on a voluntary gluten-free claim must be evaluated on a case-by-case basis per the applicable US FDA policy.

③ Food uses

Maltodextrin is sometimes used in beer brewing to increase the specific gravity of the final product. This improves the mouthfeel of the beer, increases head retention and reduces the dryness of the drink. Maltodextrin is not fermented by yeast, so it does not increase the alcohol content of the brew. It is also used in some snacks such as potato chips and jerky. It is used in "light" peanut butter to reduce the fat content, but keep the texture. Maltodextrin is also sometimes taken as a supplement by bodybuilders and other athletes in powder form or in gel packets.

Maltodextrin is used as an inexpensive additive to thicken food products such as infant formula. It is also used as a filler in sugar substitutes and other products. Maltodextrin has a glycemic index ranging from 85 to 105.

④ Other uses

Maltodextrin is used as an horticultural insecticide both in the field and in greenhouses. It has no physiological or biochemical action. Its efficacy is based upon spraying a 50% solution upon the pest insects whereupon the solution dries, blocks the insects' spiracles and causes death by asphyxiation.

(2) Cyclodextrin

Cyclodextrins (sometimes called cycloamyloses) are a family of compounds made up of sugar molecules bound together in a ring (cyclic oligosaccharides).

Cyclodextrins are produced from starch by means of enzymatic conversion. They are used in food, pharmaceutical, drug delivery, and chemical industries, as well as agriculture and environmental engineering.

Cyclodextrins are composed of 5 or more α-D-glucopyranoside units linked 1→4, as in amylose (a fragment of starch). The 5-membered macrocycle is not natural. Recently, the largest well-characterized cyclodextrin contains 32 1,4-anhydroglucopyranoside units, while as a poorly characterized mixture, at least 150-membered cyclic oligosaccharides are also known. Typical cyclodextrins contain a number of glucose monomers ranging from six to eight units in a ring, creating a cone shape, Such as

α (alpha)-cyclodextrin: 6-membered sugar ring molecule [Fig.6.12(a)]

β (beta)-cyclodextrin: 7-membered sugar ring molecule [Fig.6.12(b)]

γ (gamma)-cyclodextrin: 8-membered sugar ring molecule [Fig.6.12(c)]

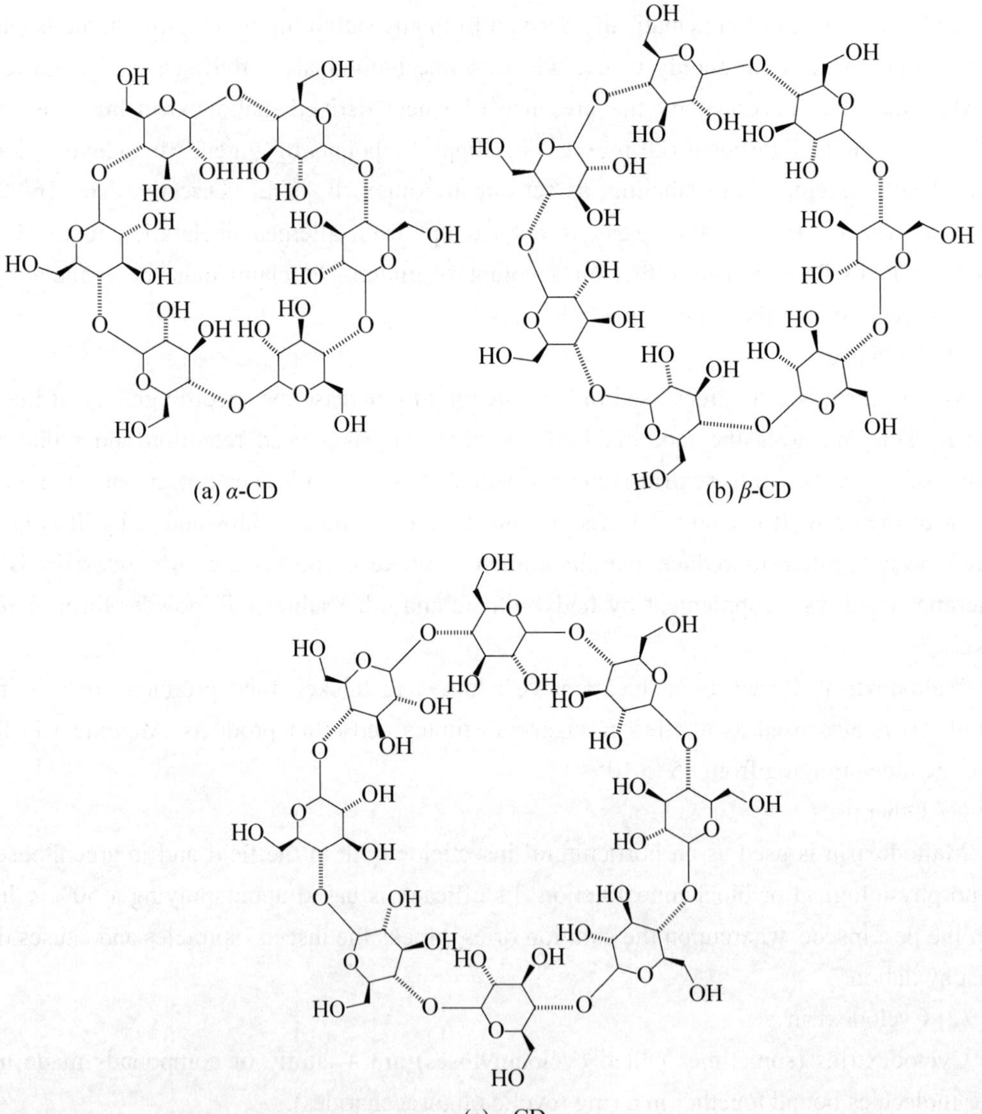

(a) α-CD

(b) β-CD

(c) γ-CD

Fig.6.12 Chemical structure of the three main types of cyclodextrins

① Application of cyclodextrin

α- and γ-cyclodextrin are being used in the food industry. As α-cyclodextrin is a soluble dietary fiber, it can be found as Alpha Cyclodextrin (soluble fiber) on the list of ingredients of commercial products.

a. Increasing bioavailability

Because cyclodextrins are hydrophobic inside and hydrophilic outside, they can form complexes with hydrophobic compounds. Thus they can enhance the solubility and bioavailability of such compounds. Cyclodextrins can also enhance drug permeability through mucosal tissues. This is of high interest for pharmaceutical as well as dietary supplement applications in which hydrophobic compounds shall be delivered. Alpha-, beta-, and gamma-cyclodextrin are all generally recognized as safe by the FDA.

b. Cholesterol free products

In the food industry, cyclodextrins are employed for the preparation of cholesterol free products: the bulky and hydrophobic cholesterol molecule is easily lodged inside cyclodextrin rings that are then removed.

c. Multifunctional dietary fiber

α-Cyclodextrin has been authorized for use as a dietary fiber in the European Union since 2008. In 2013 the EU commission has verified a health claim for alpha-cyclodextrin. The EU assessment report confirms that consumption of alpha-cyclodextrin can reduce blood sugar peaks following a high-starch meal. Weight loss supplements are marketed from alpha-cyclodextrin which claim to bind to fat and be an alternative to other anti-obesity medications.

Due to its surface-active properties, α-cyclodextrin can also be used as emulsifying fiber, for example, in mayonnaise as well as a whipping aid, for example, in desserts and confectionery applications.

d. Other food applications

Applications further include the ability to stabilize volatile or unstable compounds and the reduction of unwanted tastes and odour. Beta-cyclodextrin complexes with certain carotenoid food colorants have been shown to intensify color, increase water solubility and improve light stability. The strong ability of complexing fragrances can also be used for another purpose: first dry, solid cyclodextrin microparticles are exposed to a controlled contact with fumes of active compounds, then they are added to fabric or paper products. Such devices are capable of releasing fragrances during ironing or when heated by human body. Such a device commonly used is a typical "dryer sheet". The heat from a clothes dryer releases the fragrance into the clothing.

Cyclodextrins are also used to produce alcohol powder by encapsulating ethanol. The powder produces an alcoholic beverage when mixed with water.

e. Aerosols

Aqueous cyclodextrin solutions can generate aerosols in particle size ranges suitable for pulmonary deposition. Large quantities of aerosol can be nebulized in acceptable nebulization times. The cyclodextrin concentration does not modify nebulization efficiency in the range tested.

f. Clinical

In April 2009, compassionate use investigational new drug (IND) applications were filed by Benioff Children's Hospital Oakland and Mr. Hugh and Chris Hempel, parents of Addison and Cassidy Hempel, identical twin girls suffering from Niemann-Pick disease, type C. This was the second time in history that cyclodextrin alone was proposed to treat a fatal disease. In 1987, cyclodextrin was used in a medical case involving a boy suffering from a fatal form of hypervitaminosis A.

On January 23, 2013, a formal clinical trial to evaluate HPβCD cyclodextrin therapy as a treatment for Niemann-Pick disease, type C was announced by scientists from the NIH's National Center for Advancing Translational Sciences (NCATS) and the Eunice Kennedy Shriver National Institute of Child Health and Human Development (NICHD). A Phase Ⅰ clinical trial is currently being conducted at the NIH Clinical Center.

② Structure

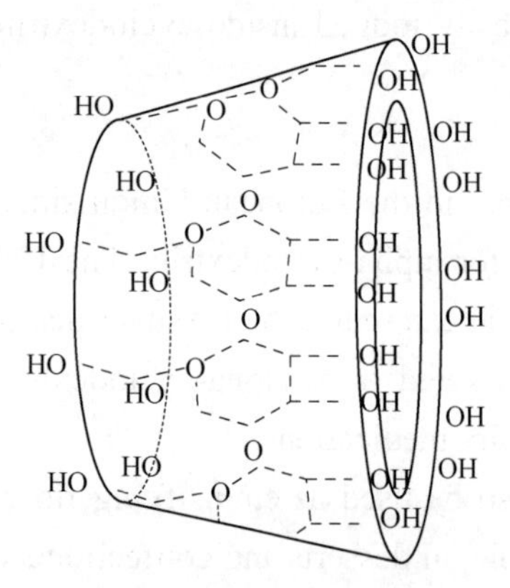

Fig.6.13 γ-CD toroid structure showing spatial arrangement.

Typical cyclodextrins are constituted by 6~8 glucopyranoside units, can be topologically represented as toroids with the larger and the smaller openings of the toroid exposing to the solvent secondary and primary hydroxyl groups respectively (Fig.6.13). Because of this arrangement, the interior of the toroids is not hydrophobic, but considerably less hydrophilic than the aqueous environment and thus able to host other hydrophobic molecules. In contrast, the exterior is sufficiently hydrophilic to impart cyclodextrins (or their complexes) water solubility.

The formation of the inclusion compounds greatly modifies the physical and chemical properties of the guest molecule, mostly in terms of water solubility. This is the reason why cyclodextrins have attracted much interest in many fields, especially pharmaceutical applications: because inclusion compounds of cyclodextrins with hydrophobic molecules are able to penetrate body tissues, these can be used to release biologically active compounds under specific conditions. In most cases the mechanism of controlled degradation of such complexes is based on pH change of water solutions, leading to the loss of hydrogen or ionic bonds between the host and the guest molecules. Alternative means for the disruption of the complexes take advantage of heating or action of enzymes able to cleave α-1,4 linkages between glucose monomers.

③ Synthesis

The production of cyclodextrins is relatively simple and involves treatment of ordinary starch with a set of easily available enzymes. Commonly cyclodextrin glycosyltransferase (CGTase) is employed along with α-amylase. First starch is liquified either by heat treatment or using α-amylase, then CGTase is added for the enzymatic conversion. CGTases can synthesize all forms of cyclodextrins, thus the product of the conversion results in a mixture of the three main types of cyclic molecules, in ratios that are strictly dependent on the enzyme used: each CGTase has its own characteristic $\alpha{:}\beta{:}\gamma$ synthesis ratio. Purification of the three types of cyclodextrins takes advantage of the different water solubility of the molecules: β-CD which is very poorly water-soluble (18.5 g•L^{-1} or 16.3mmol•L^{-1}) (at 25℃) can be easily retrieved through crystallization while the more soluble α- and γ-CDs (145 g•L^{-1} and 232 g•L^{-1} respectively) are usually purified by means of expensive and time consuming chromatography techniques. As an alternative a "complexing agent" can be added during the enzymatic conversion step: such agents (usually organic solvents like toluene, acetone or ethanol) form a complex with the desired cyclodextrin which subsequently precipitates. The complex formation drives the conversion of starch towards the synthesis of the precipitated cyclodextrin, thus enriching its content in the final mixture of products. Wacker Chemie AG uses dedicated enzymes, that can produce alpha-, beta- or gamma-cyclodextrin specifically. This is very valuable especially for the food industry, as only alpha- and gamma-cyclodextrin can be consumed without a daily intake limit.

④ Uses

Cyclodextrins are able to form host-guest complexes with hydrophobic molecules given the unique nature imparted by their structure. As a result, these molecules have found a number of applications in a wide range of fields.

Cyclodextrins can solubilize hydrophobic drugs in pharmaceutical applications, and crosslink to form polymers used for drug delivery. One example is Sugammadex, a modified γ-cyclodextrin which reverses neuromuscular blockade by binding the drug rocuronium. Other than the above-mentioned pharmaceutical applications, cyclodextrins can be employed in environmental protection: these molecules can effectively immobilise inside their rings toxic compounds, like trichloroethane or heavy metals, or can form complexes with stable substances, like trichlorfon (an organophosphorus insecticide) or sewage sludge, enhancing their decomposition.

This ability of forming complexes with hydrophobic molecules has led to their usage in supramolecular chemistry. In particular they have been used to synthesize certain mechanically interlocked molecular architectures, such as rotaxanes (Fig.6.14) and catenanes, by reacting the ends of the threaded guest. The photodimerization of substituted stilbazoles has been demonstrated using g-cyclodextrin as a host. Based on the photodimer obtained, it is established that the halogen-halogen interactions, which play an interesting role in solid state, can be observed in solution. Existence of such interactions in solution has been proved by selective photodimerization of dichloro substituted stiblazoles in Cyclodextrin and Cucurbiturils.

The application of cyclodextrin as supramolecular carrier is also possible in organometallic reactions. The mechanism of action probably takes place in the interfacial region. Wipff also demonstrated by computational study that the reaction occurs in the interfacial layer. The application of cyclodextrins as supramolecular carrier is possible in various organometallic catalysis.

In 2013, α-cyclodextrin is found to be able to selectively form second-sphere coordination complex with tetrabromoaurate anion ($[AuBr_4]^-$) from transition-metal anion mixtures, and thus is used to selectively recover gold from various gold-bearing materials in an environmentally benign manner.

β-cyclodextrins are used to produce HPLC columns allowing chiral enantiomers separation, and are also the main ingredient in P&G's product Febreze which claims that the β-cyclodextrins "trap" odor causing compounds, thereby reducing the odor.

⑤ Derivatives

Both β-cyclodextrin and methyl-β-cyclodextrin (MβCD) remove cholesterol from cultured cells. The methylated form MβCD was found to be more efficient than β-cyclodextrin. The water-soluble MβCD is known to form soluble inclusion complexes with cholesterol, thereby enhancing its solubility in aqueous solution. MβCD is employed for the preparation of cholesterol-free products: the bulky and hydrophobic cholesterol molecule is easily lodged inside cyclodextrin rings (Fig.6.15) that are then removed. MβCD is also employed in research to disrupt lipid rafts by removing cholesterol from membranes.

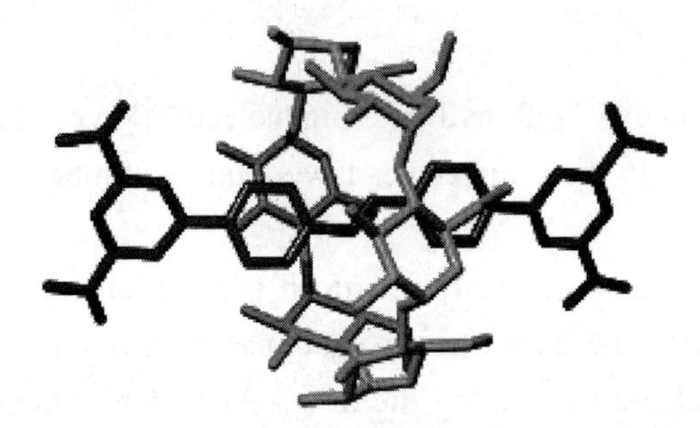

Fig.6.14 Crystal structure of a rotaxane with an α-cyclodextrin macrocycle.

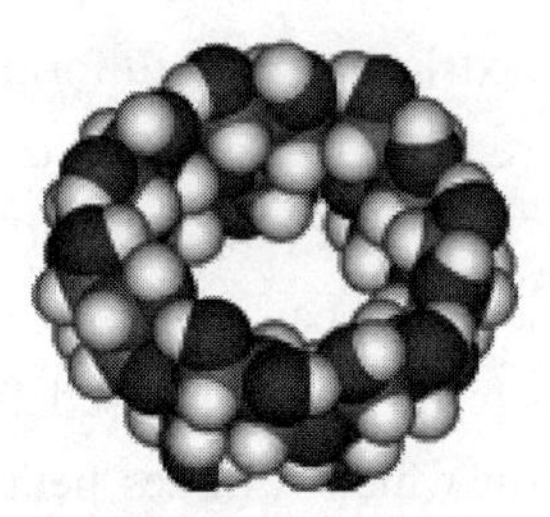

Fig.6.15 Space filling model of β-cyclodextrin.

6.3.1.6 Natural Anionic Polysaccharides

Anionic polysaccharides extracted from seaweeds (alginate, agar, carrageenans; Fig.6.16) and plant cell walls (pectin) and exudates (gum arabic) are largely used as thickening and gelling agents. Alginate is a linear polysaccharide composed of β-(1→4)-D-mannosyluronic acid (M) and α-(1→4)-L-gulosyluronic acid (G) blocks.

The ratio and the distribution of the M and G blocks (Fig.6.16) notably affect to the sensitiveness of alginate to pH and calcium ions because the different relative position of the carboxylic acid group in each block. Alginate also offers a wide range of possibilities of derivatization to improve its performance.

Alginic acid and its sodium, calcium, ammonium and potassium salts have been shown suitable for oral and topical dosage forms, acting as diluent and disintegrant in solid forms and as thickener, emulsifier and taste-masking agent in liquid and semisolid. Alginic acid is particularly suitable for preparing microcapsules, pellets and microspheres that are easily cross-linked in the presence of calcium ions.

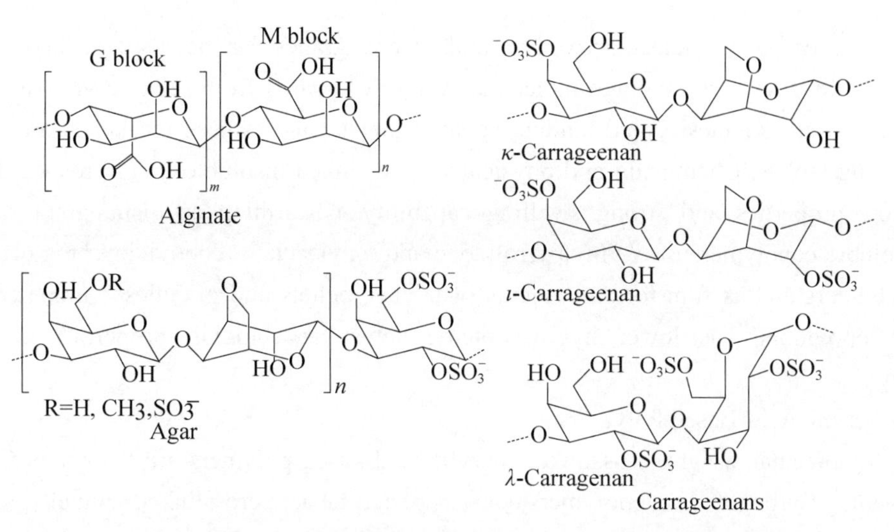

Fig.6.16　Structure of some anionic polysaccharides used as pharmaceutical excipients.

Carrageenans or carrageenins are hydrocolloids extracted from some red seaweeds. They mainly consists of linear β-D-galactose and 3,6-anhydro-α-dgalactose copolymers with a variable density in sulfated groups, which leads to three different families: kappa (κ, 1 sulfate group per dimer) which has a helical tertiary structure and strong gelling capability; iota (ι, 2 sulfate groups per dimer); and lambda (λ, 3 sulfate groups per dimer) which does not form gels (Fig.6.16). Only κ-carrageenan hydrogels exhibit pH- and temperature-sensitiveness.

At low concentration, carrageenans are added to emulsions to increase the physical stability. They are also used as pelletizer agents for extrusion-spheronization, binder agents in tableting, and as components of hard and soft capsule shells. Agar or agar-agar is obtained from the cell walls of some species of red algae or seaweeds and is widely employed as emulsifying and suspending agent, thickener, and tablet binder. Agar is a heterogeneous mixture of two unbranched polysaccharides: agaropectin and agarose, which share the same galactose-based backbone (Fig.6.16). Agaropectin is heavily modified with acidic side-groups, such as sulphate and pyruvate, while agarose has neutral charge and possesses longer chains. Thus, depending on the composition, agar are designed as neutral agarose, pyruvated agarose (with little sulfation) and sulfated galactan.

(1) Polyvinylpyrrolidone

Polyvinylpyrrolidone (PVP), also known as povidone [Fig.6.17(a)] and commercialized as Kollidon®, is a synthetic linear polymer widely used in pharmaceutical industry. It is highly soluble in water, but also in some organic solvents such as butanol. The thickening ability

(a) Povidone　　(b) Copovidone

Fig.6.17　Repeating units of povidone and copovidone

of PVP depends on its molecular weight, and several grades can be distinguished which are identified with a *K*-value (average molecular weight estimated from relative viscosity values). PVP is also used as adhesive and binding agent in tablets, and as taste masker in oral solutions and chewing tablets. Chemically and physically cross-linked insoluble PVP (crospovidone) has good flow properties and strong swelling capability, performing as disintegrant of tablets. Water-soluble copolymers of *N*-vinylpyrrolidone and vinylacetate [copovidone; Fig.6.17(b)] are used as binders and as film-forming agents of tablets, pellets and granules. Compared to PVP grades, copovidone has lower hygroscopicity, better dry binding properties, and higher plasticity .

(2) Acrylic Acid-Based Polymers

High molecular weight, crosslinked, acrylic acid-based polymers are known as Carbopol® and Noveon®. Carbopol® homopolymers consists of acrylic acid crosslinked with allyl sucrose or allylpentaerythritol. There exist also Carbopol® copolymers (with C_{10}~C_{30} alkyl acrylate) and interpolymers (which contain a block copolymer of polyethylene glycol and a long chain alkyl acid ester). Noveon® polycarbophil is a polymer of acrylic acid crosslinked with divinyl glycol.

These polymers have microgel structure and behave as thickening agents in lotions, creams and gels, oral suspensions, and in transdermal gel reservoirs. They are also useful to communicate controlled release properties to solid dosage forms. In contact with the aqueous media, acrylic acid-based polymers form gel matrices that hinder drug diffusion more efficiently than other polymers. It should be noticed that due to the presence of carboxylic acid groups, acrylic acid-based polymers undergo relevant conformational changes as a function of pH and ionic strength. At pH above the pK_a (5.5~6.5), the ionized polymer chains expand and can easily entangle with neighbor chains forming highly structured, viscoelastic gels.

(3) Polyesters

Polyesters are biocompatible and biodegradable polymers including poly(lactic acid) (PLA), poly(glycolic acid) (PGA), poly(lactic-co-glycolic acid) (PLGA), and poly-ε-caprolactone (PCL). These biocompatible polymers are widely used to encapsulate drugs in micro/nanoparticles and in macroscopic implants. Poly(α-hydroxy) esters are not biodegraded by enzymes in the body, but undergo hydrolytic degradation via surface or bulk degradation pathways. Surface degradation occurs when the rate of hydrolytic chain scission and the production of oligomers and monomers which diffuse into the surroundings is faster than the rate of water intrusion into the polymer bulk. As a consequence, the formulation erodes over time without affecting the molecular weight of the polymer bulk. Bulk degradation occurs when water penetrates the entire polymer structure, causing hydrolysis throughout the entire polymer matrix and thus overall reduction in molecular weight. In the case of PLGA, if the generated carboxyl-containing species do not diffuse out the matrix, they can act as internal autocatalyzer that accelerates the bulk degradation compared to the surface. Due to these biodegradation patterns, PLGA formulations exhibit bi/triphasic release profiles: initial burst of drug close to the surface, slow release of the entrapped drug until the molecular weight of the polymer becomes lower than a critical threshold, and rapid release when most polymer has degraded. PCL is less used mainly because it bioerodes at slower rate than other aliphatic polyesters, but has the advantage of degrading in neutral species and not in acid byproducts .

6.3.2 Functions of Polymers in Drugs Polymers for Coating

Coating of solid dosage forms allows facing up to a variety of formulation demands: taste masking, improvement of appearance, easy swallowing, prolonged stability and spatiotemporal control of drug release.

Shellac is an excellent film-forming natural polyester resin secreted by insects. It consists in a complex mixture of aliphatic and alicyclic acids, which serves as moisture barrier of tablets and pellets due to its low permeability to water and oxygen . Shellac films protect drugs from degradation in the gastric environment. However, its use is limited by stability problems.

Cellulose ethers also exhibit good filmogenic features. Low-viscosity grades are suitable for aqueous film-coating, while higher-viscosity grades are used with organic solvents. HPMC films are gastrosoluble. Oppositely, ester derivatives of HPMC such as hypromellose acetate succinate or hypromellose phthalate can provide enteric films that resist prolonged contact with the strongly acidic medium, but dissolve in the mildly acidic or neutral intestinal environment. Cellulose esters, particularly acetate and acetate phthalate (CAP), are commonly applied to solid dosage forms either by coating from organic or aqueous solvent systems, or by direct compression. CAP has the advantage of being compatible with most plasticizers.

Pectin, amylose, dextran and inulin provide coatings that degrade by enzymes of colonic flora. These polysaccharides are commonly combined with cellulose ethers or esters or synthetic polymers to obtain biphasic release profiles or colon specific release. Drugs conjugated to dextran (linear polymer backbone of α-D-glucopyranose units) show enhanced stability in the upper part of the gastrointestinal tract, but they can be absorbed in the colon .

There is a large list of methacrylate copolymers that can provide specific performances as film-coating materials (Table 6.5) . Protective coatings seal the formulation, masking unpleasant odor and taste. Gastroresistant films aim to provide protection of the drug against gastric fluid and of gastric mucosa from aggressive drugs. Films insoluble at any pH, but permeable can provide sustainedrelease along the entire gastrointestinal tract. Similarly, polyvinyl acetate films are insoluble in dilute acid and alkali media and thus act as diffusion barriers over a long period of time.

Table 6.5 Synthetic polymers commonly used as film-forming agents for coating of solid dosage forms

Function	Composition	Features	Trade names
Protective	Cationic copolymer based on dimethyl aminoethyl methacrylate, butyl meth acrylate, and methyl methacrylate	Soluble pH < 5, permeable pH >5	Eudragit® E 100, E 12.5, E PO
	Polyethylene glycol with side chains of polyvinyl alcohol	Water-soluble	Kollicoat® IR
Enteric	Poly(methacrylic acidco-ethyl acrylate) 1 : 1; poly(methacrylic acid-comethyl methacrylate) 1 : 2; poly(methyl acrylate-comethyl methacrylate-comethacrylic acid) 7 : 3 : 1	Soluble pH > 5.5~7.0	Eudragit® L 30 D-55, L 100-55, L100, L12.5, S100, S12.5, FS30 D. Acryl-EZE® 93A, MP Kollicoat® MAE

续表

Function	Composition	Features	Trade names
Sustained-release	Poly(ethyl acrylate-co-methyl methacrylate-co-trimethylammonioethyl methacrylate chloride) 1∶2∶0.2 and 1∶2∶0.1; poly(ethyl acrylate- co-methyl methacrylate) 2∶1	Insoluble, low or high permeability	Eudragit® RL 100, RL PO, RL 30D, RL 12.5, RS 100, RS PO, RS30 D, RS 12.5, NE 30D, NE 40 D, NM 30D
			Kollicoat® EMM
	Polyvinylacetate		Kollicoat® SR

6.3.3 Polymers That Enhance Drug Solubility or Dissolution Rate

A large majority of active substances are poorly soluble in water, which limits the feasibility of dissolution of the therapeutic dose. Moreover, certain treatments require that the drug is rapidly bioavailable when administered in oral solid dosage forms. Among other technological strategies, the use of polymers as solubilising and release-rate enhancer excipients has received a strong attention.

6.3.3.1 Self-Assembly Polymers

Amphiphilic polymers, mainly block copolymers, may self-assembly in aqueous medium rendering structures similar to micelles of conventional surfactants (polymeric micelles) or vesicles that resemble liposomes (polymersomes). The hydrophobic regions of the polymer chains form the core, while the hydrophilic blocks extend towards the aqueous phase as a shell. The resultant polymeric micelles can host drugs of diverse polarity in the core or in the core-shell interface, enhancing the apparent solubility of the drug up to several orders of magnitude. A variety of amphiphilic polymers have been synthetized. Common examples of hydrophilic blocks are poly(ethylene oxide) (PEO), poly(*N*-vinyl pyrrolidone), poly(*N*-isopropylacrylamide) or poly(acrylic acid) (PAA). Suitable hydrophobic blocks may be PLA, PCL, poly(propylene oxide) (PPO), poly(trimethylene carbonate), polyethers, polypeptides, and poly(β-aminoester)s.

Commercially available grades of sequential PEO-PPO-PEO or reverse PPOPEO- PPO triblock copolymers of Pluronic® family (also known as poloxamers) are already components of topical, oral and parenteral formulations approved by US Food and Drug Administration (FDA) and European Medicines Agency (EMA). Related X-shape copolymers with a central ethylenediamine group and four branches of PEO-PPO are commercialized as Tetronic® (poloxamine). These block copolymers are available in a number of varieties differing in length of hydrophilic PEO and hydrophobic PPO blocks and hydrophilic-lipophilic balance (HLB), which in turn enables the preparation of micelles with cores displaying a variety of sizes and hydrophobic environment. The choice of the core-forming segment is a critical issue for a variety of properties of polymeric micelles such as stability, drug-loading capacity, and drug-release profile. In general, polymeric micelles exhibit better thermodynamic (lower critical micellar concentration) and kinetic (less prone to rapid disassembly) stability than conventional micelles. The main advantage is that once diluted in the physiological fluids, drug-loaded polymeric micelles maintain their stability for minutes to hours even the polymer concentration in the medium is well below the critical micellar concentration. This feature endows polymeric micelles with ability to circulate for a prolonged time in the blood stream if the surface is endowed with stealth features. Although Pluronics are non-degradable biomaterials, molecules around 10~15 KDa are filtered by kidney and eliminate in the urine. Pluronic formulations have been widely investigated demonstrating their usefulness to enhance solubilization of poorly-water soluble drugs and also prolonging their release profile in oral, rectal, topical, ophthalmic, nasal and injectable preparations.

An amphiphilic copolymer composed of PEG, polyvinylcaprolactam and polyvinylacetate side chains [Soluplus®; Fig.6.18(a)] has also shown excellent solubilising properties for class Ⅱ drugs, particularly crystalline drugs. This copolymer as well as some other block copolymers cited above can be also incorporated to solid dosage forms and create self-micellizable systems that accelerate drug release.

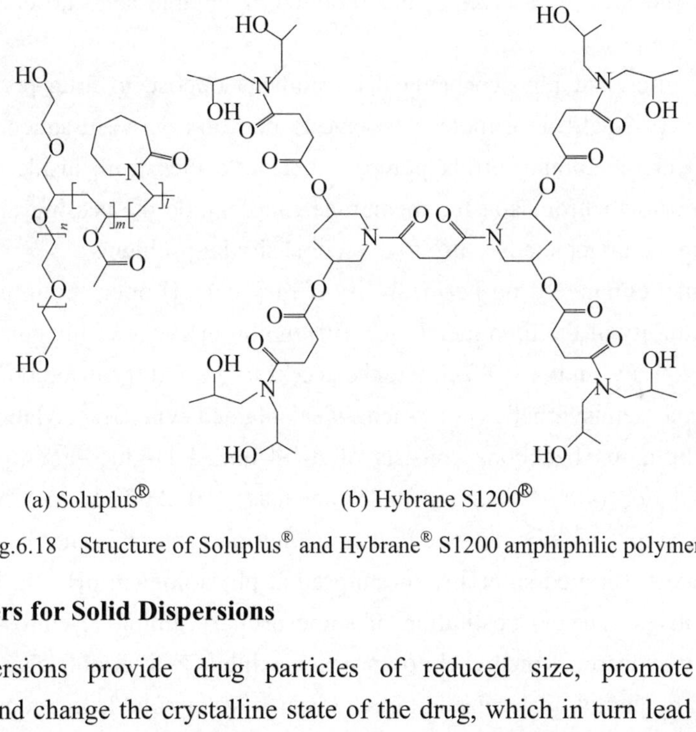

(a) Soluplus® (b) Hybrane S1200®

Fig.6.18 Structure of Soluplus® and Hybrane® S1200 amphiphilic polymers.

6.3.3.2 Polymers for Solid Dispersions

Solid dispersions provide drug particles of reduced size, promote wetting, avoid agglomeration, and change the crystalline state of the drug, which in turn lead to an increase in drug dissolution rate and even solubility. Common techniques to prepare solid dispersions require (i) dissolution of the drug and the polymer in a common solvent followed by spray-drying or freeze-drying; or (ii) mixing in solid state followed by cogrinding or melt extrusion. Suitable polymers are those that can be dissolved in a variety of organic media if the solution approach is going to be applied, or that exhibit low temperature of melting (if crystalline) or low glass transition temperature, T_g (if amorphous) for methods that require melting. In the solid dispersion, molecular interactions between drug and polymer or formation of eutectic products notably favor drug dissolution process.

Poloxamer, PVP, copovidone and PEG are suitable solubilizing agents for solid dispersions.

Soluplus® enable hot-melt extrusion to be carried out at low temperature without the addition of other excipients such as plasticizers. Some hyperbranched polymers [e.g. Hybrane®, Fig.6.18(b)] of reduced T_g are also suitable for this technique and their pharmaceutical potential is under research. Compared to other techniques, the advantage of hot-melt extrusion is that it can work in a continuous mode without the use solvents. Requirements for biocompatible polymers used in hot-melt extrusion are thermoplastic behavior, T_g in the 50~180℃, low hygroscopicity, and ability to solubilize hydrophobic drugs. For this later requirement, polymers bearing hydrogen bond acceptor or donor groups or amide moieties are particularly adequate.

6.3.4 Polymers That Enhance Drug Permeability

Non-parenteral administration requires that the drug molecules can pass through the lipid bilayer of cell membranes close to the administration site (for local effect) or through the entire physiological membranes that act as a barrier for the access to the blood stream. Once in the blood, the drug molecules themselves or incorporated in suitable nanocarriers should extravase towards the target tissues.

Several structures and physicochemical phenomena oppose to drug permeation: the gap between cells is very small; drug molecules have to partition between aqueous phase and oily phase of the bilayer membrane; efflux pumps expel molecules from inside cells to the outer cellular medium; or cytochromes at the membrane can degrade the passing through molecules. Cell penetration in the target site has also to overcome similar problems.

Polymers may enhance drug permeability if they can (i) open the intercellular spaces, (ii) increase the fluidity of the lipid membrane, (iii) inhibit or evade efflux pumps, (iv) inhibit or evade enzymatic activity, and/or (v) facilitate the access to the cell by an alternative mechanism.

Chitosan is a semisynthetic polyssacharide obtained via deacetylation or enzymatic degradation of chitin. Its backbone consists of β-1,4-linked D-glucosamines with a variable degree of N-acetylation (40%~98%) and molecular mass (50~2,000 kDa). Chitosan favors the transport of drugs across membranes due to its bioadhesion properties and the transient opening of the tight junctions. Limited solubility of chitosan at physiological pH can be overcome with the use of derivatives. The *N*-substitution of some primary amine of chitosan with carboxyl groups can yield monocarboxymethyl chitosan which exhibits zwitterionic character. *N*-trimethyl chitosan obtained by partial quaternization of chitosan is water-soluble in a range of pH between 1 and 9, and has been shown efficient for buccal penetration of high molecular weight molecules and for gene delivery. Other cationic polymers like poly-L-lysine, poly-L-arginine and polyethyleneimine (PEI) are also able to induce reversible opening of tight junctions.

Mucoadhesive polymers, such as derivatives of PAA and thiol-functionalized polymers, enable a strong attachment to cysteine groups of glycoproteins from mucus layer, resulting in an increased residence time and enhanced permeability. Functionalization with thiol groups notably increases the mucoadhesion strength of chitosan, PAA derivatives, alginate or carboxymethyl cellulose.

A variety of polymers have been shown able to inhibit efflux pumps and thus to notably increase drug accumulation inside cells. Efflux pump proteins are active membrane transporters (encoded by the ATP Binding Cassette, ABC, gene family) involved in the expelling of a broad range of structurally diverse compounds out of healthy cells, protecting them from adverse xenobiotics.

Efflux pumps affect the absorption, distribution, metabolism and elimination of endogenous substances and also of drugs, ultimately decreasing their bioavailability. P-glycoprotein (P-gp), one of the most studied mammalian ABC proteins, is ubiquitous distributed throughout the body, with the highest levels found in the epithelial cell surfaces, mainly at the apical membranes of the intestines, liver and kidney, and the blood-brain barrier. Other important drug-related efflux pumps are the multidrug resistant proteins (MRP) 1 and 2 as well as the breast cancer resistant protein (BCRP), which overexpressed in tumor cells and lead to low concentrations of anticancer agents, being responsible for multidrug resistance and, consequently, for the chemotherapy failure. Efflux pumps are polyspecific because the drug-binding site is a large and flexible region which contains multiple hydrophilic electron donor/acceptor groups, charged groups and aromatic aminoacids providing a number of subsites where drugs can bind.

The development of compounds able to selectively block or to inhibit the ATP-dependent transport function of these proteins is a very important issue for the improvement of the drug therapy, mainly in the fields of oral delivery, brain targeting and cancer therapy. Oral bioavailability of drugs substrate of efflux pumps can be improved by co-administration of efflux pump inhibitors. In the case of cancer treatment, modulators of drug efflux pumps are not expected to kill multidrug resistant cells, but to restore the cytotoxic effect of coadministered anticancer agents. The modulators or inhibitors of efflux pumps can be categorized into two groups: (i) small molecule inhibitors, and (ii) polymeric inhibitors. The polymers have the advantage of being pharmacologically inactive avoiding the toxicity problems related to the active small inhibitors. Natural polysaccharides, such as dextran, anionic gums, or sodium alginates have been proved to inhibit efflux pumps at concentration above 0.05 % . Synthetic polymers based on PEO block copolymers and dendritic and thiolated polymers are effective even at lower concentration. Moreover, hydrophilic polymeric inhibitors are not absorbed from the gastrointestinal tract and thus act locally without causing systemic adverse effects.

In general terms, drug-polymers formulations can overcome multidrug resistance phenomena following two different approaches: (i) circumventing efflux pump transport or (ii) inhibiting efflux transporter proteins. Polymeric micelles and polymer-drug conjugates take advantage of the first approach since the micelles or the conjugates can be uptaken via endocytosis, escaping from membrane diffusion affected by efflux pumps. Differently, unimers of block copolymers, e.g. Pluronics, enhance in vivo intracellular drug accumulation due to the inhibition of P-gp, MRP1 and MRP2 pumps. Pluronic® P85 (4,600 Da) is so far one of the most potent inhibitor of efflux pumps. The unimers rapidly bind cell membrane, penetrate into the cells and co-localizes with the mitochondria. This leads to inhibition of the respiratory chain, decreases oxygen consumption and causes ATP depletion. The inhibition of the ATPase and depletion of ATP and the fluidization of the membrane hinders the activity of the efflux pumps, enabling a more efficient entry of drugs into the cells. In the multidrug resistant cancer cells, the production of reactive oxygen species (ROS) and the release of cytochrome C, due to the impairment of mitochondrial respiration, enhances drug-induced apoptosis and prevents the activation of the anti-apoptotic cellular defense.

6.3.5 Polymers for Gene Delivery

Gene material cannot easily enter into cells and is extremely unstable in the extracellular medium. Therefore, systemic administration of gene material requires nanocarriers that can deliver it in the appropriate cells. Most polymers intended as non-viral gene vectors are cationically charged in order to form complexes with DNA which is anionic at physiological pH. The resultant colloidal polyplexes should exhibit removable stealth surface to enable the transport through biological barriers without preventing the entrance into the target cells. The polycation should also perform as endosomolytic component to facilitate the escape of DNA from endosomes in order to avoid degradation, traverse the cytoplasm intact and enter into the nucleus.

Although very efficient regarding DNA complexation, highly charged cationic polymers, such as PEI, poly(L-lysine) and polyamidoamine dendrimers (PAMAM), exhibit toxicity problems. Thus, combination of cationic moieties with hydrophilic non-ionic components (e.g. grafted with Pluronic, carbohydrates or cyclodextrins) or even anionic groups (e.g. glycolic acid, polyacrylic acid) is under investigation.

References

[1] L.V. Allen, N. G. Popovich, H. C. Ansel. (2011) Ansel's Pharmaceutical Dosage Forms and Drug Delivery Systems. Lippincott Williams & Wilkins, Baltimore.

[2] L. Kroewczynski. (1985) The development of pharmaceutical technology (chronological tabulated facts). Pharmazie **40**, 346.

[3] C. Alvarez-Lorenzo, A. Concheiro. (2013) From drug dosage forms to intelligent drug-delivery systems: a change of paradigm. In: C. Alvarez-Lorenzo, A. Concheiro, (eds.) Smart materials for drug delivery, Vol. 1, p. 1. Royal Society of Chemistry Publishing, Cambridge.

[4] A. Dokoumetzidis, P. Macheras. (2006) A century of dissolution research: from Noyes and Whitney to the biopharmaceutics classification system. Int. J. Pharm. **321**, 1.

[5] A. S. Hoffman. (2008) The origins and evolution of "controlled" drug delivery systems. J. Control. Release **18**, 153.

[6] C. Alvarez-Lorenzo, A. Concheiro. (2014) Smart drug delivery systems: from fundamentals to the clinic. Chem. Comm. **50**, 7743.

[7] D. N. S. Hon. (1996) Cellulose and its derivatives: structures, reactions and medical uses. In: S. Dumitriu, (ed.) Polysaccharides in Medicinal Applications, p. 87. Marcel Dekker, New York.

[8] R.C. Rowe, P. J. Sheskey, W. G. Cook, M. E. Fenton. (2012) Handbook of Pharmaceutical Excipients. Pharmaceutical Press, London.

[9] Y. Zhang, Y. Law, S. hakrabarti. (2003) Physical properties and compact analysis of commonly used direct compression binders. AAPS PharmSciTech **4**, 489.

[10] C. Alvarez-Lorenzo, J. L. Gomez-Amoza, R. Martinez-Pachecho, C. Souto, A. Concheiro. (2000) Evaluation of low-substituted hydroxypropylcellulose as filler-binders for direct compression. Int. J. Pharm. **197**, 107.

[11] L. Chen, X. Li, L. Li, S. Guo. (2007) Acetylated starch-based biodegradable materials with potential biomedical applications as drug delivery systems. Curr. Appl. Phys. **7**(s1), 90.

[12] I. Rashid, M. M. H. AlOmari, A. A. Badwan. (2013) From native to multifunctional starch-based excipients designed for direct compression formulation. Starch **65**, 552.

[13] Y. K. Lee, D. J. Mooney. (2012) Alginate: properties and biomedical applications. Prog. Polym. Sci. **37**, 106.

[14] S. N. Pawar, K. J. Edgar. (2012) Alginate derivatization: a review of chemistry, properties and applications. Biomaterials **33**, 3279.

[15] H. H. Tonnesen, J. Karlsen. (2002) Alginate in drug delivery systems. Drug. Dev. Ind. Pharm. **28**, 621.

[16] T. Coviello, P. Matricardi, C. Marianecci, F. Alhaique. (2007) Polysaccharide hydrogels for modified release formulations. J. Control. Rel. **119**, 5.

[17] Y. Freile-Pelegrín, E. Murano. (2005) Agars from three species of Gracilaria (Rhodophyta) from Yucatán Peninsula. Bioresour. Technol. **96**, 295.

[18] V. Bühler. (2008) Kollidon®, polyvinylpyrrolidone excipients for the pharmaceutical industry, 9th edn. BASF, Ludwigshafen.

Chapter 7

Pharmaceutical Nanotechnology: Overcoming Drug Delivery Challenges in Contemporary Medicine

Poor biopharmaceutical characteristics of drug and biological barriers in the body affect the drug molecules reaching the intended disease site. For instance, solubility and permeability of a drug molecule affect its transport through the cellular membranes, while their stability in the biological environment dictates residence time and efficacy. Nanomedicine, an evaluation of nanotechnology, ferry the payload safely and effectively through several anatomical and physiological barriers to the target site. Besides, nanomedicine could be engineered to provide compound effect through ligand-mediated targeting and image guided drug delivery at disease site. With illustrative examples from scientific literature, the versatility of different nanosystems and their utility in disease therapy spanning from preclinical development to approved products is emphasized. Specific issues in drug approval including quality-by-design and regulatory aspects are discussed. Based on the advances in drug delivery and nanomaterial synthesis, there is a great future for nanomedicine in diagnosis and treatment of several complex diseases.

7.1 Challenges in Delivery of Contemporary Therapeutics

Drug discovery process has been in forefront utilizing recent advances in molecular biology, together with medicinal chemistry, protein structure based screening, and computational analysis, as part of rational approach to discovering drug molecules that will address unmet clinical needs.

For example, proteins identified from structural biology platform can serve as targets for discovering new drug molecules. The discovery of antisense oligonucleotides (ASN), plasmid DNA (pDNA), peptides and protein therapeutics has also shown a greater potential in treating several complex diseases. A recent development in drug discovery is RNA interference (RNAi) which uses small stretches of double stranded RNA with 21~23 nucleotides in length, to inhibit the expression of a gene of interest bearing its complementary sequence .

Small interfering RNA (siRNA) can induce RNAi in human cells. This RNAi technology has many advantages over other posttranscriptional gene silencing methods, such as gene knockouts and antisense technologies. In addition, only a few molecules of siRNA need to enter a cell to inactivate a gene at almost any stage of development. MicroRNA (miRNA), advancement from siRNA, is a new class of drugs still in the investigative stage based on nucleic acid chemistry. miRNA with 19~25 nucleotides in length, interfere pathways that involve in disease process. In general, all these recent drugs have shown a great potential in the clinical management of several complex diseases like cancer, metabolic diseases, auto-immune diseases, cardiovascular diseases, eye diseases, neurodegenerative disorders and other illness.

Despite the diversity and size of therapeutic libraries are continually increasing, delivering them to the disease sites has been hampered by physico-chemical attributes of drugs and physiological barriers of the body. For example, many small and macromolecular drugs (ASN, pDNA, peptides, proteins, siRNA and miRNA) often fail to reach cellular targets because of several chemical and anatomical barriers that limit their entry into the cells. Therefore, the outcome of therapy with that contemporary therapeutics is often unpredictable, ranging from beneficial effects to lack of efficacy to serious adverse effects. These challenges have been discussed in the following sections with an attempt to apply nanotechnology-based concepts in designing of drug delivery systems (DDS) that overcome barriers in drug delivery.

7.1.1 Chemical Challenges

Physico-chemical properties impact on both pharmacokinetics and pharmacodynamics of the drug in vivo, and must be considered when selecting a suitable delivery method. The chemical challenges faced by small and macromolecular drugs (ASN, pDNA, peptides, proteins, siRNA and miRNA) are many folds, which mainly include:

(i) Molecular size;

(ii) Charge;

(iii) Hydrophobicity;

(iv) In vivo stability;

(v) Substrate to efflux transporters.

7.1.1.1 Size, Charge and Hydrophobicity

The chemical properties that mainly affect drug permeability through anatomical barriers are molecular size and solubility. High molecular size and increased hydrophobicity are the predominant problems particularly associated with combinatorial synthesis and high throughput screening methods. These methods allow for identification of lead molecules faster based on their best fit into receptors, but shift the molecules towards high molecular size and increased hydrophobicity, resulting in poor aqueous solubility. One estimate shows that around 40 % of the newly discovered molecules are poorly aqueous soluble, thus need a suitable delivery method to achieve pharmacologically relevant concentrations in the body. Oral route for drug delivery remains popular due to ease of administration and patient compliance. However, oral absorption can be hindered by poor aqueous solubility of therapeutics in GI fluids. The rate of dissolution, which is a prerequisite for oral absorption, depends on the drug solubility in the GI fluids. In addition, drug molecule must possess adequate lipophilicity ($\log P$) in order to efficiently permeate across intestinal epithelial cells. This is one of the reasons for hydrophilic macromolecules such as proteins, peptides, and nucleic acid constructs do not show any oral bioavailability and resulting in limited clinical success.

Drug transport mechanisms involving in intestinal epithelium are transcellular and paracellular transport. Transcellular mechanism involving in transport of drug molecules across the cell membrane, which occurs by ① passive diffusion, ② facilitated diffusion, ③ active transport, and ④ transcytosis. Lipophilic drug molecules can diffuse freely across the epithelial membrane barrier while hydrophilic and charged molecules need specialized transport carriers to facilitate cellular uptake. Transcytosis process involving endocytosis and exocytosis mechanisms is mainly for macromolecular (proteins, peptides) drugs. Recent studies show that orally given nanoparticles can pass through the epithelial membranes in GI tract through the endocytosis process, and this can be a potential route for transport of macromolecular therapeutics.

Paracellular route, on the other hand, involves diffusion of hydrophilic drugs between the cells of epithelial or endothelial membrane through sieving mechanism. The formation of tight junction between the epithelial and endothelial cells strictly limits the paracellular drug transport. Molecular cut-off for the paracellular transport is approximately 400~500 Da . Molecular mass less than the cell junction can easily pass through paracellular route regardless of polarity, for example, water and ions. It has been observed that the diffusion of drugs with molecular size <300 Da is not significantly affected by the physicochemical properties of the drug, and which will mostly pass through aqueous channels of the membrane. However, the rate of permeation is highly dependent on molecular size for compounds with M_w > 300 Da. The Lipinski rule of five suggest that an upper limit of 500 Da as being the limit for orally administered drugs. Numerous studies are focused on identifying the nature of cellular tight junctions and the signaling molecules involved in preserving the barrier function in order to find right approach to promote oral drug absorption.

Increased hydrophobicity of a molecule also causes greater protein binding. Protein binding is both help and hindrance to the disposition of drugs in the body. Elimination and metabolism may be delayed because of highly protein bind. Therefore, protein binding affects both the duration and intensity of drug action in the body.

7.1.1.2 Stability

In vivo stability is also a critical chemical property of the drug that affects drug levels in the body. For example, the extent of drug ionization, stability in the acidic environment of the stomach or stability in the presence of gut enzymes, as well as presence of food and gastric emptying can reduce oral bioavailability of many small and macromolecular drugs. On the other hand, drugs are subjected to metabolism in the body by different sequential and competitive chemical mechanisms involving oxidation, reduction, hydrolysis (phase Ⅰ reactions) and glucouronidation, sulfation, acetylation and methylation (phase Ⅱ reactions).

Cytochorme P450 enzymes which catalyze oxidation reaction are mainly responsible for first-pass biotransformation of majority of the drugs in the body, thus limiting oral absorption and systemic availability of the drugs. Cytochrome P450 abundantly present in the intestinal epithelium and liver tissue, and metabolizes several chemically unrelated drugs from major therapeutic classes.

Besides this, macromolecular drugs such as proteins and peptides, ASN, pDNA, siRNA and miRNA have poor biological stability and a short half-life resulting in unpredictable pharmacokinetics and pharmacodynamics. Proteins and peptidal drugs are highly prone to enzymatic cleavage in the blood circulation and tissues, whereas nucleic acid therapeutics are highly susceptible to degradation by intra-and extra-cellular nucleases, leading to degradation and a short biological half-life. DDS have the potential to overcome the challenges of degradation and short biological half-life, and can provide safe and efficient delivery of macromolecular therapeutics.

7.1.1.3 Membrane-Bound Drug Efflux Pumps

If the drug molecules are substrates to efflux pumps, their transport through cellular membranes is severely restricted. The ATP-binding cassette (ABC) efflux pumps are transmembrane proteins present at various organ sites within the body, and use ATP as a source of energy to actively transport drug molecules across the lipid cell membranes. Among the ABC family of efflux pumps, P-glycoprotein (P-gp) is highly expressed in epithelial cells of the small intestine, which is the primary site of absorption for the majority of the orally given drugs. Efflux pumps also present on the luminal side of the endothelial cells of BBB, and restrict entry of hydrophobic molecules into the brain. The multi-drug resistance in many cancers is linked to the ABC efflux transporters which express on cell membranes and produce intracellular drug levels below the effective concentrations necessary for cytotoxicity. All these efflux transporters preset a broad overlap in substrate specificities and act as a formidable barrier to drug absorption and availability at target sites.

DDS can be employed to overcome most of these chemical challenges. For example, paclitaxel is a potent anticancer drug, is poorly absorbed after oral administration and its bioavailability is <6 % . The obvious reason for its low bioavailability are high molecular weight, poor aqueous solubility, the affinity to drug efflux pumps, and rapid metabolism by cytochrome P450 enzymes in the gut. Nanoemulsions and self-emulsifying DDS have been employed recently for the successful oral and parenteral delivery of paclitaxel. Similarly, to protect RNA based therapeutics from enzymatic cleavage, several DDS have been proposed and they are at different stages of preclinical and clinical development.

7.1.2 Remote Disease Targets

Anatomical and physiological barriers involved in the body restrict the direct entry of small and macromolecular drugs into the target extracellular or intracellular tissue locations resulting in sub-optimal doses at target site and reduced efficacy. However, cytotoxic drugs and RNA therapeutics have their target sites inside the cells, therefore need to be delivered intracellular in sufficient doses to produce therapeutic effect. The first limiting anatomical barrier for orally administered drugs is epithelial lining of gut walls, where from drugs will permeate through either by transcellular or paracellular transport. This transport is in turn dictated by the chemical properties of drugs as alluded above. Therefore, altering the chemical properties by making the drugs in salt form, encapsulating in DDS based on cyclodextrins, lipid or polymeric carriers, or using permeability enhancers could promote bioavailability of drugs. Cytochrome P450 and efflux transporters present in the enterocytes of intestinal walls also forms as another limiting barrier to drug permeability. Use of cytochrome P450 and efflux pump inhibitors can promote oral drug absorption. For example, pretreatment with curcumin results in inhibition of P-gp and cytochrome P450 expression in the GI tract, leading to increased oral bioavailability and efficacy of drugs.

For the drug molecules given intravenously, the limiting anatomical barrier is that of vascular endothelium and basement membrane. In addition, blood serum proteins, proteolytic enzymes, RNases etc. limit the effective drug delivery to the target sites. CNS disease are likely to rise to 14 % by 2020 mainly due to the ageing population, however, many newly discovered small and macromolecular therapeutics do not cross into the brain after systemic delivery. Because brain is protected by blood-brain-barrier (BBB), which is composed of very tight endothelial cell junction and presence of several efflux transporters, resulting in formation of dynamic formidable barrier to drug transport. However, once at the BBB, hydrophobic drugs with the size below <500 Da generally do transport through lipid cell membranes by passive diffusion, but if they are substrates to drug efflux pumps, they will be pumped out from the brain. Hydrophilic molecules also cannot transport efficiently as there is very limited paracellular transport present in the tight junctions of the BBB.

Cancer mass is another complex anatomical barrier in drug delivery. For example, solid tumor microenvironment is heterogeneous and structurally complex and presents a challenging barrier in drug delivery. The cytotoxic drugs which are intended to kill a large proportion of tumor cells in a solid tumor, must uniformly distribute through the vascular network, pass through capillary walls, and traverse the tumor tissue. Nevertheless, the drug distribution in tumors is not uniform, and only a fraction of tumor cells is exposed to lethal doses of cytotoxic agents. Tumor microenvironment is composed of tumor cells with varying proliferation rate and stromal cells (fibroblasts and inflammatory cells) that are surrounded in an extracellular matrix and nourished by a vascular network, and regions of hypoxia and acidity. Each of these components may different from one site to another in the same tumor mass, and all of these factors effects tumor cell sensitivity to drug treatment. In addition, stromal components in tumors contribute to an increase in interstitial fluid pressure, which limit the penetration of macromolecular drugs. Furthermore, the three-dimensional nature of solid tumor tissue itself affects the sensitivity of constituent cells to chemo and radiation treatments. For instance, the tumor cells grown as spheroids in cell culture or tumors grown in animals, are more resistant to cisplatin and alkylating agents than the corresponding cell dispersions.

In addition, certain intracellular infections, like leishmaniasis and listeria, where macrophages are directly involved in the disease are not accessible to drug delivery, thus necessitating specific drug delivery strategies. To overcome all these challenges, it is highly important to develop DDS that render protection to the drug from biodegradation in the body, while allowing their transport through the anatomic and physiological barriers to increase their bioavailability at the target tissue.

7.2 Nanotechnology Solutions

The science of nanotechnology has begun just in the last decade, but in this short time, it has been successfully applied in several fields ranging from electronics to engineering to medicine. Recent understanding of cellular barriers and molecular profile of diseases, and controlled manipulations of material at the nanometer length scale, nanotechnology offers great potential in the disease prevention, diagnosis, and treatment. Nanotechnology has also allowed for challenging innovations in drug delivery, which are in the process of transforming the delivery of drugs. Nanosystems fabricated using controlled manipulation of material are exploited for carrying the drug in a controlled manner from the site of administration to the target site in the body. They are colloidal carriers with dimensions <1,000 nm and can traverse through the small capillaries into a targeted organ down to target cell and intracellular compartments, which represent the most challenging barrier in drug targeting. The critical attributes of any nanoparticle DDS are to

(1) protect a labile drug molecule from both in vitro and in vivo degradation;
(2) maintain the effective pharmacokinetic and biodistribution pattern;
(3) promote drug diffusion through the epithelium;
(4) enhance intracellular distribution.

However, the specificity, sensitivity and simplicity are very important for any nanosystem to be clinically successful as a DDS. Several types of nanoparticle DDS have been evaluated for their potential drug delivery applications in various stages of clinical development.

7.2.1 Enhancing Solubility and Permeability

Solubility and permeability are two of the most critical biopharmaceutical characteristics impacting the successful delivery of drug molecules through anatomical membranes in the body. If the drug molecule is not a substrate to efflux transporters and metabolizing enzymes, then the solubility (hydrophilic and hydrophobic) plays a major role in determining oral intestinal permeability.

Biopharmaceutical Classification System (BCS) is proposed based on the solubility and permeability properties of the drugs which classifies drugs into one of four classes. Class Ⅰ drugs are highly soluble and permeable in the GI tract, therefore, bioavailability is not an issue with Class Ⅰ drugs. Class Ⅱ drugs are poorly aqueous soluble but highly lipophilic. They are well permeable across the GI tract due to high lipophilicity, but the bioavailability is likely to dissolution rate limited due to low aqueous solubility. Class Ⅲ drugs are highly soluble but have low permeability due to their low lipophilicity. In both Class Ⅱ and Class Ⅲ examples, DDS plays a critical role in overcoming poor solubility and permeability. On the other hand, Class Ⅳ drugs show low solubility and low permeability, and exhibit poor and variable bioavailability. Methods to enhance both solubility and permeability should be adopted for these drugs.

To improve solubility and permeability, several methods have been employed over the years. Such as preparation of prodrugs, use of chemical or physical permeability enhancers to transient openings of the tight junctions, or direct administration to the target site. However, formulation efforts can best exemplify in improving poor solubility and permeation profiles of both small and macromolecular drugs. Nanoparticle DDs like, liposomes, nanoemulsions, nanosuspensions, solid-lipid nanoparticles (SLN), micelles or polymeric nanoparticles are highly useful over the current methods to deliver the highly hydrophilic or highly lipophilic molecules across the intestines and BBB. For example, drug nanocrystal suspensions (nanosuspension) allow for increased dissolution velocity and saturation solubility of poorly aqueous soluble drugs, which is accompanied of an increase in oral bioavailability. In addition, nanocrystals can be delivered intravenously for controlled drug release, and their surface can be tailored for both passive and active targeting. On the other hand, lipid-based systems like nanoemulsions and SLN could allow for the delivery of lipophilic drugs, by incorporating them in the lipid core of the formulation. These DDS can enable direct transfer of drug to the intestinal membranes and excluding the dissolution of drugs in aqueous fluids in GI tract. In once such study, we have formulated highly lipophilic paclitaxel into deoxycholic acid modified nanoemulsion, which showed increased oral bioavailability compared to paclitaxel solution. In another example, saquinavir, an anti-HIV protease inhibitor incorporated in nanoemulsion, showed enhanced oral absorption.

Recent studies show that nanoemulsions made using oils rich in omega-3 and omega-6 polyunsaturated fatty acids (PUFA) can promote drug delivery to the brain. This is some extent attributed to the presence of PUFA transporters on the abluminal membrane side of the endothelial cells of BBB. Tissue and cell permeability also altered by surface modification of the nanoparticles with targeting ligands which can facilitate the nanoparticle uptake along with its payload into the cells.

7.2.2 Targeted Delivery to Disease Sites

Targeted delivery exploiting the structural changes and cellular markers of a given pathophysiology can potentially reduce the toxicity and increase the efficacy of drugs. This is highly important in case of diseases like cancer, where dose-limiting toxicities and drug resistance constitute major barriers to drug success. General targeting mechanisms consists of passive and active targeting.

Upon parenteral delivery, passive targeting depends on the size of the DDS and the disease vascular pathophysiology in order to preferentially accumulate the drug at the site of interest and avoid distribution to normal tissue. For example, nanosystems escape from the blood circulation and accumulate in sites where the blood capillaries have open fenestrations as in the sinus endothelium of the liver or when the integrity of the vascular endothelial membrane is perturbed by inflammation due to infections, rheumatoid arthritis or infarction or by tumors. In the liver, the size of capillary fenestrae can be as large as 150 nm and liposomal nanocarriers showed extravasation to hepatic parenchyma nanosystems in the size range of 50~200 in size can extravasate and accumulated the tumor tissue and inflammatory sites. Therefore, the nanomedicine in the size range is expected to provide therapeutic benefits for treating the diseases. In case of solid tumors, passive targeting involves in the transport nanosystems through a newly formed leaky tumor microvasculature into the tumor interstium and cells (Fig.7.1). This phenomenon has named as “enhancepermeability and retention” (EPR) effect, first discovered in murine tumors for macromolecules accumulation by Maeda and Matsumura. EPR effect observed in many human solid tumors with the exception of hypovascular tumo (prostate or pancreatic cancer). This effect will be optimal if nanosystem can escape reticulo-endothelial system (RES) and show longer circulation half-life in the blood. Poly(ethylene glycol) (PEG) grafting on nanosystems will evade RESuptake, allow for prolonged circulation in the blood and enhance tumor accumulation through EPR. Besides, the RES uptake of non-PEG grafted nanosystems also offer an opportunity for passive targeting against intracellular infections such aleishmaniasis, candiasis, and listeria which reside in macrophages.

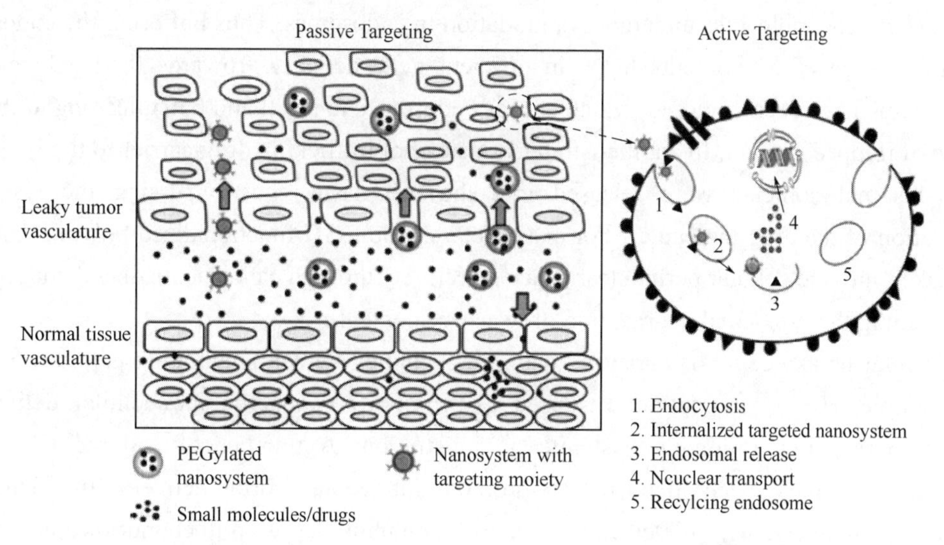

Fig.7.1 Schematic illustration of passive and active targeting strategies in tumor drug delivery.

The specificity of passive targeting can be remarkably improved when the targeting ligands are used with nanosystems, which selectively bind to cellumarkers overexpressed on the disease site termed as active targeting (Fig.7.1). For example, folic acid-nanoparticles can be used to target tumor cells that over express folate receptors, such particles internalize via folate receptor mediate endocytosis. In another example, arginine-glycine-aspartic acid (RGDsequence containing peptides can be conjugated to nanoparticle to target $\alpha5\beta5$ or $\alpha5\beta3$ integrin receptors overexpressed on endothelial cells of the newly formed angiogenic blood vessels and also on tumor cells. Furthermore, the targeting ligand anchored to nanosystems will allow for carrying of many drug molecules compare to direct conjugation of targeting ligands with drug molecules.

7.2.3 Intracellular and Subcellular Delivery

The nanosystems once in the disease vicinity, they need to enter the cells and transfer the payload to sub-cellular organelles. There are two mechanisms playing a role in intracellular and subcellular delivery are non-specific or specific uptake of nanosystems by cells. In case of non-specific uptake, cells surround the nanosystems and forms a vesicle in the cell called an endosome. The endosomes then fuse with the highly acidic organelles called lysosome, which are rich in degrading enzymes. Endosomes usually travel in a specific direction and join at the nuclear membrane. Specific uptake on the other hand, involves receptor mediated endocytosis, where the actively targeted nanosystem binds to the cell-surface receptor, resulting in internalization of the entire nanoparticle-receptor complex and vesicular transport through the endosomes. The receptor can be re-cycled back to the cell surface following dissociation of complex. After the cellular internalization, stability of the payload in the cytosol and delivery to specific organelles, such as mitochondria, nucleus etc, is also essential for therapeutic activity. However, many drugs do not survive in the lysosomal environment. For example, 99 % of the internalized gene molecules undergoes degradation in endosomes. Thus buffering the endosomes for safe release of its contents helps in efficient gene delivery. Towards this, poly cationic nanosystems have been explored, which causes endosomes to swell and burst, leading to the safe release of trapped content. In another strategy, instead of trafficking drug carrier to the lysosome, the endosomal contents were released into the cytoplasm, thus bypassing the lysosomal degradation of the drug molecules. For example, a cyclic RGD functionalized polyplex micelles were taken into the cellular perinuclear space selectively through caveolae mediated endocytosis, thus escaping the lysosomal degradation of its active content .

Cellular uptake could be enhanced using of arginine rich cell penetrating peptides (CPP's). For example, HIV-1 tat peptide was used to promote non-specific intracellular delivery of various therapeutics following systemic administration. A number of cationic CPP's like penetratin also have been identified to promote intracellular drug delivery. In addition to intracellular delivery, use of delocalized cationic amphiphiles or mitochondriotropic nanosystems can promote mitochondrial drug delivery.

7.2.4 Enabling Non-invasive Delivery

Non-invasive delivery is an alternate to systemic delivery of drugs, and mainly includes drug delivery via intranasal, pulmonary, transdermal, buccal/sublingual, oral and transocular routes. Patient compliance has been found to be much higher when drugs given by non-invasive routes and therefore they are considered to be a preferred route of drug delivery. However, the preferred route of administration for a given drug selected based on several factors, such as biopharmaceutical properties (solubility, permeability and stability) of a drug molecule, disease state, onset of action, dose frequency and adverse effects. For example, sumatriptan and zolmitriptan administered via intranasal route provide rapid-onset of relief from migraine related pain in minutes compared to oral tablet in hours. Similarly, potent peptidal drugs like calcitonin, desmopressin allows therapeutic blood levels that are not achieved with oral route of administration. In another example, selegiline and fentanyl transdermal products eliminate GI related adverse effects. In addition, noninvasive insulin products for inhalation and buccal administration improve patient compliance by reducing multiple daily injections.

In general, oral route is much convenient for high doses of administration. However, macromolecular drugs are not stable in the GI fluids, where intranasal, buccal/sublingual or pulmonary offers a non-invasive route of choice. These routes also favor treatments that need faster absorption of drug and where a rapid systemic exposure is well tolerated. Transdermal delivery is useful in chronically administered treatments (chronic pain, depression, Parkinson's, dementia, attention deficit hyperactivity disorder and hormonal therapies), where sustained plasma profiles and low C_{max} to C_{min} ratio are required.

7.3 Illustrative Examples of Nanotechnology Products

Nanotechnology based concepts have been extensively applied in engineering of nanosystems for delivery of contemporary therapeutics in a controlled manner from the site of administration to the target disease in the body. The history of nanosystems reaches back to 1950s when the first polymer-drug conjugate was reported with *N*-vinyl pyrrolidine conjugated to glycyl-L-leucine-mescaline. However, the most relevant nanosystems were conceptualized only after the first report of liposomal preparations in 1964 and their subsequent use as vehicle for drug delivery application. Soon after, synthesis of albumin nanoparticle was reported in early 1970s with a subsequent early attempt of exploiting them as the first protein based DDS. The pharmacological effects of polymer-based nanoparticles were studied and their application as DDS envisioned around the same time. As alluded earlier, ground breaking discovery of EPR effect in tumors by Matsumura and Maeda further emphasized on relevance of the size of delivery vehicle. These seminal works drew tremendous attention on nanosystems for a sustained and controlled delivery of drugs. It was realized that for an optimized delivery system, the size of the payload vehicle should be between 10 nm and 100 nm. Kidneys easily clear off particles smaller than

10 nm while the particles larger than 100 nm are removed by the RES. Since then, several different types of nano systems have been researched and much focus has specifically been on tailoring the size, physical properties and surface functionality of the delivery systems for varying therapeutic applications. The collective research input on the nanotechnology based improvement of DDS has enabled several products in to the market in the past two decades (Table 7.1).

Table 7.1 Nanotechnology-based products in clinical application

Nanotechnology Platform	Trade Name	Active Agent	Indication(s)	Approval Year
Liposomes	Abelcet	Amphotericin B	Fungal infection	1995
	AmBisome	Amphotericin B	Fungal infection	1997
	Amphotec	Amphotericin B	Fungal infection	1996
	Daunoxome	Daunorubicin	Antineoplastic	1996
	DepoCyt	Cytarabin	Lymphomatous meningitis	1999
	Doxil/Caelyx	Doxorubicin	Antineoplastic	1995
	Myocet	Doxorubicin	Antineoplastic	2000
	OncoTCS	Vincristine	Non-Hodgkin's lymphoma	2004
Micelles	Estrasorb	Estradiol	Vasomotor symptoms	2003
Nanocrystal	Emend	Aprepitant	Antiemetic	2003
	Tricor	Fenofibrate	Hypercholesterolemia and hypertriglyceridemia	2004
	Triglide	Fenofibrate	Hypercholesterolemia and hypertriglyceridemia	2005
	Megace ES	Magesterol acetate	Anorexia, cachexia or an unexplained significant weight loss in AIDS patients	2005
	Rapamune	Sirolimus	Immunosuppressant	2000
Nanoemulsion	Tocosol	Paclitaxel	Nonsuperficial urothelial cancer	2003
Nanoparticle	Abraxane	Paclitaxel	Metastatic breast cancer	2005
Nanotube	Somatuline depot	Lanreotide	Acromegaly	2007
Superparamagnetic iron oxide	Feraheme injection	Ferumoxytol	Treatment of iron deficiency anemia in patients with chronic kidney disease	2009
	Feridex	Ferumoxide	MRI contrast agent	1996
	GastroMARK	Ferumoxsil	Imaging of abdominal structures	1996

Sandimmune® and Taxol® are US Food and Drug Administration (FDA) approved dosage forms of cyclosporine and paclitaxel respectively, formulated using Cremophor®EL as solubilizing nonionic surfactant. However, due to hyper sensitivity reactions associated with these products, Cremophor®-free formulations based on nanosystems have been developed and commercialized. Genexol PM is one such example of Cremophor-free polymeric micelles formulated paclitaxel where poly (ethylene glycol) is used as a nonimmunogenic carrier while biodegradable poly (*dl*-Lactic acid) forms the drug solubilizing hydrophobic core. Several such DDS including liposomes, nanoemulsions, polymeric nanoparticles, micelles and nanocrystals (Fig.7.2) have been developed, granted regulatory approval and have been marketed since then. The following section will focus on each of such DDS with illustrative examples of commercialized products.

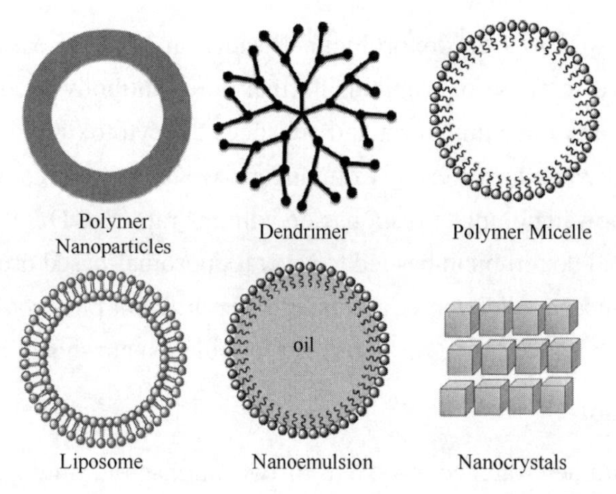

Fig.7.2 Different types of pharmaceutical nanosystems used in drug and gene delivery.

7.3.1 Lipid-Based Nanosystems

Lipid-based carriers are extremely popular since they facilitate a controlled administration of both small and macromolecular drugs at therapeutically relevant doses. Liposomes and nanoemulsions are two most commonly used lipid-based nanosystems for drug delivery application.

7.3.1.1 Liposomes

Liposomes are vesicles formed of a lipid bi-layer, first developed by Alec Bangham in 1961, and their lipid bi-layer membrane is similar to that of cellular membranes. The lipid bi-layer of liposomes is composed of phospholipids with a hydrophilic head and a hydrophobic long-chain tail. The hydrophilic core of the liposomes facilitates in compartmentalizing water-soluble drugs into the aqueous core while the hydrophobic bi-lipid membrane has been exploited to load water-insoluble drugs. Initial attempts using liposomes as nanosystems focused largely on improving their circulation time in the blood and targeting efficiency. PEG-modification of liposomes, first reported in 1990 has by far been the most promising approach to achieve longer circulation of the liposomes in the blood. There has been a plethora of literature since then on the application of PEG-modified liposomes to achieve a selective delivery of drugs post-administration. However, several other surface modifications of liposomes such as poly[*N*-(2-hydroxypropyl) methacrylamide], poly-*N*-vinylpyrrolidones, polyvinyl alcohol and amino acid-based polymer-lipid conjugates have been explored. Many studies showed that the opsonization of the liposomes might be dependent on the hydrophobicity of the surface, charge of the lipid and the molecular weight of the modifying polymer. Antibody, folate and peptide mediated surface receptor targeting has been primarily enabled directing the liposome based drug delivery to the target organ.

The first liposome based formulation, PEG-liposome encapsulated doxorubicin was approved in 1995 (Doxil™, Orthobiotech) initially for the treatment of HIV related *Kaposi sarcoma* and later for ovarian cancer and myeloma. Doxil has remarkably reduced the cardiotoxicity by lowering cardiac exposure to free doxorubicin. Besides, it also increased half-life and tumor accumulation compared to free doxorubicin. Furthermore, antibody modification of Doxil has shown a much higher tumor accumulation and enhances the cytotoxicity of the doxorubicin. In a study conducted on 53 patients suffering from advanced Kaposi's sarcoma, 19 patients showed partial and 1 patient showed complete response on administration of Doxil™ once every 3 weeks. The success of liposomal doxorubicin has led to several liposomal-based drug formulations that are either approved for clinical application or are undergoing different phases of clinical trial. Some of the key drugs that have been exploited for liposomal formulation are shown in Table 7.1.

7.3.1.2 Nanoemulsions

Nanoemulsions are heterogeneous system of two immiscible liquids; typically oil-in-water (O/W) or water-in-oil (W/O) with a droplet size in the range of 50~200 nm. These kinetically stabilized nano-sized droplets have several advantages over macroemulsions such as higher surface area and hence more free energy, higher stability with lower creaming effects, coalescence, flocculation and sedimentation. The formation of nanoemulsions however requires an external shear force to decrease the droplet size to desired range and their productions methods are broadly classified as high-energy and low-energy methods. The high energy methods could include laboratory or industrial scale high-pressure homogenization, micro-fluidization or laboratory scale ultrasonication. However, these methods may not be conducive for applications involving thermolabile drugs, nucleic acids and proteins. Low-energy methods such as spontaneous emulsification, the solvent-diffusion method and the phase-inversion temperature (PIT) method are used for such payloads. The nanoemulsions serve as an excellent vehicle for solubilizing lipophilic drugs into the oil phase or hydrophilic drugs in the aqueous phase. The application of nanoemulsions as DDS has been envisaged only in the past decade and several attempts have been realized to increase their stability, circulation time and achieve a targeting efficiency.

For example, propofol was first formulated in Cremophor® EL by Imperial Chemical Industries as ICI35868, and went into clinical use. However, due to the toxicity of Cremophor®, it was withdrawn from the market, reformulated in oil-in-water emulsion and launched by the trade name Diprivan® (ICI, now AstraZeneca). Apart from propofol as active pharmaceutical ingredient, the formulation contains generally regarded as safe grade excipients (GRAS) such as soyabean oil, glycerol, egg lecithin and disodium edetate. Diprivan® is used as a short acting, intravenous sedative used in intensive care medicine. It is known to have low toxicity, controlled sedation effect, rapid onset, a short duration of action and quick recovery despite prolonged usage. TOCOSOL is another Cremophor® EL-free nanoemulsion formulation of paclitaxel that was approved by FDA in 2003 for the treatment of nonsuperficial urothelial cancer. Dexamethasone (Limethason®, Mitsubishi Pharmaceuticals), alprostadil palmitate (Liple®, Mitsubishi Pharmaceuticals), flurbiprofen axetil (Ropion®, Kaken Pharmaceuticals) and Vit A, D, E, K (Vitalipid®, Fresenius Kabi) are some other examples of therapeutically relevant compounds that have been formulated in nanoemulsions for clinical applications. Recently, NanoBio Corporation has formulated an emulsion-based antiviral drug NB 001 that shows potent activity against HSV-1 virus and antifungal drug NB 002 for the treatment of distal subungual onychomycosis (DSO). Both these formulations are currently in phase Ⅱ/Ⅲ trails.

7.3.2 Polymer-Based Nanosystems

Polymeric nanoparticles clearly are the most studied system for drug delivery applications. Different polymeric materials, natural, semi-synthetic and synthetic, have been exploited as polymer-drug conjugate or polymer-based nanoparticle for drug encapsulation to facilitate therapeutic applications. It is important to realize that while polymer-drug conjugate is a system which involves a single polymer chain conjugate to the drug, polymer-based nanoparticles are actually made up of several polymer chains which encapsulate the drug of interest.

Polymer-drug Conjugate Polymer-drug conjugates preparation date back to early 1950 and the field has rapidly evolved since then. Most drug molecules suffer from permeability through biological membranes, short half-life, non-specific distribution and dose dependent toxicities. Polymer conjugates on the contrary not only tremendously improves the in vivo circulation time of the drug but also facilitates passive delivery of these conjugates through leaky vasculature in diseases like cancer and inflammation. They however also suffer from certain drawbacks such as polymer dependent toxicity, immunogenicity, rapid drug release, conjugate instability and poor drug loading. Several endeavors have been taken to overcome some of these shortcomings with much success. Besides, many bio-inspired polymers such as proteins (albumin, antibodies etc.) have also been looked upon as promising candidates for drug delivery applications.

The first polymer conjugate to undergo clinical trial was SMANCS where anti-tumor protein neocarzinostatin was (NCS) was covalently conjugated to two styrene maleic anhydride (SMA) . SMANCS was approved subsequently in Japan in 1994 to treat advanced and recurrent hepatocellular carcinoma. PEG-cconjugate were the first candidate to get US FDA approval when PEG-L-asparaginase conjugate (Oncaspar) was accepted to treat acute lymphoblastic leukaemia. Several other PEG -conjugates of drugs such as Neulasta (PEG-G-CSF; neutropaenia associated with cancer chemotherapy), PPEG-Intron (PEG-IFN$\alpha 2\beta$; Hepatitis C) EG-asys (PEG-IFN$\alpha 2\alpha$; Hepatitis B and C), have been approved to clinical treatment while several others are under various preclinical development. Besides, several other polymers (or their derivatives) conjugates (products names) such as polyglutamate (CT-2103, CT-2106), dextran (DOX-OXD, DE-310), *N*-(2-hydroxypropyl) methacrylamide (PK1, PK2, MAG-CPT, AP-5280, AP-4346) are being looked upon as promising candidates in their preclinical trial stages.

Though first protein nanoparticle based drug conjugation was reported in as early as 1974 the first approved conjugate was realized only in 2005 when paclitaxel bound to albumin (Abraxane, AstraZeneca) was approved by FDA for treatment of metastatic breast cancer. It is a non-targeted formulation with particle size around 130 nm, which is localized into the tumor partly through EPR effect and partly through albumin-binding protein. Clinical studies have demonstrated that Abraxane increases the therapeutic response, reduces the rate of disease progression and improves the survival rate among the patients. Antibodies have also been explored for drug conjugation and some examples of products from this class of nanovector includes Gemtuzumab (Mylotarg), Tositumomab and ibritumomab tiuxetan.

Micellar Delivery Systems Micelles are submicroscopic structures formed in an aqueous phase by amphiphilic surfactants or polymers that have a polar and a nonpolar group. The typical size of these structures for delivery application ranges from 10 nm to 100 nm. These structures have a hydrophobic core, which facilitates the solubility of a lipophilic therapeutic agent and a hydrophilic corona that is exploited for surface functionalization to improve their tumor accumulation. These properties render them an attractive choice as carriers for drug delivery applications. Conventional surfactants however have a very high critical micellar concentration, and therefore are prone to disintegration on dilution in the blood stream. Alternatively, polymeric micelles are usually prepared by self-assembly of a copolymer having hydrophobic moiety forming the biodegradable core while hydrophilic component for the surface. These polymers form micelles in aqueous media but at a much lower concentration compared to conventional surfactants. Such polymeric micelles have been extensively researched for drug encapsulation, enhanced tumor targeting and longer in vivo circulation to aid an improved delivery system. Various approaches have been utilized to prepare polymeric micelles of desired properties using block copolymer, their lipid or cyclodextrin derivatives, diblock copolymers, triblock copolymers, pluronic polymer and graft polymers.

Genexol-PM, a cremophor-free PLA-PEG copolymer-based micellar formulation completed its preclinical Phase Ⅰ trial in 2004. Currently, the formulation is in its Phase Ⅱ trial for the treatment of the patients suffering from taxane-pretreated recurrent breast cancer. SP1049C is another doxorubicin encapsulated pluronic polymer micelle based formulation that is under Phase Ⅱ preclinical trial for the treatment of advanced level inoperable adenocarcinoma of esophagus . NK911 is yet another example of a micellar formulation of PEG and doxorubicin conjugated poly (aspartic acid) which is under preclinical development.

Dendrimer Delivery System Dendrimers are roughly spherical nanoparticles made of several monomers, which branch out radially from the center. The advantages associated with dendrimers such as their controlled size, multiple valency, water solubility, modifiable functionality and an internal core render them a promising choice as drug carriers. They are therefore applied as delivery vehicles in several administration routes such as intra-venous, ocular, dermal and oral. Their biocompatibility and immunogenicity has been studied in vitro as well as in vivo and similar to cationic macromolecules like liposomes and micelles, cationic surface groups render dendrimers cytotoxic to cells. Surface functionalization of dendrimers with PEG or fetal calf serum however has shown to reduce the cytotoxicity effects. The drug could be loaded on the dendrimers mainly by physical interaction or by covalent attachment. Physical adsorption of drug could suffer from poor drug loading and less control on drug release kinetics. Alternatively, the pro-drug approach is far more viable where the drug is chemically attached to the dendrimer directly or using a linker giving a much better pharmacokinetic and pharmacodynamic profile.

The field of dendrimer-based DDS has evolved greatly in the last decade and several dendrimer-drug conjugates are in their preclinical testing. One of the key examples is conjugation of PEO modified 2,2-bis(hydroxymethyl) propionic acid based biodegradable dendrimer to doxorubicin, which shows 9-fold higher tumor accumulation and 10-fold less cytotoxicity than free drug. The intravenous administration of prodrug to doxorubicin-nonresponsive tumor showed a rapid tumor regression in a single dose. Poly(glycerol-succinic acid) dendrimer (PGLSA)-camptothecin prodrug similarly has shown an enhanced solubility, cellular uptake and retention. Since these initial success reports, several drugs such as artemether, cisplatin, diclofenac, mefenamic acid, dimethoxycurcumin, diflunisal, etoposide, ibuprofen, 5-florouracil, indomethacin and many more have been conjugated to dendrimer and are undergoing preclinical/clinical trials.

7.3.3 Nano-sized Drug Crystals

Poor aqueous solubility is one of the key problems with many small drug molecules, which affects their delivery and therapeutic applications. It is a well-established fact that with size reduction to nanometer scale, the properties of a material is governed by quantum laws and entirely different from its macro/micro size counterpart. A drug nanocrystal is therefore drug particle with its size in the nano-range i.e. 10~100 nm, and a suspension of such nanocrystals is popularly known as nanosuspension. The suspension of these nanocrystals can be achieved in aqueous solutions as well as non-aqueous medium (liquid PEG, oil) with help of stabilizers like amphiphilic surfactants (poloxamers, PVP, phospholipids, polysorbate 80) or polymeric (hydroxypropyl methyl cellulose) materials. The hallmark of drug nanocrystals is that these crystals are pure drug particles with no carrier system. Similar to typical nanoparticle preparation, drug nano crystals could be prepared by a "bottom-up approach" (molecular level to nanocrystals) such as precipitation method or "top-down approach" (macro/micro level to nanocrystals) such as pearl milling (technology owned by Elan Nanosystems), high-pressure homogenization in water (technology owned by Skyepharma as well as Baxter) and in non-aqueous medium (technology owned by Pharmasol). Sometimes, a combination of the two approaches is used for nanocrystal production e.g. Nanoedge® (Baxter) that uses precipitation followed by homogenization. The major advantages of nanocrystallized drug are increased rate of drug dissolution and saturation solubility, improved oral bioavailability, reduced dose variations and general applicability to all routes of administration.

Rapamune® was the first nanocrystalline drug to obtain FDA approval in 2000 and was licensed to Wyeth Pharmaceuticals. It was produced by pearl milling method developed by Elan Nanosystems and contains rapamycin as the active drug. The formulation is marketed in two forms as tablets and oral suspensions. Soon after, Emend® was approved in 2003, which contains Aprepitant and is marketed by Merck. The production process has been developed by Elan Nanosystems and it is used for the treatment of emesis. Tricor® (drug Fenofibrate), Megace ES® (drug Megestrol acetate) and Theralux® (drug Thymectacin) are three other drugs which have been developed by Elan Nanosystems and have been licensed to Abbott, Par Pharmaceuticals and Celmed respectively. Several other products have however been introduced by other companies which include Semapimod® (Guanylhydrazone, Cytokine Pharmasciences), Paxceed® (Paclitaxel, Angiotech) and Nucryst® (Silver, Nucryst Pharmaceuticals).

7.4 Multifunctional Nanotechnology

As detailed in previous sections, biological system presents several barriers to effective drug delivery. It is therefore germane to develop drug delivery strategies to circumvent these barriers. This could be achieved by making the right choice of material as delivery vehicle, surface modification to increase targeting and intracellular availability of the drug and improving the functionality of the delivery system to achieve the diagnostic applications. The nanosystem with these multifunctional abilities (Fig.7.3) offer new possibilities in diagnosis, treatment and disease monitoring. The following sections will provide an in depth discussion on these aspects of drug delivery systems.

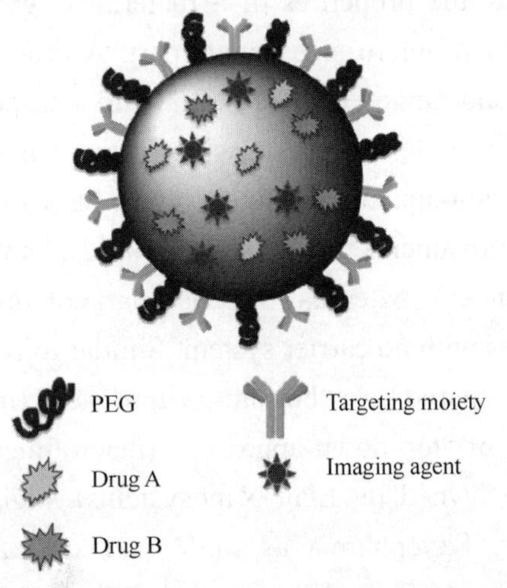

Fig.7.3 A conceptual model of multifunctional nanomedicine with targeting ability, imaging capability, and drug/gene delivery in a single platform.

7.4.1 Choice of Materials for Nanotechnology

The material property of the delivery system is essentially the most important factor that governs the biocompatibility of formulation, stability and bioavailability of the drug and its clearance from the body. It is also equally important to understand the microenvironment of the target where the drug has to be delivered to achieve an effective therapeutic concentration. Design of nanosystems governed by microenvironment of the disease site results into a class of delivery systems that are popularly known as stimuli-responsive DDS. The delivery payload, route of administration and material safety profile, would also govern the components of such delivery vehicles.

7.4.1.1 pH Responsive Delivery Systems

The physiological profile of an infected, cancerous or inflammatory body tissues differ drastically compared to the normal tissue. It has also been noted that various cellular compartments maintain their own characteristic pH levels; such as a lysosomal pH is around 4.5 where as in a mitochondria, the pH is around 8. These physiological differences result into a transmembrane pH gradient within the cellular compartments in a cell as well as among the cells. Such subtle differences in physiological environment could be actively exploited to design a pH responsive delivery system, which would be stable at physiological pH of 7.4 but actively degrade to release the drug under other conditions. For example, a tumor is composed of rapidly dividing and metabolizing cells that are always short of the desired food and oxygen supply and thus rely on glycolytic pathways for harvesting energy to sustain. The lack of oxygen in the tissue results in development of acidic condition within the tumor cells that could be exploited to achieve the delivery of a desired payload. The physicochemical properties of the delivery vehicles in response to difference in pH therefore are important characteristic, which has been actively focused in the past two decades.

Poly(β-amino ester) (PbAE) is a biodegradable cationic polymer which has been used for pH stimuli responsive delivery of drug. The polymer rapidly degrades under acidic environment with pH levels below 6.5 to release its payload into the cells. Significantly enhanced accumulation of drugs in the tumors has been demonstrated using PbAE polymer, leveraging pH stimuli-responsive delivery compared to a non-responsive polymer based delivery. It has also been shown that the pH sensitivity of the polymeric delivery systems can be tailored by altering the length of the hydrophobic carbon chain length. The pH responsiveness of poly (alkyl acrylic acid) polymer can be controlled by the choice of the monomer as well as the ratio of carboxylated to non-carboxylated alkylacrylate monomer. This polymeric system has been used for enhanced and effective in vitro transfection of lipoplex formulations. In yet another study, pullulan acetate- sulfadimethoxine polymer conjugate has been utilized to develop pH responsive, self-assembled hydrogels for an enhanced delivery of doxorubicin.

Polymers have also been directly conjugated to the target drug using pH responsive spacers, which would degrade under the low pH environment inside the tumors or lysosomes/endosomes to release the drug. In one such attempt, poly (vinylpyrrolidone-co-dimethyl maleicanhydride) (PVD) was conjugated to doxorubicin and its pH responsive controlled release increased the accumulation of the drug in to the tumor site. Similarly, copolymer *N*-(2-hydroxypropyl) methacrylamide (HPMA) and linear PEG based nanosystems are other candidates which have shown promise in delivery of drug to the tumor targets. Hydrolytically labile hydrazone linkage has been used for the drug release by enzymatic action in the lysosomes/endosomes from the polymeric or protein-based conjugate. Serum albumin conjugates of anticancer drugs such as chlorambucil and anthracyclines have shown an enhanced antiproliferative activity compared to free drug. Polyacetals are other pH labile candidates, which have been exploited for developing polymer based pH-responsive DDS.

Liposomes have similarly been suitably modified to achieve pH stimuli and controlled drug delivery. The intact pH-sensitive liposomes are internalized into the cells by endocytosis and fuse to the endosomes to deliver its contents inside the cytoplasm. The desired modification of the liposomes is mainly achieved by using new lipid candidates, which provides acid sensitivity to liposomes or by conjugation of pH sensitive polymers on liposome surface to render them prone to pH sensitive degradation. Mildly acidic amphiphiles have been used to design such phosphotidylethanolamine based liposomes where at physiological pH, these amphiphiles act as stabilizers but get protonated under acidic conditions causing a destabilization of the liposome and facilitating the delivery of the payload. These delivery systems have been successfully researched to show in vitro delivery of antitumor drugs, toxins, DNA, antisense oligonucleotides and antigens. Other lipids such as cholesteryl hemisuccinate (CHEMS), poly(organophosphazenes) and dioleoyl phosphatidyl ethanolamine (DOPE) have also been used for pH-sensitive liposomal formulations.

Micelles are yet another class of nanocarriers which have been extensively investigated to develop pH-responsive delivery. One approach to realize this aim has been the employment of titratable amines or carboxylic groups on the copolymer surface such that the micelle formation relies on the protonation of these groups. In certain cases, water-soluble block copolymers exist in different forms depending on the pH of their aqueous solution and thus have been manipulated for drug delivery applications. Besides, several other water-soluble copolymers have been extensively used to develop long circulating, pH responsive micelles. Some of the common examples include block copolymers based on poly[4-vinyl benzoic acid (VBA) and 2-*N*-(morpholino) ethyl methacrylate (MEMA), poly(acrylic acid)-b-polystyrene-b-poly(4-vinyl pyridine) (PAA-b-PS-b-P4VP), Poly[2-(dimethylamino) ethylmethacrylate]-block-poly [2-(*N*-morpholino) ethyl methacrylate](DEA-MEMA), poly(L-lactide)-b-poly(2-ethyl-2- oxazoline)-b-poly(L-lactide) (PLLA-PEOz-PLLA) ABA triblock copolymers and diblock copoly- mers (PEOz PLLA) etc., have been used for such applications.

Dendrimers are relatively new class of materials that are being investigated to develop pH-responsive delivery systems. One promising report has been the use of dendrimer composed of 2,2-bis(hydroxymethyl)propanoic acid monomer which has been conjugated to doxorubicin to produce a pH responsive delivery. In another recent attempt, the terminal ends of core-forming PEO dendrimers have been modified with hydrophobic groups using acid-sensitive acetal groups. The hydrophobic groups are cleaved off the dendrimer in acidic environment resulting in the release of the drug.

7.4.1.2 Thermo-responsive Delivery Systems

The cancerous cells are known to be highly fragile and sensitive to heat-induced damage (compared to normal cells) largely due to their rapid dividing nature. Incorporation of components that facilitate heat induction in presence of external stimuli such as magnetic field has therefore been looked upon as attractive choices to pursue. These facts have led to the development of hyperthermia as an adjunct to the radiation and chemotherapy for treatment of cancer cells. Several recent research efforts have shown that loading of superparamagnetic iron oxide particles to a delivery system leads to hyperthermia induced cell death at tumor sites. Use of drug delivery vehicle to localize these magnetic particles in the tumor sites ensure that only cancerous cells are subjected to elevated temperatures without affecting the normal cells. The tumor ablation by hyperthermia coupled with incorporation of an antitumor drug in the formulation leads to enhanced efficacy and accumulation of the drug .

The thermo-sensitive polymers display a low critical solution temperature (LCST) in aqueous solution, below which they are water-soluble but become insoluble above it. This interesting property makes them an exciting choice as thermo-responsive DDS. One such example has been the accumulation of rhodamine——poly(*N*-isopropyl acrylamide-co-acrylamide) conjugate at the tumor site using targeted hyperthermia. Certain amphiphilic polymers exhibit thermosensitivity where they have a temperature sensitive hydrophilic component and a hydrophobic component. Poly (*N*-isopropylacrylamide) (NIPAAm) and its other copolymers have been the most researched thermo-sensitive amphiphilic polymers. In an interesting report, gold nanoparticles coated cross-linked Pluronic® (poloxamer) micelles that showed a thermo-sensitive reversible swelling-shrinking behavior caused by hydrophobic interactions of copolymer chains in the micells. Several other illustrations of such polymer based thermoresponsive nanocarriers have been accounted in details in literature for further reading.

Fabrication of temperature-sensitive liposomes has been an area of tremendous interest to the researchers due to the simple known fact that the membranes of different phospholipids are known to undergo phase-transition from gel-to-liquid crystalline and lamellar-to-hexagonal transition and are release small water-soluble components during such transitions. One popular example is use of dipalmitoyl phosphatidylcholine as primary lipid for liposome formation. It shows a leaky behavior at gel-to-liquid transition at 41 °C and this transition can be tailored by adding distearoyl phosphatidylcholine as a co-lipid. Polymers have also been employed to design thermo-sensitive liposomes that also show LCST. These polymer chains exhibit a coil-to-globule transition with a change in temperature and thus impart temperature-regulated functionality to the liposomes. Such polymers stabilize the liposomes in their hydrated form below the LCST but their dehydrated form destabilizes the liposomal structural integrity resulting in delivery of the drug. Several reports exploit the modifi cation of liposomes with NIPAAm copolymers for the fabrication of thermo-responsive substitutes.

7.4.1.3 Redox-Responsive Delivery Systems

Nucleic acid based therapeutics has acquired considerable interest lately and numerous attempts have been made to deliver ASN, pDNA, siRNA and miRNA, peptides and proteins for treatment of many genetic diseases. However, successful delivery of these biomolecules to the target cells is an important challenge considering the fact that these agents are highly prone to degradation. A stimuli-responsive system will be of tremendous application as DDS for these biomolecules to ascertain their structural integrity and therefore the therapeutic functionality. It has been established that there is a redox potential difference between the reducing extracellular space and the oxidizing intracellular compartment, which can be potentially exploited to guide the DDS into the cells.

Redox-sensitive delivery systems largely rely on components containing disulfide linkage that are taken up in the cell by endocytosis and the disulfide linkage is disrupted in the lysosomes to facilitate payload delivery. The glutathione pathway plays a key role in reduction of the disulfide linkage in the reducing intracellular environment by maintaining an elevated level of reducing glutathione. Besides, the disulfide crosslinking also ensures more stable and robust structural integrity of the nanosystem that decreases the chances of early release of the payload.

One of the strategies to exploit the redox stimuli has been the use of polyaspartamide that uses positively charged groups in the polymer to electrostatically entrap DNA while the thiol groups on the polymer chain form the disulfide linkage resulting in formation of thiopolyplexes. Thiolated gelatin particles have also been shown to form gelatin thiopolyplexes and have been used as potential redox responsive nanosystem for pDNA delivery. Thiolated polyethylene imine has been directly conjugate to DNA to form polyplexes or have been used with a crosslinking agent to successfully delivery DNA into the cells with high transfection effi ciency. In yet another report, glutathione sensitive polymer coated chitosan particles were used for designing of nanosystems stabilized by disulfide bond to provide gene delivery. FDA has recently approved redox responsive anti-DC33 antibody conjugate (Mylotarg®) for the treatment of acute myeloid leukemia.

Disulfide bond based redox-responsive liposomes have also been explored to enhance liposomal stability and delivery efficiency. Such liposomes are formed by a standard phospholipid along with a small chain lipid of which the hydrophobic and hydrophilic ends are linked by disulfide bond. These liposomes show tremendous structural stability until they reach the reducing environment inside the cells where the disulfide bond cleavage results in destabilization and delivery of the gene. Thiocholesterol lipid based liposomes have been shown to successfully delivery gene into the cell in the reducing environment of the cells. Mitomycin C conjugate with a cleavable disulfide bond incorporated into liposomes has shown lesser toxicity and better therapeutic potential than the free drug .

7.4.2 Surface Modification to Increase Availability at Tissue and Cell Levels

A careful designing of the nanosystems will enable them to deliver the drugs successfully to the target disease through active or passive targeting. However, to do so successfully, the DDS should be available in the blood stream for longer period of time by avoiding recognition by the components of immune system, circumventing the process of opsonization and preventing subsequent clearance by the RES. The longevity of nanosystem in the circulation not only allows their deposition at the target site through EPR effect but also improves targeting ligand to interact to its receptor. Suitable surface modifications of the nanocarriers for a prolonged and sustained presence in the body have therefore garnered tremendous interest.

Water-soluble polymers have been most commonly used to improve the retention time of the nanosystem in the blood and PEG is found to be most efficient in this regard. The PEG coating on the nanosystem surface provides a steric hindrance that prevents the interaction and binding of blood proteins to nanoparticle surface. The fact that RES recognition of a foreign object in the body largely depends on the binding on these plasma proteins to the surface, the sterically stabilized nanocarriers successfully escape body clearance. This property to evade the immune system is popularly known as the "stealth" effect of the polymer. PEG is an excellent choice as surface protection moiety due to its high solubility in aqueous medium, flexibility of chain length, low immunogenicity and low toxicity. Besides, it does not interfere with the biological performance of the drug loaded in the delivery vehicle. PEG therefore by far is the most studied surface modifying agent to improve the residence time of the pharmaceutically relevant nanosystems. It has also been observed that while the particles modified with brush-like PEG effectively escape the immune response, surfaces modified with mushroom-like PEG molecules seem to activate the immune system against the particles. Literature serves several derivatives of PEG that have actively been used to enable the surface functionalization of the delivery vehicles .

Besides PEG alone, copolymers of PEG have also been explored for surface modification of drug delivery constructs. Block copolymer of PEG-polylactide glycolide (PLGA) forms a hydrophobic core of PLGA and a hydrophilic shell of PEG that shows a longer residence time in the blood circulation. Such polymeric preformed particles of PLGA could also be functionalized by PEG derivatives to prevent recognition by the immune system and therefore an enhanced retention time in the body. For example, the PLGA particles functionalized with polylysine-PEG copolymers shows a considerably reduced opsonization while PEG modified poly (cyanoacrylate) particles provided longer-circulation as well as permeation into the brain tissue. In a similar attempt, surface modification of polystyrene nanosystems by hydrophobically-modified dextran and PEG-dextran was studied to show that the stability of construct could be tailored by the density and also the nature of the surface modifying polymer. Lipid derivatives of PEG have similarly been used to prepare PEG modified liposomes for enhanced circulation and improved performance of the delivery system.

Even though use of PEG has largely dominated the surface modification of DDS to increase retention time, several other alternatives have also been explored. The prerequisite for a substitute of PEG has to be a water-soluble, biocompatible and non-immunogenic material. Polyoxomers, polyoxamines, polysorbate 80 and many more polymers have been used to modify the surface of nanoparticles to improve the bioavailability inside body. Lipid derivatives of poly(acryl amide) and poly(vinyl pyrrolidone) as well as other amphiphilic polymers such as poly(acryloyl morpholine) (PAcM), phospholipid (PE)-modified poly(2-methyl-2-oxazoline) or poly(2-ethyl-2-oxazoline), phosphatidyl poly glycerols, and polyvinyl alcohol have been successfully employed for surface modification of the liposomes.

7.4.3 Image-Guided Therapy

Imaging is an indispensable component of therapy and has been routinely used in hospitals and clinics for diagnosis of diseases and defects in the body. Conventional methods such as computerized axial tomography (CAT), magnetic resonance imaging (MRI), X-Ray imaging etc. have been employed in medical science for past several decades. Therefore, it was only fitting that with the advent of nanotechnology and more specifically nano-pharmaceutics, the concept of "molecular imaging" has been envisioned. Ability to image a DDS has therefore been an integral aspect of drug delivery application since it provides a visual feature to locate the site and extend of a disease in the body. Besides, it also enables a real-time assessment of the site of localization of a delivery vehicle in the body, its extent of sequestration in a particular organ and more specifically within a cell in question. For instance, presence of an imaging modality in a delivery vehicle customized to target a metastatic tumor could be essentially tracked to the end site of its localization providing a direct visual evidence of the efficiency of a targeted or non-targeted system as well as the location of the tumor in the patient. Owing to the versatility of such a delivery system, extensive endeavors have been exercised to develop multifunctional nanosystem (Fig.7.3) comprising of targeting ligands, therapeutic agent(s) as well as imaging agents. To this date, several organic and inorganic imaging agents have been explored including liposomes, dye-conjugated silica, quantum dots, gold nanoparticle and nano-shells magnetic nanoparticles and many other contrast enhancing agents. Along with advances in conventional techniques like CAT and MRI scan, many new molecular imaging approaches such as radioactivity-based imaging [gamma scintigraphy, positron emission tomography (PET), single-photon emission computed tomography (SPECT)], surface enhanced raman scattering (SERS), optical coherence tomography (OCT), near-infrared fluorescence imaging etc., are been actively researched.

Radiolabelled probes are the most commonly used imaging agents in the drug delivery systems. Gamma scintigraphy provides a 2-dimension imaging ability while SPECT and PET enable a 3-D scanning. These techniques have their own advantages and disadvantages. However, radioactivity based imaging systems are plagued by difficulties such as handling radioactive material, regulations concerned with their administration, their residence and clearance time from the body. Alternatively, improvement in MRI by the use of magnetic nanoparticles or contrast enhancing agents in the delivery system has been explored with vigor because of the non-invasive nature of the technique. Complexes of gadolinium, manganese, ferrofluids as well as superparamagnetic iron oxide are some of the most commonly applied contrast enhancing agents in MRI scans. Other popular imaging modalities include application of fluorescent dyes and quantum dots, SERS agents such as gold and silver nanoparticles.

7.4.4 Combination Therapeutics

Reports of multiples drug resistance (MDR) against antibacterial, antiviral, antifungal and anticancer drugs have become regularity in the previous decade. Numerous research endeavors have been applied to understand the origin of MDR and design therapeutic agents against them. However, the more we strive to overcome the medical enigmas by new drug discovery, the more complex the problem of MDR becomes. The gravity of the situation can be envisaged by a fact that the probability of MDR tuberculosis infection in acquired immunodeficiency syndrome (AIDS) patient is many folds more than a normal person. The inception of drug resistance has triggered the use of combination of drugs targeting a disease causing organism/process. The components of combination therapy may impact different independent targets, complement each other effect on the same target or bind independent of each other to give a combined effect for containment of the disease. Such combination therapy has successfully been realized in the treatment of cancer, diabetes, bacterial and viral infections and asthma.

Co-administration of paclitaxel and ceramide using nanosystems has been proven to be extremely effective against MDR ovarian cancer as well as brain tumor cells compared to the effect of individual drugs. Similarly, the use of a combination of paclitaxel and curcumin as well as doxorubicin and curcumin enables to overcome the MDR in cancer cells. Several commercialized drugs such as Vytorin®, Caduet®, Lotrel®, Glucovance®, Avandamet®, Truvada®, Kaletra®, Rebetron®, Bactrim® and Advair® are actually a combination of two drugs. Celetor Pharmaceuticals have developed CombiPlex® technology to launch combination chemotherapies for treatment of cancer. The technology uses high throughput screening, mathematical algorithm for synergy analysis and advanced nanosystems to predict right drug combination for therapy. This platform is meant to design chemotherapies so as to maintain an optimized ratio of the drugs in the body for enhanced efficacy. Their formulation CPX-1 is a fixed ratio combination of irinotecan and floxuridine that has shown positive results in its Phase-Ⅰ trial and is currently under Phase-Ⅱ trial for treatment against colorectal cancer. CPX-351 similarly is a combination of cytarabine and daunorubicin and is under Phase-Ⅰ trial for the treatment of acute myeloid leukemia.

7.5 Regulatory Issues in Nano-pharmaceuticals

7.5.1 Approval of Pharmaceutical Products

Despite the advances in nanomaterial application in disease diagnosis and drug delivery, significant amount of work still to be done in terms of characterizing nano medicine safety and long term effects on biological system. Currently, all nano medicine go through the FDA's traditional regulatory pathway within the Center for Drug Evaluation and Research (CDER) or Center for Devices and Radiological Health (CDRH). This pathway includes the following general requirements prior to approval.

(i) CDER reviews applications for new drugs.

(ii) Prior to clinical testing, laboratory and animal testing is performed to determine pharmacokinetic and pharmacodynamic attributes of the drug to determine a likely safety and toxicology profile in humans.

(iii) Clinical trials are performed in stages to determine if the drug is safe in healthy, then sick patients, and whether it provides a significant health benefit.

(iv) A team of FDA physicians, chemists, toxicologists, pharmacologists, and other pertinent scientists evaluates clinical data, and if safety and efficacy are established, the drug is approved for marketing.

Prior to the initiation of clinical trials, pre-clinical testing and manufacturing are regulated by several levels of regulation or guidance. These are FDA internally generated guidance documents, codified regulations listed in Title 21 Code of Federal Regulations (CFR) and International Conference on Harmonization (ICH) guidelines. Guidance documents are not codified law, but represent the Agency's current thinking on a particular subject. They do not create or confer any rights for, or on any person and do not operate to bind FDA or the public. An alternative approach may be used if such approach satisfies the requirements of the applicable statute, regulations, or both.

Title 21 is the portion of the CFR that governs food and drugs within the United States for the FDA. It is divided into three chapters: Chapter Ⅰ——Food and Drug Administration; Chapter Ⅱ——Drug Enforcement Administration; and Chapter Ⅲ——Office of National Drug Control Policy.

Most of the Chapter Ⅰ regulations are based on the Federal Food, Drug, and Cosmetic Act. Notable sections in Chapter Ⅰ are:

(a) 11 Electronic records and electronic signature related;

(b) 50 Protection of human subjects in clinical trials;

(c) 54 Financial Disclosure by Clinical Investigators;

(d) 56 Institutional Review Boards that oversee clinical trials;

(e) 58 Good Laboratory Practices (GLP) for nonclinical studies.

The 200 and 300 series sections are regulations pertaining to pharmaceuticals:

(a) 202~203 Drug advertising and marketing;

(b) 210 cGMP's for pharmaceuticals;

(c) 310 Requirements for new drugs;

(d) 328 Specific requirements for over-the-counter (OTC) drugs.

The 600 series covers biological products (e.g. vaccines, blood):

(a) 601 Licensing under section 351 of the Public Health Service Act;

(b) 606 cGMP's for human blood and blood products.

The 700 series includes the limited regulations on cosmetics:

701 Labeling requirements.

The 800 series are for medical devices:

(a) 803 Medical Device Reporting;

(b) 814 Premarket Approval of Medical Devices;

(c) 820 Quality system regulations (analogous to cGMP, but structured like ISO);

(d) 860 Listing of specific approved devices and how they are classified.

ICH guidelines are the result of The International Conference on Harmonization of Technical Requirements for Registration of Pharmaceuticals for Human Use and are unique in bringing together the regulatory authorities and pharmaceutical industry of Europe, Japan and the US to discuss scientific and technical aspects of drug registration. Since its inception in 1990, ICH has evolved, through its ICH Global Cooperation Group, to respond to the global face of drug development, so that the benefits of international harmonization for better global health can be realized worldwide. The FDA has adopted ICH guidance within four main categories as described below.

(1) ICH: Efficacy

(a) Clinical Safety E1–E2F

(b) Clinical Study Reports E3

(c) Dose-response Studies E4

(d) Ethic Factors E5

(e) Good Clinical Practice E6

(f) Clinical Trials E7–E11

(g) Clinical Evaluation by therapeutic Category E12

(h) Clinical Evaluation E14

(i) Pharmacogenomics E15–E16

(2) ICH: Joint Safety/Efficacy (Multidisciplinary)

(a) MedDRA Terminology M1

(b) Electronic Standards M2

(c) Nonclinical Safety Studies M3

(d) Common Technical Document M4

(e) Data Elements and Standards for Drug Dictionaries M5

(f) Gene Therapy M6

(g) Genotoxic Impurities M7

(h) Electronic Common Technical Document (eCTD) M8

(3) ICH: Quality

(a) Stability Q1A–Q1F

(b) Analytical Validation Q2

(c) Impurities Q3A–Q3D

(d) Pharmacopoeias Q4–Q4B

(e) Quality of Biotechnological Products Q5A–Q5E

(f) Specifications Q6A–Q6B

(g) Good Manufacturing Practice Q7

(h) Pharmaceutical Development Q8

(i) Quality Risk Management Q9

(j) Pharmaceutical Quality System Q10

(k) Development and Manufacture of Drug substance Q11

(4) ICH: Safety

(a) Carcinogenicity Studies S1A–S1C

(b) Genotoxicity Studies S2

(c) Toxicokinetics and Pharmacokinetics S3A–S3B

(d) Toxicity Testing S4

(e) Reproductive Toxicology S5

(f) Biotechnology Products S6

(g) Pharmacology Studies S7A–S7B

(h) Immunotoxicology Studies S8

(i) Nonclinical Evaluation for Anticancer Pharmaceuticals S9

(j) Photo-safety Evaluations S10

The most relevant FDA regulatory document associate with nanomedicine manufacturing is the "Liposome Drug Products" guidance document proposed in August of 2002. This document currently guides development of liposomal based drugs, which generally fall into the definition of nanomedicine based on particle size. The guidance provides recommendations for drug development applicants on chemistry, manufacturing and controls (CMC), human pharmacokinetics and bioavailability; and labeling documentation for liposome drug products submitted in new drug applications (NDAs). The guidance recommendations are segmented as follows:

(1) Chemistry, Manufacturing, and Controls

(a) Description and composition

(b) Physiochemical Properties

(c) Description of Manufacturing Processes and Controls

(d) Control of excipients: Lipid Components

(e) Control of Drug Product Specifications

(f) Stability

(g) Changes in Manufacturing

(2) Human Pharmacokinetics and Bioavailability

(a) Bioanalytical Methods

(b) In Vivo Integrity (Stability) Considerations

(c) Protein Binding

(d) In Vitro Stability

(e) Pharmacokinetics and Bioavailability

(3) Labeling

(a) Product Name

(b) Cautionary Notes and Warnings

(c) Dosage Administration

Nanomedicine platforms have a number of common issues that are related to regulatory oversight. Some of these include functional qualities such as significantly different chemical properties than corresponding small or large molecules, different PK/PD/ADMET properties, delivery, targeting, release, stabilization, and bioavailability. Characterization, in terms of physiochemical attributes and general CMC issues (stability, sterility, etc.), are also common to many of the nanomedicine platforms, but differ greatly from the traditional small/large molecule drug.

While nanomedicine are becoming more prevalent in the areas of cancer, AIDS, and brain disorders, there are concerns that the unique properties of nanoparticles, such as size, shape, affinity, and surface chemistry may not fit the traditional safety and quality evaluation protocol proposed under current regulations.

The FDA and European Medicines Agency (EMA) have begun to address the lack of a more comprehensive regulatory framework for nanomedicine through the establishment of international scientific workshops such as the EMA 1st International Workshop on Nanomedicine in September of 2010. The FDA has also recognized the need for specific nanomedicine guidance, and is working toward that goal. In August 2006, the FDA established a Nanotechnology Task Force to determine the regulatory framework needed to develop safe and effective FDA-regulated products that use nanotechnology materials. The resulting Nanotechnology Task Force Report recommended that the FDA pursue the development of nanotechnology guidance for manufacturers and researchers, and that because of the emerging and uncertain nature of nanotechnology and the potential for multiple medical applications, there was a requirement for transparent, consistent and predictable regulatory pathways.

Current FDA recommendations, until specific guidance documents are developed, are to follow current FDA guidance including all normal testing procedures, normal drug stability testing, and those associated with CMC, in vivo, and in vitro analysis. Though understanding specific technical and scientific aspects of the drug product, tests should be designed accordingly. All parts of the drug product should be tested for stability, both individually and formulated. It will be critical for nanomedicine drug companies to communicate and develop acceptable procedures in concert with the FDA as early in the product development process as possible.

7.5.2 Preclinical and Clinical Development

There are more than twenty FDA approved products that contain nanomaterials (Table 7.1). To date, all of these products have been approved through the traditional regulatory pathway. As previously described nanomedicines are becoming more prevalent in the areas of cancer, AIDS, and brain disorders. There are currently hundreds of nanotechnology companies and research facilities trying to benefit from the emerging nanomedicine marketplace. Within the life sciences industry sector, funding has been primarily focused on those companies that apply nanotechnologies to ‘conventional’ therapeutics (i.e. drugs as either chemicals or biologics) to increase or extend their application; for example, targeted drug delivery systems (Nemucore Medical Innovations, BioDelivery Sciences International, CytImmune Sciences Inc., NanoBioMagnetics Inc., Nanobiotix, Nanotherapeutics Inc.), diagnostics (Nanosphere Inc., Oxonica Ltd) and medical imaging systems (Life Technologies Inc.- Qdots). These products and applications have a relatively well-defined route to commercialization (subject to the regulatory hurdles facing nanotechnologies in general).

Most notable of the nanomedicine products are the combinatorial drugs that combine targeting, drug delivery, stability, protection, and imaging. Fig.7.3 illustrates a typical combinatorial nanomedicine unit. The multifunctional nanoparticle is by nature a complex mixture of hydrophobic/hydrophilic molecules, inorganic components, peptides, and/or small molecule organic drug molecules.

Many issues, regarding in vivo and in vitro assays need to be developed to segregate different properties of a multifunctional drug product. Some of these are:

(a) Synergies or interactions between the nanoparticle components;

(b) Biocompatibility;

(c) Long-term/chronic exposure assays/data;

(d) General toxicology assays and analytics;

(e) Animal models;

(f) Molecular weight;

(g) Particle size;

(h) Charge distribution;

(i) Purity;

(j) Contaminants;

(k) Stability——individual components and formulated;

(l) Consistency in manufacturing;

(m) PK/PD/ADMET assays/profiles;

(n) Aseptic processing/sterilization;

(o) Immunogenicity.

7.5.3 Knowledge Management, Manufacturing and Scale-Up

Process development and manufacturing of nanomedicine is at its early stages of development and thus is also in its seminal stages of preparing to respond to the guidance of the FDA. With FDA's push to move from quality by testing to quality-by-design (QbD) (Fig.7.4) for nanomedicine community to succeed in this new environment it is imperative to develop robust

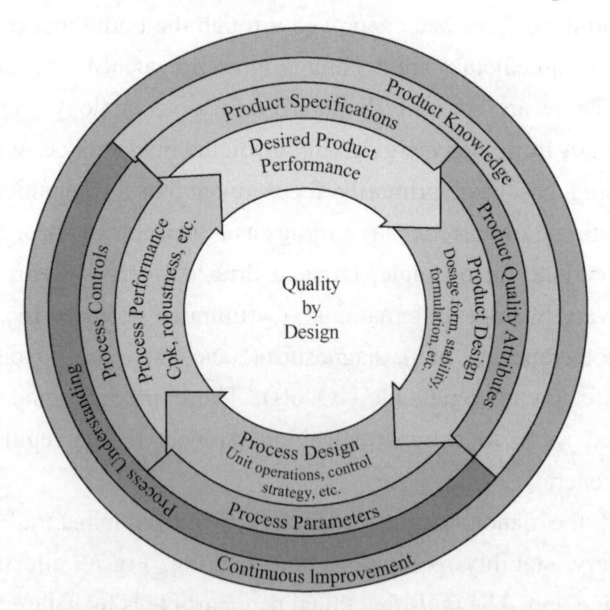

Fig.7.4 The United States Food and Drug Administration recommended quality-by-design (QbD) approach to link product knowledge with process knowledge and create a continuous improvement product development environment.

documented, process knowledge for the fabrication of nanomedicine. Acquisition and development of process knowledge will enable practitioners to bring novel therapies to the clinic with unique multifunctional capabilities. Articulation of the key variables (equipment, materials, idiosyncratic protocols etc.) at an early stage (i.e. the discovery lab) involved in production process will lead to a better understanding of how to translate good lab scale synthesis into scale processes for future clinical translation and assist manufacturing partners to produce material according to FDA's QbD principles.

QbD first implemented in pilot capacity by the FDA in 2005 has been formally adopted as a way to harmonize the development life cycle of biopharmaceuticals and move away from sampling to find product defects to an environment were "in control" validated processes drive a data rich environment where variations within specification are acceptable. QbD is based on the underlying principle that quality, safety and efficacy must be designed into a product and that quality specifically cannot be tested or inspected into a product. Officially defined as "a systematic approach to development that begins with predefined objectives and emphasizes product and process understanding and process control, based on sound science and quality risk management" . We consider QbD to be essential for the development of manufacturing processes for nanomedicine. QbD creates a continuous knowledge cycle, an important concept for advancing beyond the seminal steps for identifying innovative means to scale production of complex nanomedicine products. The FDA has come out with guidance that covers Pharmaceutical Development, Quality Risk Management, and Quality System with a predisposition that the future state of biopharmaceutical manufacturing, of which nanomedicine will be a part, will be an environment governed by QbD. Table 7.2 illustrates the differences in approach and the information requirements of QbD over traditional biopharmaceutical manufacturing which is dependent upon inspection, testing, locked processes and reproducibility.

Table 7.2 Quality-by-design (QbD) approach in manufacturing

Aspect	Traditional approach	Enhanced QbD approach	Informatics requirements
Overall pharmaceutical development	a. Mainly empirical	a. Systematic, relating mechanistic understanding of input material attributes and process parameters to drug product CQAs	a. Knowledge management across entire life cycle
	b. Developmental research often conducted one variable at a time	b. Multivariate experiments to understand product and process	b. Process traceability and change management from development through manufacturing
		c. Establishment of design space	c. One-point access to all phases and all levels of data
		d. Process Analytical Technology (PAT) tools utilized	
Manufacturing process	a. Fixed	a. Adjustable within design space	a. Full documentation of process analysis and verification decisions
	b. Validation primarily based on initial full-scale batches	b. Lifecycle approach to validation and, ideally, continuous process verifycation	b. Integration of these decisions with PAT tool configuration setups
	c. Focus on optimization and reproducibility	c. Focus on control strategy and robustness	
		d. Use of statistical process control methods	

续表

Aspect	Traditional approach	Enhanced QbD approach	Informatics requirements
Process controls	a. In-process tests primarily for go/no go decisions	a. PAT tools utilized with appropriate feed forward and feedback controls	a. Web-based "Digital Dashboard" providing remote process monitoring
	b. Off-line analysis	b. Process operations tracked and trended to support continual improvement efforts post-approval	b. Record of all significant parameter variances and trends
Product specifications	a. Primary means of control	a. Part of the overall quality control strategy	All lifecycle documents (from URS though PBR) interlinked, with traceability of changes
	b. Based on batch data available at time of registration	b. Based on desired product performance with relevant supportive data	
Control strategy	Drug product quality controlled primarily by intermediate and end product testing	a. Drug product quality ensured by risk-based control strategy for well understood product and process	a. Inventories and risk assessments of all systems, equipment and processes
		b. Quality controls shifted upstream, with the possibility of real-time release or reduced end-product testing	b. Decisions on parameter prioritization and acceptable variances
			c. Integration with PAT records, for refinement analysis
Lifecycle management	Reactive (i.e., problem solving and corrective action)	a. Preventive action	Change management integrated across all lifecycle phases
		b. Continual improvement facilitated	

It is important that nanomedicine manufacturers understand that QbD is knowledge rich environment dependent upon user definition of critical quality attributes (CQA), such that the physical, chemical, or biological property or characteristic of the intended nanomedicine should be within a proper range or distribution to ensure product quality. Linking CQA to process inputs (raw materials, chemicals, biologics etc), and process parameters (temperature, pressure, pH, etc) is performed in the early stage experimentation defined as the "design space" which is defined as the range of input variables or parameters for a single operation or it can span multiple operations.

Early articulation of the design space, CQA and process inputs can provide a very flexible operational environment with the desired attributes for scale-up manufacturing.

7.5.3.1 Importance of Knowledge Management in Nanomedicine

Nanomedicine holds the promise to cure complex diseases like cancer and save lives. Today, academic scientists lead the development of the complex multifunctional nanomedicine, but for all their promise, there is a striking lag in clinical translation. This lag rests on the fact that nanomedicine investigators under appreciate the value of target product profiles (TPP), a key component of QbD, for ensuring that processes used in the laboratory are compatible with commercial scale-up processes and regulatory guidance. A solution to this problem is at very early stage, put information into the hands of investigators to guide efforts towards nanomedicine that will have a chance to make it to the clinic. Innovation in informatics is another essential area and is complementary to the NIH's proposed investment to create National Center for Advancing Translational Sciences.

Nanomedicine translation faces substantial challenges related to managing the complex data streams emerging from the work at the bench, from process development work, and from preclinical studies all with important attributes required to drafting a TPP. The critical information developed during these activities is required to navigate a complex regulatory environment. Without effective data capture solutions and subsequent translation of large quantities of data into shared information, it will be "challenging" to coordinate the bench level process with scale-up process development, risk management and regulatory compliance. We are currently developing a software package, Fig.7.5, designed to assist academics in overcoming this translational bottleneck for nanomedicine by consolidating existing drug development best practices into a single package for use as a guide to further advance nanomedicine development."Nanolytics", developed by Nemucore Medical Innovations, Inc. (NMI) is a knowledge management system for information pertinent to development of TPP, processes development plans, validation plans and risk management assessment needed to support effective nanomedicine translation. Nanolytics allows academic investigators early in research to contextualize how a nanomedicine could move to the clinic. Unlike either small molecule or biologic development the creation of nanomedicine, which are complex molecular entities, is very process and design intensive. A manual process already demonstrated value of an informatics approach to identify barriers (use of equipment not compatible with scale-up) and risks (regulatory, material, etc.) to translating these nanomedicine to the clinic. Nanolytics software consists of three suites: a TPP Suite, a Process Suite and Validation Suite. These suites and the knowledge they will manage should mitigate cost and reduce time of development of scale-up processes, lower barriers to clinical development for nanomedicine and leverage research costs more effectively. Nanolytics allows for the input of key information based on initial research and outputs documentation on how to achieve for the pilot scale production of the target nanomedicine. As always is the case, better information, begets a more realistic product development plans. This development of information "outside" of the typical areas of focus of a nanomedicine researcher will reduce risk and clarify efforts in translating nanomedicine from bench to bedside.

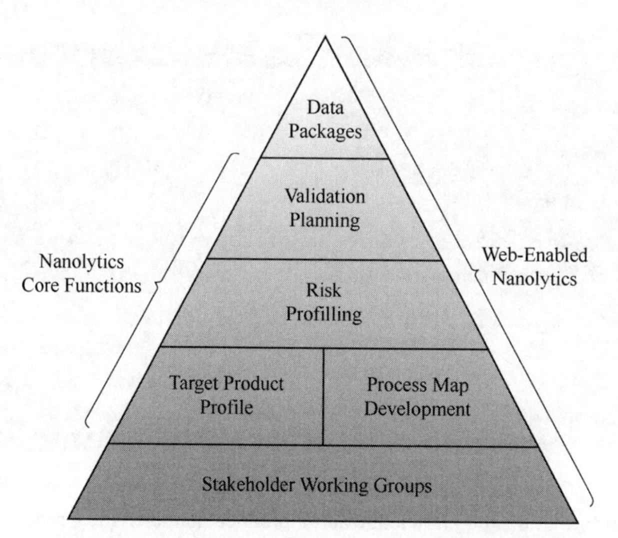

Fig.7.5 "Nanolytics": Conceptual framework for combination of informatics with processing technology for optimization of nano-pharmaceutical formulations.

7.5.3.2 Significance to Nanomanufacturing

Practices Developing manufacturing capability in the past has been capital intensive and typically relegated to a commercial responsibility. But with many of the advances happening in nanomedicine there is a discreet need to lower the barrier to access manufacturing capabilities on a molecule agnostic platform. In an effort to create such an environment, we have begun the process to establish the first in the nation FlexFactory™ nanomedicine manufacturing facility compliant with QbD principles. FlexFactory™ was developed by Xcellerex, Inc, (Marlborough, MA) to transition from single molecule manufacturing footprint to a modular, single use backbone which is agnostic to molecule.

FlexFactory™ provides the ideal manufacturing environment for nanomedicine as the controlled environmental units (CEMs) are able to maintain a single unit operation of a manufacturing process with the ability to grow with the progress of the molecule from preclinical thru commercial launch. The innovation of the FlexFactory™, briefly, is if a unit operation needs to change for the development of a new nanomedicine manufacturing process the modular CEMs can be opened a new unit operation installed, the new step and the new process validated allowing for the production of a nanomedicine that conforms to different CQA. While there are other modular platforms that can be used in a similar manner they often have to be pieced together. The FlexFactory™ system has withstood numerous FDA audits, inspections, and license applications for a variety of biologics. The sophistication required for biologic therapeutic manufacturing is suspected to be similar to the complexity required for multifunctional nanomedicines. This level of complexity and novelty of scaling nanomedicine production is why we have taken a two-step approach to aggregate knowledge using Nanolytics and the molecule agnostic manufacturing platform FlexFactory™, Fig.7.6.

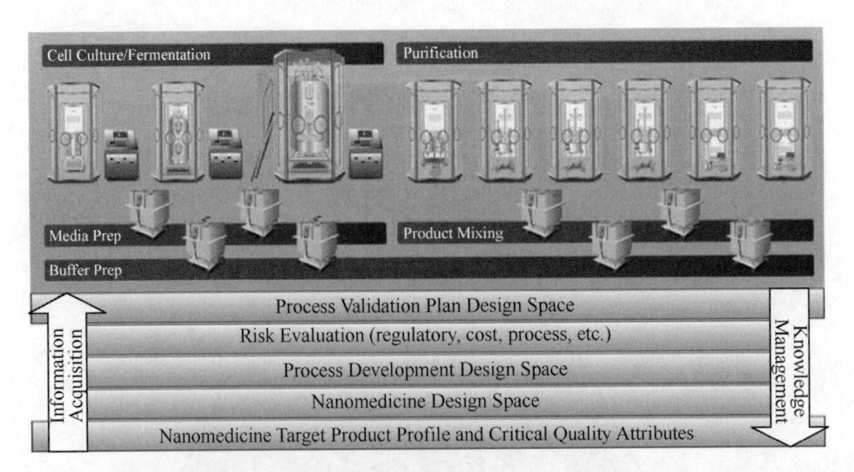

Fig.7.6 NMI FlexFactory™ footprint shown to illustrate that data captured in Nanolytics serves as foundation for manufacturing Information and knowledge about product characteristics, process, and systems drive manufacturing design to optimize manufacturing of nanomedicine.

7.6 Conclusions and Future Outlook

With greater understanding of chemical and physiological barriers associated in drug delivery and advances in nanomedicine design, there is an opportunity to efficient delivery of small and macromolecular drugs to complex diseases. Along these lines, the nanosystems have been engineered with specific attributes such as biocompatibility, suitable size and charge, longevity in blood circulation, targeting ability and image guided therapeutics, which can deliver the drug/imaging agent to the specific site of interest, based on active and passive targeting mechanisms. These systems cannot only improve the drug delivery to the target disease, but also the resolution of detection at cellular and sub-cellular levels.

To fully realize the potential of nanosystems for delivery of contemporary therapeutics in clinical setting, it is imperative that researchers also address the material safety, scale-up and quality control issues. Scale-up and quality control becomes extremely challenging especially when dealing with nanosystem designed to carry multiple drugs, imaging agents and targeting moieties. Furthermore, in vivo fate of nanomedicine engineered using novel nanomaterials are need to be fully assessed before being used in clinical application.

References

[1] P. N. Pushparaj, J. J. Aarthi, J. Manikandan, S. D. Kumar. (2008) siRNA, miRNA, and shRNA: in vivo applications. J Dent Res 87:992~1003.

[2] K. Baumann. (2014) Gene expression: RNAi as a global transcriptional activator. Nat Rev Mol Cell Biol 15(5):298~299.

[3] M. Ha, V. N. Kim. (2014) Regulation of microRNA biogenesis. Nat Rev Mol Cell Biol 15: 509~524.

[4] M. J. Alonso. (2004) Nanomedicines for overcoming biological barriers. Biomed Pharmacother 58: 168~ 172.

[5] C. V. Pecot, G. A. Calin, R. L. Coleman, G. Lopez-Berestein, A. K. Sood. (2011) RNA interference in the clinic: challenges and future directions. Nat Rev Cancer 11:59~67.

[6] S. Stegemann, F. Leveiller, D. Franchi, H. de Jong, H. Linden. (2007) When poor solubility becomes an issue: from early stage to proof of concept. Eur J Pharm Sci 31:249~261.

[7] C. A. Lipinski. (2000) Drug-like properties and the causes of poor solubility and poor permeability. J Pharmacol Toxicol Methods 44:235~249.

[8] E. M. Merisko-Liversidge, G. G. Liversidge. (2008) Drug nanoparticles: formulating poorly watersoluble compounds. Toxicol Pathol 36:43~48.

[9] B. J. Aungst. (1999) P-glycoprotein, secretory transport, and other barriers to the oral delivery of anti-HIV drugs. Adv Drug Deliv Rev 39:105~116.

[10] M. Goldberg, I. Gomez-Orellana. (2003) Challenges for the oral delivery of macromolecules. Nat Rev Drug Discov 2:289~295.

[11] N. Salama, N. Eddington, A. Fasano. (2006) Tight junction modulation and its relationship to drug delivery. Adv Drug Deliv Rev 58:15~28.

[12] A. T. Florence. (2005) Nanoparticle uptake by the oral route: fulfilling its potential? Drug Discov Today 2:75~81.

[13] S. C. Yang, S. Benita. (2000) Enhanced absorption and drug targeting by positively charged submicron emulsions. Drug Dev Res 50:476~486.

[14] P. Artursson, A. L. Ungell, J. E. Lofroth. (1993) Selective paracellular permeability in two models of intestinal absorption: cultured monolayers of human intestinal epithelial cells and rat intestinal segments. Pharm Res 10:1123~1129.

[15] C. A. Lipinski, F. Lombardo, B. W. Dominy, P. J. Feeney. (2001) Experimental and computational approaches to estimate solubility and permeability in drug discovery and development settings. Adv Drug Deliv Rev 46:3~26.

Chapter 8

Thermal Analysis of Pharmaceutical Polymers

Thermal methods have been widely used and are well-established routine methods for pharmaceutical raw material and dosage form characterization. The conventional thermal methods all involve measuring a response from a material (usually in the form of energy/temperature or mass changes) as a result of applying heat to the sample. In this chapter, the most widely used thermal analytical methods along with some more recently developed local thermal analysis and thermally based imaging methods are reviewed with regards to their working principle and applications in pharmaceutical product development. In the recent years, with the addition of the newly developed thermal imaging techniques, the capability of thermal analysis has broadened from conventional bulk sample analysis to also allowing more localized micron to sub-micron scale distribution and compositional analysis. The limitations of thermal methods for different applications are also discussed in relation to other characterization methods.

8.1 Introduction

The thermal properties of pharmaceutical raw materials and products are extremely important for understanding their processability and stability. Indeed, in many cases thermal measurements can be used to predict the physical and chemical stabilities of the materials. Thermal analytical methods include both classic thermal techniques, such as differential scanning calorimetry, for which the measurement of energy/temperature changes on heating are used to characterize a material's behavior and more recently developed techniques where thermal methods are combined with other analytical techniques such as scanning atomic force microscopy. These have extended the capability of thermal analysis to allow imaging and compositional identification. In this chapter, both conventional thermal analytical techniques as well as new developments in local thermal analysis and thermally based imaging technique are reviewed and the applications of these methods in the pharmaceutical field are discussed.

8.2 Theoretical Background of Thermal Analysis

8.2.1 Differential Scanning Calorimetry (DSC)

8.2.1.1 Conventional DSC

DSC is a thermal analysis technique, which was first commercially introduced in 1963. It provides qualitative and quantitative information as a function of time and temperature regarding thermal transitions in materials that involve endothermic or exothermic processes, or changes in heat capacity. In terms of instrumentation, there are two main types of DSC instruments: power compensation and heat flux. Power compensation DSC involves two separate furnaces for the reference and for the sample (Fig.8.1). The common principle of power compensation DSC is to heat both the reference and the sample simultaneously in such a way that the temperature of the two is kept identical, and the difference in power required to maintain the temperature is measured. Unlike power compensation DSC, which uses two furnaces, heat flux DSC uses two crucibles for the sample and for the reference within one furnace (Fig.8.1). They are both heated from the same source and the temperature difference between the sample and the reference over the heating profile is measured.

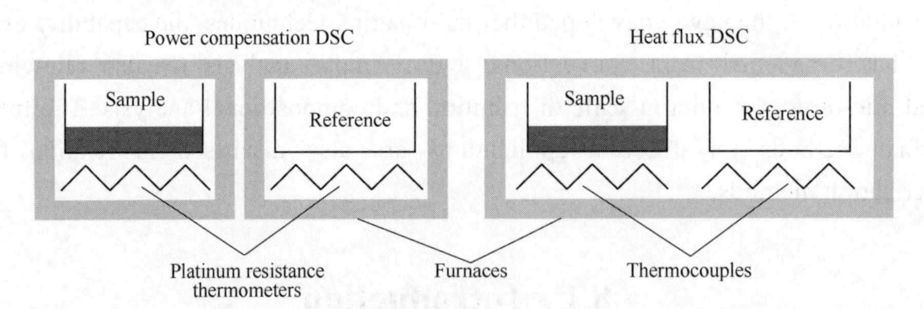

Fig.8.1 Schematic illustrations of power compensation and heat flux DSC

For heat flux DSC, the output signal is then converted into the power difference as shown in the following equation:

$$dQ/dt=\Delta T/R \tag{8-1}$$

where Q is the heat, t is the time, ΔT is the temperature difference between the sample and the reference and R is the thermal resistance of the heat paths between the furnace and the crucible. As described by Eq. (8-1), if the furnace and heat paths are truly symmetrical, the temperature difference between samples and reference is a measure of the difference in heat flow of the sample and reference.

The total heat content of a material has a linear relationship to its heat capacity (C_p, J•g^{-1}•℃$^{-1}$), which is defined as the quantity of heat required to change the temperature of the material by 1 K:

$$C_p = \mathrm{d}Q/\mathrm{d}T \tag{8-2}$$

Rearranging this equation with time:

$$\mathrm{d}Q/\mathrm{d}t = C_p(\mathrm{d}T/\mathrm{d}t) \tag{8-3}$$

where $\mathrm{d}Q/\mathrm{d}t$ is the heat flow and $\mathrm{d}T/\mathrm{d}t$ is the heating rate. With this equation, the differential heat flow provides a measure of the sample heat capacity. DSC data is normally expressed as the heat flow as a function of temperature. Typical transitions can be measured using DSC including a range of first and second order phase transitions, such as melting, crystallisation, glass transition (T_g) and relaxation. As an example, Fig.8.2 shows the thermal behaviour of amorphous paracetamol examined by DSC at different heat rates. Due to the low T_g and physical instability of amorphous paracetamol on heating, glass transition, recrystallization, polymorphic form transformation and crystalline melting occurred as a sequence of responses of the sample heated in DSC. It can be clearly seen that the crystallisation process is significantly affected by the heating rate of the DSC experiment, which highlighted the strong kinetic nature of the crystallization behavior.

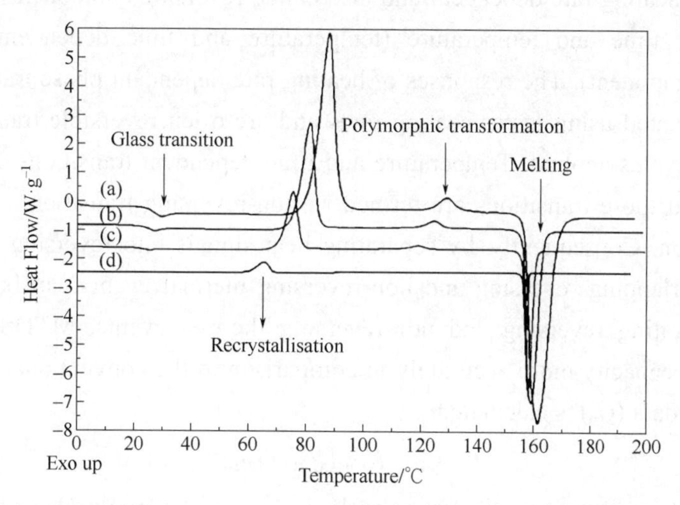

Fig.8.2 DSC responses of amorphous paracetamol examined at (a) 20℃•min^{-1}; (b) 10℃•min^{-1}; (c) 5℃•min^{-1}; (d) 1℃•min^{-1}.

8.2.1.2 Modulated Temperature Differential Scanning Calorimetry (MTDSC)

Conventional DSC mentioned above is a powerful tool to measure a wide range of thermal events such as melting accurately. However it often struggles to distinguish overlapping thermal

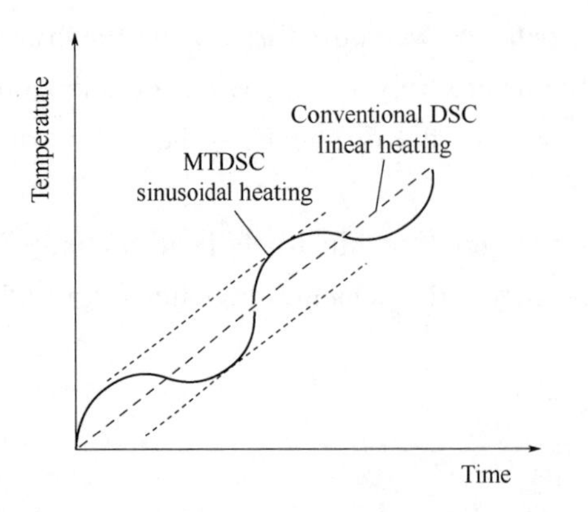

Fig.8.3 Different scanning mode of conventional (linear) DSC and MTDSC (sine wave).

events such as overlapped glass transitions and endothermic relaxation events, which can occur within the similar temperature range for many amorphous drugs and polymers. MTDSC was designed to separate overlapping thermal events. Compared with conventional DSC, where a linear heating rate is applied, in MTDSC the sample follows a heating rate commonly with a sinusoidal modulated wave as seen in Fig.8.3, and the uses of square and saw tooth modulated waves have also been reported. Sinusoidal modulation is overlaid on the linear ramp, namely, a perturbed heating or cooling process. As a result, there are three important parameters related to the MTDSC methodology, the average heating rate (the underlying heating rate), the amplitude of the modulation and the frequency of the modulation. Taking into account the contributions of these parameters, the heat flow signal of a MTDSC run can be described as follows:

$$dQ/dt = C_p \cdot dT/dt + f(t, T) \qquad (8\text{-}4)$$

where dQ/dt is the total heat flow, dT/dt is the heating rate, C_p is the heat capacity, and $f(t,T)$ is a function of time and temperature that governs the response associated with the phase transition. This equation shows that total heat flow is comprised of two contributions: one is a function of the heating rate (heating rate dependent and also called reversing component) and the other one is a function of time and temperature (temperature and time dependent and also called non-reversing component). The responses of heating rate dependent phase transitions tend to be larger when evaluated using faster heating rates and are often reversible transitions if multiple heating-cooling cycles applied. Temperature and time dependent transitions cannot be reversed once initiated and these transitions are termed as non-reversing transitions, such as relaxation and decomposition. Consequently, by separating heat signals into reversing and non-reversing components, overlapping reversing and non-reversing thermal events can be distinguished. In addition to separating reversing and non-reversing thermal events, MTDSC is also able to measure the heat capacity more accurately in comparison to the conventional DSC. In MTDSC, the heat capacity data (C_p) is calculated:

$$C_p = K_{Cp} A_{MHF} / A_{MHR} \qquad (8\text{-}5)$$

where K_{C_p} is the heat capacity calibration constant, A_{MHF} is the amplitude of the modulated heat flow and A_{MHR} is the modulated heating rate. Therefore, the application of large amplitude in MTDSC can increase the heat capacity precision by reducing noise. In practice, adjusting the combination of amplitude, heating rate and period to suit the sample is extremely important for the effective use of MTDSC. As a rule of thumb, slow heating rate ($<5℃\cdot min^{-1}$) should be used and the combination of all three parameters should allow at least six modulations over the course of the transition of interest.

8.2.1.3 Hyper DSC

High speed or high performance conventional DSC, also known as hyper DSC, operates at extremely fast heating rates from 200℃•min^{-1} up to 750℃•min^{-1}. Conventional DSC using slow (linear) heating rates (typically heating rate below 100℃•min^{-1}) can result in good resolution but poor sensitivity particularly for phase transitions that strongly affected by kinetic factors, whilst fast heating rates can result in poor resolution but good sensitivity. Fast heating rates have the same total heat flow signal as in a DSC or MTDSC experiment. However as transitions occur over a shorter time period, the signal response to the thermal event appears larger. One issue that can occur with conventional DSC with slow heating is that the heating process may alter the sample, before the thermal transition of interest is reached. Using fast heating rates these effects can be eliminated or reduced, allowing for the characterisation of samples in their "as received" state. This technique is also of particular advantage for materials possessing properties that may change upon prolonged exposure to increased temperatures like amorphous products or formulations of biological molecules. Fox example, detecting the amorphous content of spray dried lactose increased from 10% (*W*/*W*) for conventional DSC to 1.5% (*W*/*W*) for hyper DSC with 500℃•min^{-1}.

8.2.2 Thermogravimetric Analysis (TGA)

TGA is one of the oldest thermal analytical procedures and has been used extensively in the study of material science. The technique involves monitoring the weight change of the sample in a chosen atmosphere (usually nitrogen or air) as a function of temperature. The measurement is operated by applying a temperature programme to a closed sample furnace containing an electronic microbalance (for holding the sample), which allows the sample to be simultaneously weighed and heated in a controlled manner, and the mass, time and temperature to be captured. In the pharmaceutical industry, TGA is routinely used for thermal stability and volatile components analysis of pharmaceutical materials.

In research and development phase of a pharmaceutical formulation, TGA can also be used to facilitate the evaluation of processing temperatures of thermally based manufacturing processes such as hot melt extrusion. It can determine the thermal stability of the drugs and polymers upon heating to assist with selecting the operation temperature in hot melt extrusion to avoid thermal degradation occurred in process. As an empirical method, TGA is often used to measure the moisture or residue solvent contents in processed materials. However, the drawback is that it has no capability to identify the chemical composition of lost solvent. Therefore in some studies TGA has been coupled with spectroscopic detection methods such as gas chromatography (GC and GC-MS) to allow the chemical identification of the volatile material liberated from the sample.

8.2.3 Scanning Probe Based Thermal Analysis

The coupling of thermal analysis with atomic force microscopy (AFM) gives the new generation of thermal analysis with the capability to allow thermal measurement to be performed at the selected point of interest. Such technique is often known as localized thermal analysis (LTA). LTA is an extension of conventional AFM, replacing conventional probes with thermal probes. The tips of these probes are composed of material with high resistance to electrical current. When an electrical current is applied, it causes the tip of the probe to become heated. The probes have a unique advantage in that not only can they be used as normal probes to generate topographical images but they can also be used as a localised heating source. The first LTA probe (called a Wollaston probe) was developed in 1994 and first used in 1996. The probe consists of a 75 μm diameter silver wire with a 5 μm platinum filament core. The wire is manipulated to form a sharp point at which the silver is electrochemically removed to expose the filament (Fig.8.4). This V-shaped wire is then used as the sensor to perform the functions mentioned above. As a result of the micron spatial resolution of the technique, the technique is known as micro-thermal analysis (μTA). There are several modes in which heated probes can be used but one of the most common is to carry out the thermal analysis on specific locations on a sample surface. Following the generation of a topographic image, the probe is placed in contact with a point of interest. The system monitors the position of the probe above the surface as the voltage is applied. Once the temperature at the end of the tip reach the softening temperature of the tested sample, such as a glass transition or melting transition, penetration of the tip into the sample will occur, which is reflected as a decreased deflection of the probe. Other applications of heated probe have also been reported. For example, the LTA probe can be used at a constant (heated) temperature as it is scanning across the sample, which is often known as heated-tip AFM. This temperature is normally selected as being above the softening temperature of one the components of interest in a multi-component sample. The image generated often can reveal the distribution of the components within the sample.

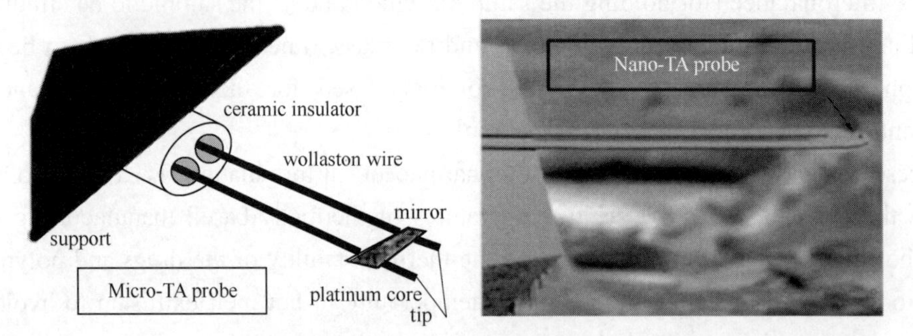

Fig.8.4 Comparison of micro- and nano-thermal analysis probes: schematic design of micron size Wollaston wire thermal probe on the left and nano-TA probe on the right.

Recently, thermal probes with the similar dimensions to conventional tapping mode AFM probes have been invented, which allows the application of LTA studies at a increased spatial resolution (sub-micron) in comparison to the Wollaston probe. The new generation of thermal probe has a length of circa 200 μm and the height of the tip is below 1 μm. This provides the probe with a topographic spatial resolution of 5 nm. The probe is made of highly doped silica with boron or phosphorus which has a high electrical resistance. This allows the tip to be heated when an electrical current is applied. With a sub-micron resolution, the technique is named nano-thermal analysis (nano-TA). A typical nano-TA tip is shown in Fig.8.4.

A recent major development of LTA is the transition temperature mapping (TTM) technique. Instead of manually selecting the tested locations at the sample surface in conventional LTA, a defined region of interest at the surface of the sample can be selected. With in the selected region, each LTA tested point is a pixel of the final thermal image of the entire region produced. The LTA test at each point (pixel in the image) will stop once a thermal transition is detected and the probe will move on to the next adjacent point to continue the same process. The distance between the two consecutive pixels can be as close as 1 μm. As with other imaging techniques, the detected transition temperatures within that area are assigned a colour based on a particular palette and hence the coloured image is assembled based on different transition temperatures.

As an imaging method, TTM can distinguish and reveal distribution of different materials or phases within a small area (i.e. 50 μm×50 μm). It can also "draw" the borderline between different phases in the same sample. Therefore, it can be useful to investigate the distribution of different materials within the sample at a sub-micron level. Figure 8.5 shows the possible identification crystallised region at the surface of an amorphous felodipine sample.

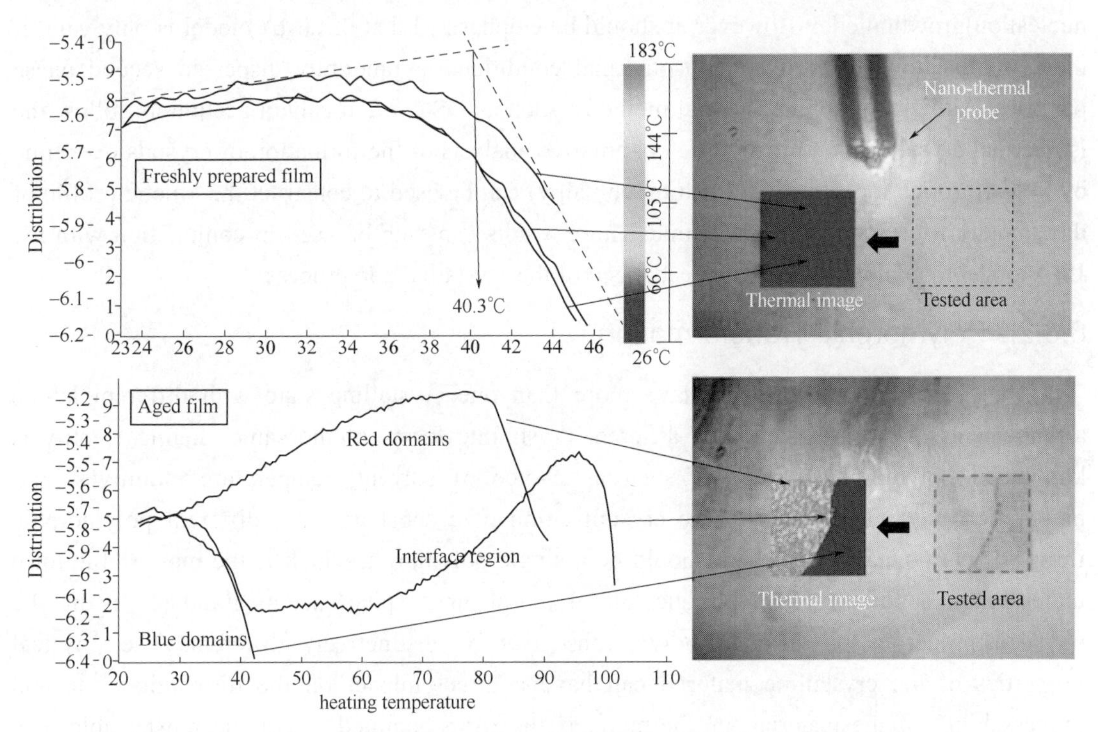

Fig.8.5 Nano-TA profiles of the thin films of amorphous felodipine before and after aging (left hand side panel) and the optical images showing the tested areas on the film and the corresponding TTM images (right hand side panel).

8.3 Physical and Chemical Phenomena Commonly Investigated Using Thermal Approaches

8.3.1 Crystallisation

Crystallisation is one of the most common processes that occur (both intentionally and unintentionally) for both active pharmaceutical ingredients and many excipient materials. It can take place in materials with low molecular weights, such as many drug molecules, lipids and salts, as well macromolecules including therapeutic proteins and polymers. The kinetic behaviour of crystallisation is often of significant importance in pharmaceutical formulation development and processing design in order to produce physically and chemically stable and effective medicines.

Thermal analysis has been used to study the crystallisation behaviour of pharmaceutical materials in solution and in solid state. The most commonly used approaches for studying crystallisation kinetics using thermal methods is applying Johnson-Mechl-Avrami (JMA) nucleation-growth model. However, it should be emphasized that the JMA model is only valid if the crystallisation occurred under isothermal conditions at randomly dispersed second phase particles. Although scanning thermal methods, such as DSC, often cannot accurately follow the isothermal crystallisation process the quantitative analysis of the formation of crystals over time by DSC (through measuring the melting enthalpy) can be used to construct the kinetic profile of the amount of crystallisation over time. These results can then be used in conjunction with the JMA model to explore the kinetic mechanism of the crystallisation process.

8.3.2 Polymorphic Transformation

Many pharmaceutical solids have more than one crystalline state with different lattice arrangements. The co-existence of different crystalline forms of the same chemical entity is known as polymorphism. Factors such as choice of solvent, temperature, additives, and preparation method can lead to the crystallisation of a substance into different polymorphic forms. Thermodynamically, there should be a single polymorph, which is the most stable form under a fixed condition. The importance of polymorphs in the pharmaceutical industry is that the physical properties of different polymorphs may be distinctively different. The physical properties of the crystalline material can have a direct impact on the dissolution rate and processability of the material. Furthermore, if the form obtained is not the most stable one, polymorphic transition may occur spontaneously. Thermal methods are the most frequently used characterization approach for determining the stability relationship between the polymorphs either being enantiotropic or monotropic. As seen in Fig.8.6, for enantiotropically related polymorphs, their free energies become equal at the transition temperature T_0. Because the enthalpy of form A is lower than that of form B at T_0, the transition is often detected as an endothermic transition by DSC. However, sometimes the transition temperature may be higher than the free energy crossing point which may lead to the melting of form A without polymorphic transformation. In this case, the enthalpy of fusion of form A is usually larger than that of form B. For monotropically related polymorphs, the stability relationship of the two polymorphs does not depend on the temperature below their melting points (no free energy crossing point). Therefore polymorphic transformation from metastable form B to form A may or may not occur on heating. If it occurs, this transition is an exothermic process. If it does not occur, metastable form B usually melts with a smaller enthalpy of fusion than that of the stable form A.

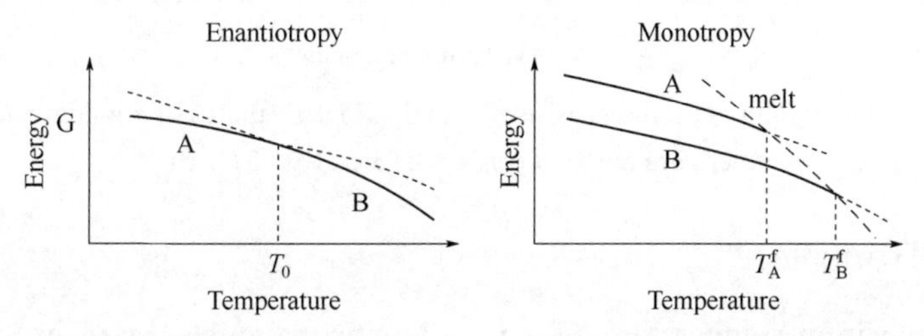

Fig.8.6 Gibbs free energy diagrams of enantiotropic and monotropic polymorphs.

However, the kinetic influence from the conventional thermal scanning method often makes the polymorphic transformation data difficult to interpret. Recently, the development of modulated, quasi-isothermal and ultraslow thermal method as a means of characterizing the α to γ indomethacin polymorphic transition was reported. By using both scanning and quasi- isothermal MTDSC to measure subtle heat capacity changes through the transformation, they were able to use Lissajous analysis to study the modulated heating signal in order to identify the poorly resolved polymorphic transformation (as seen in Fig.8.7). Ultraslow heating rates (down to 0.04℃•min^{-1}) were also used to minimize the kinetic interference from the method. By combining these thermal methods, both kinetic and thermodynamic analyses of the polymorphic transformation via melt and recrystallization process between polymorphs of indomethacin was achieved.

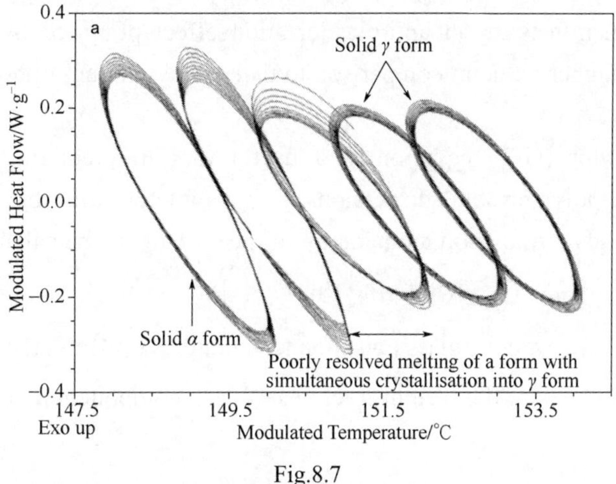

Fig.8.7

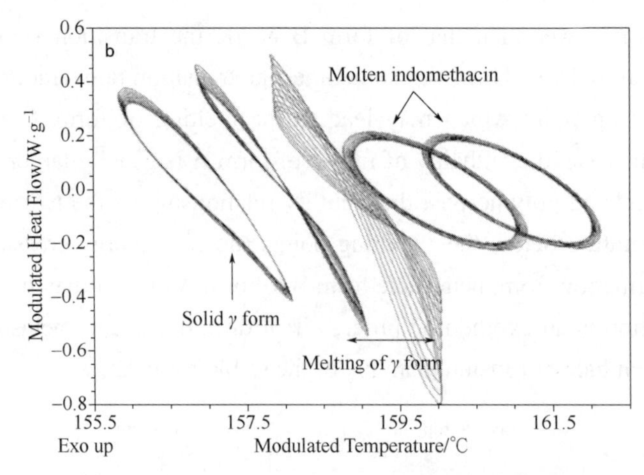

Fig.8.7 Lissajous figures for the quasi-isothermal MTDSC studies of α indomethacin at the temperature range of (a) 148~153℃; (b) 157~161℃.

8.3.3 Glass Transition

The glass transition temperature (T_g) is a kinetic parameter associated with molecular motion in an amorphous state, which can be measured using DSC and MTDSC. Below T_g the amorphous materials are "kinetically frozen" into the glassy state, and any further reduction in temperature only causes a small decrease in molecular motions of amorphous solids. Whereas above the T_g, amorphous solids will enter a rubbery state with significantly increased molecular motion. Molecular motion has been related to the occurrence of the recrystallisation of amorphous solids and phase separation in amorphous solid dispersions.

For multicomponent amorphous systems, such as amorphous solid dispersions, glass transition temperatures can also be used as indicators to describe their physical state. If a binary drug-polymer system is miscible, only a single T_g can be observed using DSC. The presences of two glass transitions indicate the presence of phase separation, which often is due to the poor miscibility between components. For miscible drug-polymer systems, if the T_g of the drug is lower than the T_g of the polymer, drugs can act as a plasticizer and reduces the T_g of the system. However this also manifests as an anti-plasticization effect provided by the polymer, which increases the T_g to a higher value in comparison to that of the pure amorphous drug, and stabilise the amorphous drug.

The Gordon-Taylor (G-T) equation is a useful tool in predicting the glass transition temperatures of drug-polymer solid dispersions. For completely miscible binary drug-polymer solid dispersions, the glass transition temperature of the system can be calculated by:

$$T_{g_{mix}} // [(w_1 T_{g_1}) + (K w_2 T_{g_2})] / [w_1 + (K w_2)] \tag{8-6}$$

where T_{g_1}, T_{g_2} and $T_{g_{mix}}$ are the glass transition temperatures (in Kelvin) of component 1,2 and the mixture, and w_1 and w_2 are the weight fractions of each component and K is a constant which can be calculated by:

$$K \approx (\rho_1 T_{g_1}) / (\rho_2 T_{g_2}) \tag{8-7}$$

where ρ_1 and ρ_2 are the true densities of each component. The application of the equation in predicting the miscibility of solid dispersions lies in the comparison between the theoretical and experimental $T_{g_{mix}}$ values of the system. The Gordon-Taylor equation relies on the assumption that the different components form and ideal mixture. Therefore, if the calculated and experimental $T_{g_{mix}}$ values are consistent, it may indicate that the system is miscible. However, exceptions to this have been reported. The discrepancies of T_gs between the calculations from Gordon-Taylor equation and experimental data may be caused by two main reasons. Firstly, interactions between drugs and polymers in amorphous solid dispersions can affect the values obtained. Positive deviation whereby the experimental value is higher than the G-T predicted value could occur if the interaction between drugs and polymers are stronger that what occurs between two drug molecules, whereas negative deviation may be observed if the drug-polymer interaction is weaker than that between two drug molecules. Secondly, water sorption in amorphous solid dispersions can decrease the Tg value of the system via the strong plasticization effect of water sorption.

8.3.4 Molecular Mobility

The molecular mobility of drugs is commonly considered to be a key factor associated with the stability of amorphous solids and solid dispersions, since a high molecular mobility can lead to high tendency of recrystallisation and phase separation on aging. On aging non-equilibrium amorphous materials tend to approach the equilibrium state by releasing the extra enthalpy and configurational entropy. This reducing extra energy process is termed as structural relaxation and the length of time over which the structural relaxation occurs is known as the relaxation time. Molecular mobility has a reciprocal relationship with the relaxation time constant (τ). In the Adam-Gibbs model, it is calculated as:

$$\tau = \tau_0 \exp[C/(TS_c)] \tag{8-8}$$

where τ is the molecular relaxation time constant, τ_0 is a constant, T is the absolute temperature, S_c is the configurational entropy, and C is a material dependent constant. This equation was further modified to the Adam-Gibbs-Vogel equation:

$$\tau = \tau_0 \exp(DT_0/[T(1-T_0/T_f)]) \tag{8-9}$$

where D is the strength parameter, T_0 is the temperature of zero molecular mobility, and T_f is the fictive temperature. The fictive temperature is defined as the temperature of intersection between the equilibrium liquid line and the non-equilibrium glass line. In most cases, T_f values are very close to T_g values, and therefore in calculations the T_f value can be replaced by the T_g value. By measuring the T_g and configurational entropy using thermal methods, the molecular mobility of amorphous solids can be estimated.

8.3.5 Structural Relaxation

Two types of relaxations of amorphous materials have been defined at temperatures below and above the glass transition temperatures. For molecules with low M_w, at temperatures below

T_g, the dominant relaxation procedure is β-structural relaxation (also known as local molecular mobility), which can take place by means of the atomic movements within the molecular structure. For amorphous polymers, at temperatures below the T_g, β-structural relaxation refers to the vibrating of the side chains of the polymer. At temperatures higher than the T_g, the dominant relaxation is α-structural relaxation (also known as global molecular mobility), which could occur for both amorphous drug and polymer whereby an entire molecule will be mobilised.

These two types of relaxation can be detected by techniques such as DSC, dynamic mechanical analysis (DMA) and dielectric spectroscopy. For instance, when stored under ambient conditions for a certain time period, the relaxation enthalpy of amorphous indomethacin at the glass transition region can be detected on heating in DSC, and the relaxation enthalpy was a contribution of both α- and β- relaxations on aging. For multicomponent system, such as solid dispersions, both relaxations of drugs and polymers are responsible for physical stability of the sample on aging. As discussed earlier, the molecular mobility of an amorphous solid is associated with temperature, and will decrease with decreasing storage temperature leading to an increased physical stability. It has been used as an empirical practice that storing amorphous materials at temperatures 50 K below the T_g can sufficiently lower the molecular motions and leads to good physical stability of the amorphous material on aging.

8.3.6 Dehydration and Decomposition

Dehydration and thermal decomposition behaviour of pharmaceutical active ingredients and excipients are commonly studied using TGA, or in some cases, TGA combined with FTIR, MS or GC/MS for identification of evaporated fume. In terms of dehydration studies, weight loss over the heating period is quantitatively related to the moisture loss within the sample (assuming no further thermally related decomposition occurs). Thermal decomposition temperatures can be identified by sharp and continuous significant weight loss of the samples during heating.

8.4 Applications of Thermal Analysis for the Characterisation of Pharmaceutical Raw Materials

8.4.1 Amorphous Drugs

Thermal analysis methods have been one of key streams of analytical techniques for the characterization of pharmaceutical amorphous solids including APIs and excipients. The physical parameters associated with their amorphous nature discussed earlier can all be fully investigated using thermal methods. For example, the α- and β-relaxation processes of amorphous indomethacin were studied using DSC. The β-relaxation was detected as a small endothermic relaxation peak that occurred in the similar temperature region of the T_g. The activation of this relaxation process increased with increasing the annealing temperature of amorphous indomethacin. When the annealing temperature is well below the T_g, the β-relaxation fades significantly and leads to good physical instability of amorphous indomethacin. The crystallization kinetics of an amorphous API is another topic that has been widely studied using thermal methods such as DSC. It has been reported the rapid but incomplete recrystallization behaviour of cyro-milled amorphous Etravirine at temperatures below the T_g of the amorphous drug measured using DSC and hyper DSC. Furthermore, the effect of thermal history on the stability of amorphous material is also frequently studied using DSC. Amorphous paracetamol, as an example, shows a marked difference in recrystallization and polymorphic form conversion behaviour depending on the cooling rate that was used during sample preparation.

8.4.2 Pharmaceutical Polymers and Lipidic Excipients

Thermal analysis methods are routine tests used in research and industry for the characterization of pharmaceutical polymers and lipids, including their T_g, melting and crystalline content and relaxation behaviour. Hydroxypropylmethylcellulose (HPMC) is a classical example of such practice. Semi-crystalline polymers such as PEGs are another example of a pharmaceutical excipient where thermal analysis is considered to be key for revealing the behaviour of the material. For example DSC has been used to investigate the miscibility of PEG and crystalline drug as a fast screening method for pre-formulation development. Lipidic excipients such as TPGS, Gelucire and Poloxamer, all have distinct melting transitions which can be measured by DSC. Changes in the melting behaviour of these materials after drug incorporation or processing provide an indication of the occurrence drug-excipient interactions and the effect of processing on the excipients.

8.4.3 Polymer Blends

Pharmaceutical polymer blends have been recently valued as a class of useful drug carrier materials with the advantage of having easily adjusted drug release profiles and enhanced formulation physical stability. The phase behaviour of a polymer blend can directly impact on the physical performance of the blends as a matrix.

Thermal analysis methods such as DSC, MTDSC, LTA and TTM have been all used to characterize pharmaceutical polymer blends. For example, MTDSC was used to estimate the miscibility between Eurgadit® E PO and PVPVA by analyzing the T_g of the blend. Visualization of the phase separation of the processed polymer blend using TTM confirmed the immiscibility of the two polymers (as seen in Fig.8.8). This immiscibility was found to be advantageous, enhancing the physical stability and in vitro dissolution rate of the formulation. MTDSC was used to investigate the influence of the PEG chain length on the miscibility of PEG/HPMC 2910 E5 polymer blends and the influence of polymer on the miscibility with itraconazole. It was found that the polymer miscibility increased with decreasing chain length due to a decrease in the Gibbs free energy of mixing. Lower miscibility leads to more phase separated system indicated by a more rapid decrease of the T_g of the ternary solid dispersion and a high level of itraconazole recrystallization from the ternary solid dispersions. Scientists have used DSC to assess the miscibility of the blends and to screen a wide range of polymer blends as potential carriers for oral drug delivery. As mentioned earlier, they also used T_g of the blend as a particular indicator to evaluate the miscibility between different combinations of hydrophilic and hydrophobic polymers.

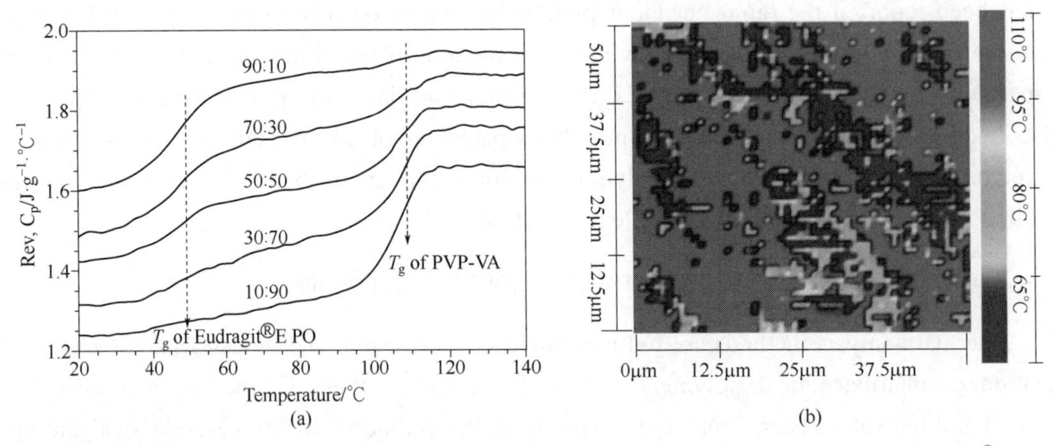

Fig.8.8 (a) DSC profiles of the immiscible polymer blends with clearly separated two T_gs of the Eudragit® E PO and Kollidon® VA 64 and (b) TTM result of the surface of hot melt extruded 0:50 (*W*/*W*) polymer blend.

8.5 Applications of Thermal Analysis for the Characterisation of Pharmaceutical Dosage Forms

8.5.1 Solid Dosage Forms

8.5.1.1 Physical Stability Evaluation

Thermal analysis has been widely used for monitoring the physical stability of solid dosage forms, such as detecting amorphous contents, polymorphic conversions of either API or crystalline excipients and excipient stabilities (lactose and crystalline lipids as typical examples). Here solid dispersion formulations are used as an example to demonstrate how thermal methods can be used for not only monitoring but also predicting the physical stability of such formulations. An amorphous drug-polymer solid dispersion is a thermodynamically unstable system since the molecularly dispersed drugs (high energy level) in the system will tend to convert back to the more stable crystalline form (low energy level), leading to a physical instability on aging. Freshly prepared amorphous solid dispersions may have the drug molecularly dispersed in the polymeric carriers. On aging the physical instability of amorphous solid dispersions can be observed in the form of phase separation and recrystallization due to the relaxation of high energy-level drug molecules and molecular mobility of drugs under accelerated storage conditions. Recrystallization could then occur within the drug-rich phase whereby high concentration amorphous drug crystallizes to form the more stable crystalline state. Factors which can potentially affect the physical stability of amorphous solid dispersions have been investigated widely, and glass transition temperatures, molecular mobility, physical stability of amorphous drugs alone, miscibility between drugs and polymers and solid solubility of drugs in polymers have been considered to be key factor influencing the physical stability of solid dispersions. Amongst these, many can be studied using thermal analysis methods.

(1) Crystallisation Tendency

Drug crystallization may take place if nuclei are presence either in form of residual crystalline drug or foreign particles from the surrounding environment. For an amorphous solid dispersion, drug crystallisation of a poorly water-soluble drug from the dispersion can significantly compromise the dissolution advantage an amorphous solid dispersion formulation can provide. Thermal analysis can be used as a tool for the evaluation of the tendency of crystallisation of amorphous drug either alone or present in a formulation, such as solid dispersions. For example, Yang used DSC and MTDSC to study the effect of milling on the physical stability of hot melt extruded solid dispersions. As seen in Fig.8.9, DSC can clearly indicate the instability of milled amorphous solid dispersions formulations by the appearance of a recrystallization transition of the phase separated amorphous drug on heating and the separation of single T_g to a double T_g.

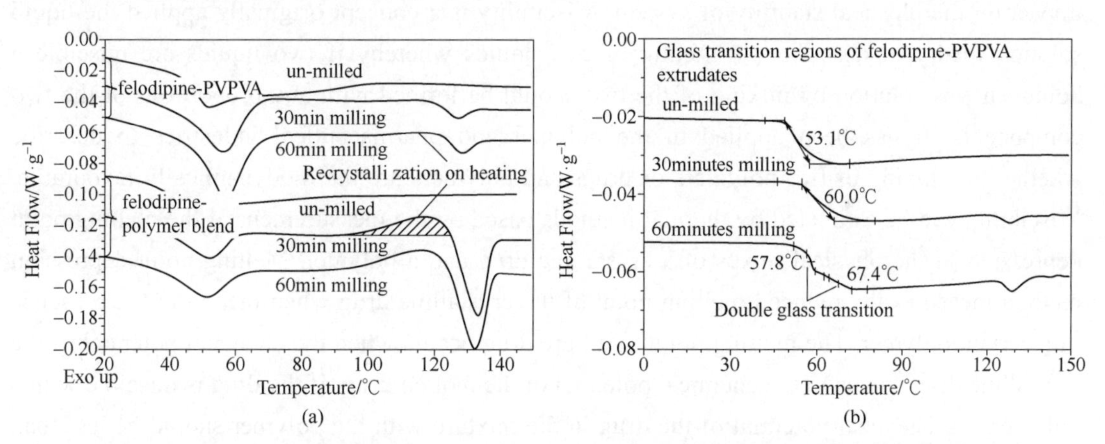

Fig.8.9 (a) Total heat flow signals of the MTDSC results of 70% (*W*/*W*) felodipine loaded melt extrudates before and after milling; (b) reversing heat flow signals of 70% (*W*/*W*) felodipine-PVPVA melt extrudates before and after milling.

(2) Quantitative Analysis of Crystallinity

The use of thermal methods including DSC and MTDSC for quantitative analysis of crystallinity of pharmaceutical materials including API, processed formulations and polymers has been widely reported. However, the absolute accuracy of the quantitative measurement of the crystalline content is still debatable. The common practice of using DSC for crystallinity measurement relies on the measurement of the enthalpy of the melting peak of the crystalline component and compares this to the melting enthalpy of the crystalline material with known purity. However, this often can only be used as an empirical approach for rough estimation of the crystallinity as the melting enthalpy of the crystalline material is affected by the presence of other excipients such as polymers and lipids. In addition, in many cases, recrystallization of the amorphous proportion of the material on heating during DSC testing can significantly affect the estimation of the crystalline content. Instead of using the melting enthalpy, heat capacity and recrystallization enthalpy can also be used to measure the amorphous content and as a means to indirectly calculate the crystalline content in a formulation with drug and excipients.

(3) Moisture-Induced Instability

Moisture can induce physical and chemical instability (such as degradation via hydrolysis reaction) in an API and in formulated pharmaceutical products. In terms of the physical instability of amorphous systems in particular, the sorption of moisture can promote the recrystallization of amorphous drug and phase separation of amorphous solid dispersions that initially containing molecularly dispersed drug in polymer matrices. The change in T_g and the presence of crystalline material in the samples after exposure to moisture, which can be measured using DSC and MTDSC, can often be used as indicators of physical instability.

8.5.1.2 Drug-Excipient Miscibility Estimation

As discussed earlier, for a solid dispersion, drug-polymer miscibility can significantly impact on the physical stability of system. Miscibility is a concept originally applied the liquid solution theory to describe the mixing of two liquids whereby if two liquids are miscible a homogeneous solution by mixing of the two should be formed with any proportions of the two components. It has been applied in the polymer and pharmaceutical industries to describe whether the mixing of two polymers or drugs and polymers are thermodynamically favourable. Miscibility can be estimated by thermal methods based on the measurement of the melting point depression of the physical mixes of a crystalline drug and a polymer. Melting point depression method measures the reduced melting point of the crystalline drug when melted in the presence of a certain polymer. The melting point of a pure drug occurs when the chemical potential of the crystalline drug equals to the chemical potential of the molten drug. If the drug is miscible with a polymer, the chemical potential of the drug in the mixture with the polymer should be less than the chemical potential of the pure crystalline drug, and thus the reduced chemical potential will result in a depressed melting point of the drug in the presence of the miscible polymer. In contrast, if the drug and polymer are immiscible, no melting point depression is expected due to the unchanged chemical potential brought by the presence of polymers.

Briefly, the melting point depression approach is a simple method to test the miscibility between drugs and polymers by running a physical mixture of the drug and the polymer in DSC on heating. The presence of a depressed melting point may indicate the drug and the polymer are miscible. Previously the melting point depression approach has been proved to be effective in predicting the miscibility between felodipine and nifedipine with PVP. However, the depression of the melting point is heating rate dependent, which can affect the accuracy of the prediction. Drug-polymer solid solubility is also a concept derived from solution theory, which considers the mixing of two liquids, solubility is defined as dissolving a solid-state material into a solvent until the equilibrium maximum concentration at a defined temperature and pressure. Several theoretical methods have been reported for the prediction of solid solubility of drugs in polymers. Melting point depression combined with Flory-Huggins lattice based theory has been reported to predict solid solubility of felodipine and nifedipine in PVP. The interaction parameter χ from Flory-Huggins theory can be calculated using the detected depressed melting points from drug-polymer physical mixtures with different ratios, and by further using the obtained interaction parathion the solid solubility of drugs in polymers can be predicted. However, the calculation used by this method is lengthy and in practice the depression of the drug melting is highly heating rate dependent. Therefore, finding the appropriate experimental conditions for measuring melting point depression is also not straightforward.

Another thermal method for estimating drug-polymer solubility based on the measurement of melting enthalpy of crystalline drugs in physical mixtures with polymers by DSC was recently developed. This model was a modification of a previous melting enthalpy based method. It assumed the dissolution of drugs in polymers on heating is endothermic and two possibilities, which involve drugs completely dissolved in polymers and vice versa were included in the model. The method was effectively used to predict the solubility of felodipine in Eudrgait® E PO, as seen in Fig.8.10. There are discrepancies between the solubility measured using this DSC based method and other prediction methods including melting point depression and solubility parameter methods. This is likely to be because of the nature of the characterization method, which is using the behaviour of the drug at temperatures close to the melting transition to predict the mixing behaviour with the polymer at ambient temperature.

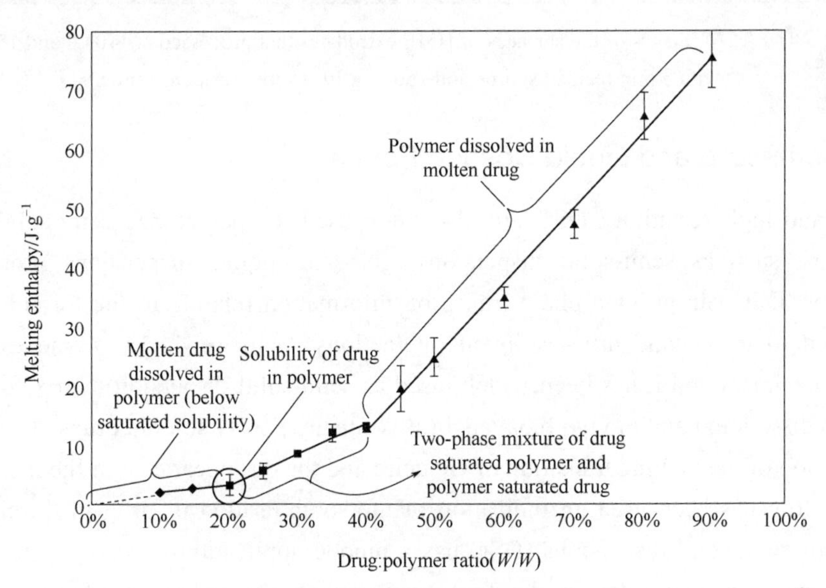

Fig.8.10 The model prediction result of the miscibility between felodipine and Eudragit® E PO.

8.5.1.3 Micro-scale Drug Distribution Uniformity Assessment

Although conventional thermal analysis methods are mostly used for characterize materials in bulk, the scanning and local thermal analysis have high precision for characterisation of materials at a micron to sub-micron scale. This offers the possibility of using these methods to study the distribution of drug or certain excipients within a formulation. This has been achieved by other imaging methods such as IR and confocal Raman imaging. However, each imaging method has advantages and limitations. The essential requirement for successful application of vibrational spectroscopic based imaging methods is to have sufficient separation of the vibrational bands used to identify the drug and excipients. If this is not available, local thermal analysis and thermal imaging method are good alternatives for studying the drug distribution within a dosage form by using the thermal transition temperature difference between the drug and other excipients. As seen in Fig.8.11, TTM was used to map out the micro-phase separation of a series of cyclosporine A containing solid dispersions processed using hot melt extrusion under different operational parameters. The presence of colourcoded domains is an indication of the separation and distribution of phases (polymer-rich and drug-rich domains) in the formulation. The results clearly demonstrated that with increasing the processing temperature and residence time, a significant improvement in the homogeneity with a narrow distribution of transition temperatures could be obtained.

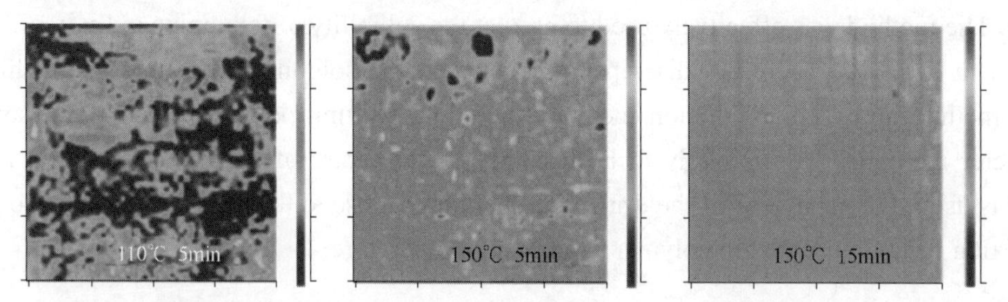

Fig.8.11 TTM images of the surfaces of HME extrudates that processed at 110℃ and 150℃ with 5 min residence time and 150℃ with 15 min residence time.

8.5.2 Semi-solid and Liquid Dosage Forms

DSC and high sensitivity DSC are also often used to characterize semi-solid and liquid formulations, such as semi-solid dispersions, gels and liquid suspensions. For semi-solid formulations, DSC can provide phase transition information relating to the formulation, which can be used to understand drug-excipient interactions. Gelucire 44/14 is one example of a semi-solid excipient and it has been widely used as semi-solid dispersion matrix for enhancing the in vitro dissolution and in vivo bioavailability of many poorly soluble drugs. The broadening and depression of the melting transition of Gelucire and the disappearance of the melting peak of crystalline drug has been used an indication of the solubilisation of drug in the lipid and the formation of semi-solid dispersions. DSC has a unique application in studying the interactive behaviour of water molecules in the formulations with the polymer/surfactant in gels and solutions. DSC can allow the identification of non-frozen water (highly interactive water bound with polymer/surfactants at a molecular level), freezable bound water, and free water by detecting the different melting transitions of different type of water. DSC can also been used to study liquid suspensions. High sensitivity DSC was used to study the interactions of saturated fatty acids in alkaline buffer solution. The in solution results confirmed the prediction of the occurrence of interactions observed in the DSC data of the dried fatty acids microspheres.

8.6 Conclusion

Thermal and mechanical analysis are extremely useful characterization methods for studying a highly diverse range of physical chemistry issues associated with drug and formulation development in the pharmaceutical industry. With the continuous development of new thermal analytical methods and thermal methods combined with spectroscopic and imaging methods, a wider range of sample forms and sizes can be studied and the application of thermal analysis in pharmaceutical product development and production is likely to be significantly extended.

References

[1] C. Ahlneck, G. Zografi. (1990) The molecular basis of moisture effects on the physical and chemical stability of drugs in the solid state. Int J Pharm 62:87~95.

[2] V. Andronis, M. Yoshioka, G. Zografi. (1997) Effects of sorbed water on the crystallization of indomethacin from the amorphous state. J Pharm Sci 86:346~351.

[3] Y. Aso, S. Yoshioka, S. Kojima. (2004) Molecular mobility-based estimation of the crystallization rates of amorphous nifedipine and phenobarbital in poly(vinylpyrrolidone) solid dispersions. J Pharm Sci 93:384~391.

[4] P. Badrinarayanan, W. Zheng, Q. Li, et al. (2007) The glass transition temperature versus the fictive temperature. J Non-Cryst Solids 353:2603~2612.

[5] C. Bhugra, R. Shmeis, S. Krill, et al. (2006) Predictions of onset of crystallization from experimental relaxation times I—Correlation of molecular mobility from temperatures above the glass transition to temperatures below the glass transition. Pharm Res 23:2277~2290.

[6] K. L. N. R. Bohmer, C. A. Angell, D. J. Plazek. (1993) Nonexponential relaxations in strong and fragile glass formers. J Chem Phys 99:9.

[7] C. D. Bruce, K. A. Fegely, A. R. Rajabi-Siahboomi, et al. (2010) The influence of heterogeneous nucleation on the surface crystallization of guaifenesin from melt extrudates containing Eudragit® L10055 or Acryl-EZE®. Eur J Pharm Biopharm 75:71~78.

[8] A. Burger, R. Ramberger. (1979) On the polymorphism of pharmaceuticals and other molecular crystals. II. Microchim Acta 72:273~316.

[9] P. Claudy, S. Jabrane, J. M. Le′toffe′. (1997) Annealing of a glycerol glass: enthalpy, fictive temperature and glass transition temperature change with annealing parameters. Thermochim Acta 293:1~11 .

[10] D. Q. M. Craig, M. Barsnes, P. Royall, et al. (2000) An evaluation of the use of modulated temperature DSC as a means of assessing the relaxation behaviour of amorphous lactose. Pharm Res 17:696~700.

[11] A. B. da Fonseca Antunes, B. G. De Geest, C. Vervaet, et al. (2013) Gelucire 44/14 based immediate release formulations for poorly water-soluble drugs. Drug Dev Ind Pharm 39:791~798.

[12] T. Felix, H. Anwar, H. Takeru. (1974) Quantitative analytical method for determination of drugs dispersed in polymers using differential scanning calorimetry. J Pharm Sci 63:427~429.

Chapter 9

Applications of AFM in Pharmaceutical Polymers

Atomic force microscopy (AFM) is a high-resolution imaging technique that uses a small probe (tip and cantilever) to provide topographical information on surfaces in air or in liquid media. By pushing the tip into the surface or by pulling it away, nanomechanical data such as compliance (stiffness, Young's Modulus) or adhesion, respectively, may be obtained and can also be presented visually in the form of maps displayed alongside topography images. This chapter outlines the principles of operation of AFM, describing some of the important imaging modes and then focuses on the use of the technique for pharmaceutical research. Areas include tablet coating and dissolution, crystal growth and polymorphism, particles and fibres, nanomedicine, nanotoxicology, drug-protein and protein-protein interactions, live cells, bacterial biofilms and viruses. Specific examples include mapping of ligand-receptor binding on cell surfaces, studies of protein-protein interactions to provide kinetic information and the potential of AFM to be used as an early diagnostic tool for cancer and other diseases.

9.1 Introduction

AFM continues to find ever wider applications in the fields of materials characterisation and life sciences. By using a small tip to scan across the sample's surface, the requirement to focus light and electrons as with light and electron microscopies is eliminated; this overcomes the Rayleigh criterion resolution limit, enabling nanometre and sometimes atomic resolution imaging, depending on the sample and/or imaging regime, to become routine. AFM can be operated without the need for conducting or stained samples and therefore can be operated in physiological media. Further, the probe can be used to push into or pull away from sample surfaces, yielding quantitative nanomechanical and adhesion data, which can also be displayed graphically. This chapter will discuss the fundamentals of AFM and highlight a few recent applications with relevance to pharmaceutical science.

9.1.1 Background

The atomic force microscope (AFM; scanning force microscope, SFM) is the principal member of a number of related scanning probe microscopes (SPM). The first of these was the scanning tunnelling microscope (STM), invented by Binnig and Rohrer, who received the Nobel Prize for Physics, in 1982 after revealing the first atomic resolution images. By applying a bias potential across a small gap between a sharp metallic tip and a conducting sample, it was possible for electrons to tunnel across the forbidden energy gap; if the probe was simultaneously scanned across the sample, the tunnelling current yielded an image related to the topography (strictly, the local density of states) of the sample. The technique necessitated the use of conducting specimens, with the exception of DNA, proteins and other small molecules, and so was restrictive for the study of thicker biological samples and other insulating materials, such as polymers. To overcome this issue, the imaging method was developed, by the same inventors, to measure forces on a cantilever, to which the tip was mounted, rather than tunnelling currents: this became the AFM instrument. In 1992, SPMs capable of both STM and AFM, became commercially available and very soon afterwards a vast array of derivative techniques and modes, almost solely AFM-based, were developed. AFM has finally come of age, with new imaging modes and related techniques continually coming to market; highlights of these are summarized later.

9.1.2 Principles of Operation

Before discussing the multitude of acquisition modes, some of the fundamentals of AFM will be presented. AFM operates by scanning a small, usually square pyramidal shaped tip, prefabricated onto a cantilever, across a sample, mounted on a stub, which is magnetically held into place on top of a piezoelectric ceramic scanner. Piezoelectric crystals change shape on the application of a voltage, the relationship being roughly proportional and accurate to 0.1 Å; it is this feature that provides AFM with its high resolution capability. The piezos are arranged in the scanner (tripod or the more usual tube design) to provide x, y and z (height) movement, the former two of which allow the scanning motion. The arrangement here describes a sample-scanning rather than tip-scanning instrument, i.e., the sample scans a stationary tip. It is often more convenient, however, to imagine the tip scanning the sample, as will be discussed here.

The cantilever is often coated with a gold layer to make it reflective, so that a laser can be reflected from the cantilever onto a photodectector, via a mirror (Fig.9.1). This laser assembly serves as a tracking system so that the position of the tip in relation to the surface can be continuously monitored. In contact mode, the simplest of operations, the probe (tip + cantilever) is brought into contact with the sample until a small deflection of the cantilever, corresponding to a repulsive force (the set-point), is detected via a displacement on the photodetector. The probe is then scanned across the surface causing the cantilever to move vertically corresponding to variations in topography (height) of the sample. As the probe begins to go over a high feature on the specimen, for example, the cantilever will start to deflect more (bending will increase) causing the laser spot on the photodetector to move. So as not to cause damage to the sample, and/or tip, a voltage signal from the photodetector is sent to the scanner, to which the sample is mounted, causing the scanner to retract. This then lowers the sample, relieving the increased deflection (above the set-point) on the cantilever. Upward movement of the piezo, and sample, is also allowed when, for example, the probe moves over a low feature, such as a pit. This cycle is repeated as the sample is scanned, so that the deflection of the cantilever is kept constant by means of this feedback loop. The potential required to maintain the set-point via the feedback circuit is used, alongside the *x* and *y* coordinates, to create the contact mode AFM image.

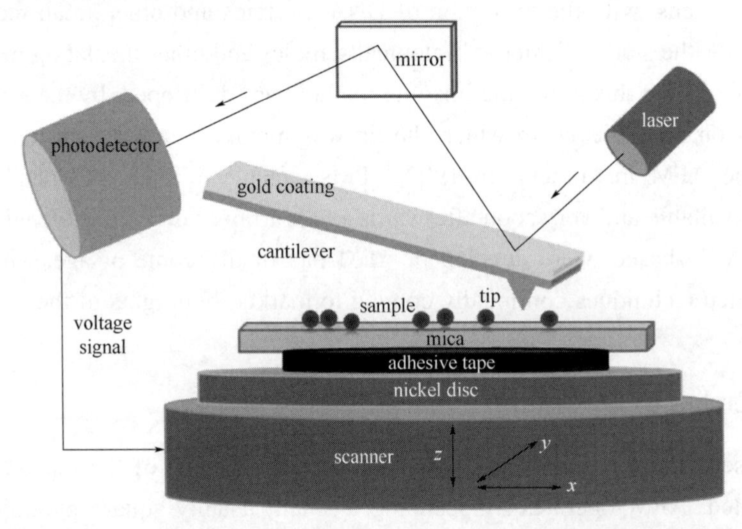

Fig.9.1 A schematic of the main components of a typical AFM, with particles on a mica surface.

9.1.3 Operating Modes

There are three operating modes in AFM: contact mode, Tapping Mode® (a trade mark of Bruker, and also known as intermittent contact mode) and non-contact mode. The last two modes are referred to as 'AC' techniques while contact mode is referred to as 'DC'. Each has their own advantages and disadvantages. Contact mode is the simplest, where the cantilever moves along the surface maintaining a set force. If too much force or too little force is detected, the height will be adjusted using the piezoelectric motors through the feedback loop. Tapping Mode® is probably as widely encountered as contact mode. Here, to minimise potentially damaging lateral forces being exerted on the sample, the cantilever is oscillated at its resonant frequency during scanning; the measurement is made only at the moment when the tip touches the specimen, and for the most part, while the scanner moves the sample, the tip is safely retracted. Tapping Mode® is often used for imaging delicate biological samples, such as DNA, and this can also be carried out in physiological media.

With most commercial AFM instruments, different image channels can be acquired at the same time as the topography (height) image. For example, in contact mode, the error signal (raw signal detected by the photodetector) and the friction response (or lateral force microscope, LFM image, derived from the horizontal component of the photodetector's signal as the cantilever twists during scanning) may be obtained. Similarly, in Tapping Mode®, in addition to the error signal, the phase image (not to be confused with other definitions of this term in microscopy) and amplitude image may be obtained. These respectively correspond to the time (phase) lag and amplitude of the driven tapping sine wave signal relative to the response wave after interacting with sample. For example, variations in viscoelastic and/or adhesion properties of the sample across the surface may modify the phase and/or amplitude signals. These modes often give more defined visual contrast than their simultaneously acquired topography images, although the exact mechanism behind their complex origin is poorly understood.

In addition to imaging, force data can be obtained by pushing the tip into or pulling the tip away from a sample's surface; this is often referred to as force spectroscopy. This can be explained with reference to a force *vs*. distance plot (force curve; Fig.9.2), where a tip approaches a surface, typically experiences a short-range attraction event and then is pushed into the

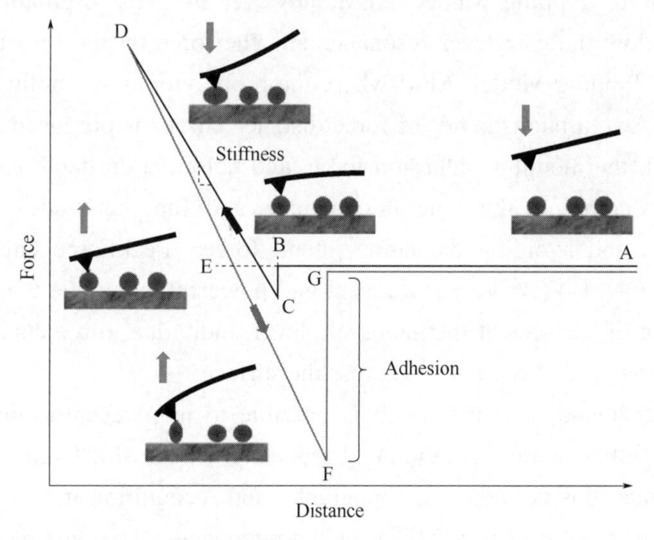

Fig.9.2 A schematic showing the some key features of an AFM force *vs*. distance plot: Upper line: approach curve, lower line: retract curve; regions (A-B) tip approaches the surface, (B-C) tip-surface attraction, (C-D) tip driven into the surface combined with cantilever bending (where stiffness/compliance information may be extracted), (D) point at which applied load is withdrawn and tip pulled away from the surface, (E) no net attractive of repulsive forces between tip and sample, (E-F) tip is adhered to sample, (F) applied load overcomes adhesion force and tip released from surface (maximum adhesion force), (F-G) adhesion force, (G) tip away from surface, (G-A) continued withdrawal of the tip.

surface. The gradient of the force curve in this region is a combination of indentation of the tip into the sample and cantilever bending, and with a suitable hard reference surface and use of mathematical models, quantitative nanoindentation (Young's modulus) data can be extracted.

After a user-defined selected total force (or deflection), the applied load on the cantilever can be reduced, and the tip may eventually become adhered to the surface. Overcoming this force leads to an adhesion event, the importance of which becomes significant when the tip is suitably chemically derivatised or biologically functionalised. Typically, only one-half of the force curve (an approach or retract curve) is relevant for any given experiment.

Nowadays, force measurements are combined with the scanning capability to produce local physical property maps, e.g., indentation, Young's modulus, adhesion and friction. These can be quantitative or qualitative depending on the extent of pre-calibration. Instrument manufacturers tend to have their own trade mark names for these modes, such as Torsional Resonance (TR) Mode® and PeakForce Tapping® (PFT).

In the TR mode, the tip is parallel rather than vertical to the surface, and the forces between the tip and the sample cause a change in resonance behaviour that can be used to track the surface at a constant distance. The TR mode has the advantage that the tip remains at a constant distance from the surface at all times.

The dynamics of the tip-cantilever system was investigated, when it is operated in TR mode, with or without tip-sample interaction, and they also described the basic methodology to extract in-plane surface properties in TR mode.

PFT is similar to Tapping Mode® AFM, however the PFT oscillation is performed at frequencies well below the cantilever resonance and the force on the tip can be kept constant which differs from Tapping Mode® AFM where the probe vibration amplitude is controlled by the feedback loop. A continuous series of force-distance curves is produced and by keeping the peak force constant, the modulus, adhesion force, and deformation depth can be calculated. In PFT, the oscillation combines the benefits of contact and Tapping Mode® imaging by having direct force control and avoiding damaging lateral forces. PeakForce Quantitative Nanomechanical Mapping (QNM®) (Bruker) is a recent and powerful technique that provides quantitative characterisation of surfaces at the nanoscale level. Individual force curves can be acquired and quantitatively analysed as the tip taps across the surface.

By attaching a ligand onto the tip, it is possible to map receptor-binding sites on cell/substrate surfaces, with a lateral resolution of a few nanometres, whilst simultaneously acquiring the topography image; this is known as topography and recognition imaging (TREC) mode or single-molecule force spectroscopy (SMFS), a single-molecule interaction method.

Outside the scope of this short chapter lie a whole host of related probe techniques (not strictly modes). These, as their names suggest, are able to obtain other local physical properties; they include electrochemical AFM (EC-AFM), magnetic force microscopy (MFM), electrostatic force microscopy (EFM), kelvin probe force microscopy (KPFM), photoconductive AFM (PC-AFM), scanning spreading resistance microscopy (SSRM), scanning thermal microscopy (SThM), scanning capacitance microscopy (SCM) and surface potential microscopy (SPoM).

Some of the latest developments include fast scanning, where the typical 1 Hz scan rate (ca. 10 min/scan) can be rapidly increased, and in some cases, for flat samples, to real-time acquisition. This is achieved through the use of small cantilevers.

9.1.4 Cantilevers and Tips

A wide variety of cantilevers and tips (often jointly named probes, although terms are used interchangeably) are available depending on their intended application. Broadly, contact mode probes tend to be made from silicon nitride, a hard material of approximate stoichiometry Si_3N_4, whereas Tapping Mode® probes are normally formed from silicon, owing to its stiffness, although Si_3N_4 is also sometimes used.

The shapes of the cantilevers also vary, for example, V-shaped (often used for contact mode), beam-shaped (rectangular) and arrow shaped. The cantilever's spring constant (k, $N \cdot m^{-1}$), a proportionality constant used to relate cantilever deflection to force, can be determined from the dimensions (typically, 10~450 μm in length and various widths) and material properties of the cantilever. There are numerous methods for measuring k, such as the thermal noise method, described elsewhere. Accurate determination of this property, for which manufacturers provide approximate ranges, is essential for all AFM studies where force measurements are required. The tip itself can be a square pyramid (approximating at its apex to a small sphere, the radius R of which may need to be determined for some force measurements), an oxide-sharpened tip (small R), a high-aspect ratio tip or a colloid probe. The latter type includes small entities, such as cells or polymer spheres that can be adhered to a tip-less cantilever to bestow biological functionality or defined contact geometries. Probes can also be coated on either side for increased laser reflection (top side: Au), chemical functionalisation (underside: Au) or to change physical properties (underside: Pt, diamond-like coatings, magnetic Co alloys). Chemically modified probes are typically prepared by grafting thiol self-assembled monolayers (SAMs) on to Au coated probes or by silanising the surface OH groups of Si/Si_3N_4 probes, which are then used to measure specific short-range intermolecular (usually adhesion) forces. This is often termed chemical force microscopy (CFM). At the earliest, functionalised AFM tips and surfaces were first used to measure forces between biotin-avidin ligand-receptor pairs; this area has now been extended to investigate many interacting pairs.

Many different approaches have been used to attach (bio)molecules to AFM probes. For example, AFM was used to investigate protein-protein (the core methylase and the HsdR subunit) interactions within the EcoR124I molecular motor. To do this, Researchers functionalised a tip with a glutathione S- transferase-HsdR complex via a NHS-PEG-MAL linker and probed the core methylase (Fig.9.3). The NHS end of the PEG linker reacts with amines on an amino-terminated, silanised Si_3N_4 tip, forming a stable amide bond, while the MAL group forms a C—S linkage with the protein. Another group used a similar linker to attach the drug rituximab to an AFM tip.

An "easy-to-use test system" for investigating single ligand-detectors with AFM was developed, demonstrating this using the avidin-biotin interaction where the AFM tip was

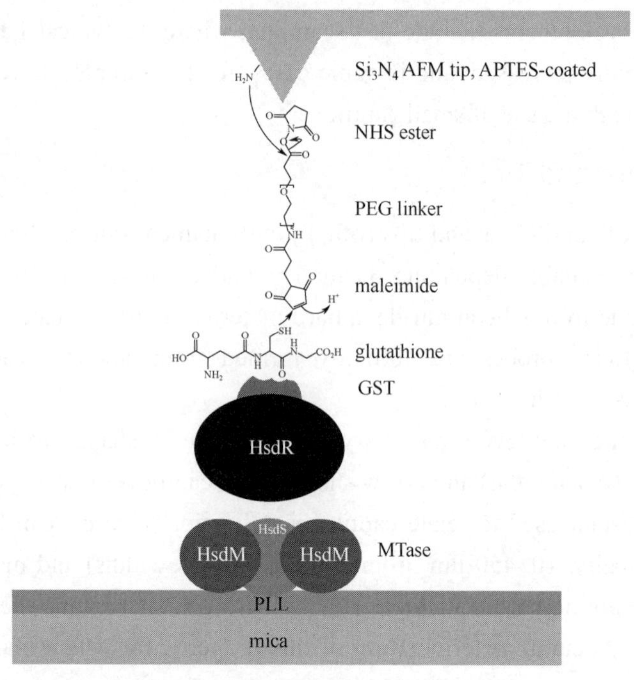

Fig.9.3 Surface chemistry used to immobilise proteins to a Si_3N_4 AFM tip and a mica surface to study protein-protein interactions, in this case different subcomponents (HsdR/HsdS/HsdM) of a molecular motor (type I restriction-modification enzyme EcoR124I). APTES = (3-aminopropyl) triethoxysilane (to form surface NH_2 groups), PLL= poly(L-lysine) (a polycation to couple proteins to negatively charged mica), GST= glutathione S-transferase. Reproduced with permission.

biotinylated via a 6 nm PEG linker. Some researchers functionalised a tip with an antibody via an 8 nm PEG linker and an amino-functionalised Si_3N_4. And another research team added a polystyrene microbead to a cantilever to investigate sticking efficiency in colloid systems; spheres were simply glued to the cantilevers with UHU 5-min epoxy glue. The procedure was carried out by placing a tip-less Si_3N_4 cantilever in the AFM apparatus, using the piezoelectric actuators to drive the cantilever into the adhesive and then onto a glass slide covered with microspheres and picking up a single microsphere.

9.1.5 The Need for AFM in Pharmaceutical Research

The potential of AFM for structural biological studies was recognised since the conception of the technique. Indeed, it was the inability of tunnelling currents in STM to penetrate through the thickness of biological specimens greater than the diameter of proteins that led to the realisation of AFM. Nowadays, with the multitude of operating modes and techniques available, including the ability for real-time imaging in physiological buffer, many questions of interest to those engaged in pharmaceutical research can be answered. For example, ligand-receptor binding on cell surfaces can be measured and mapped, protein-protein interactions can be studied to provide kinetic information and AFM shows potential for use as an early diagnostic tool for cancer and other diseases.

9.2 Use of AFM in Pharmaceutical Sciences

9.2.1 Tablet Coating and Dissolution

Pharmaceutical solid dosage forms are usually coated to control drug release, to protect active pharmaceutical ingredients (APIs) from degradation in the stomach or in humid atmospheres, to provide a barrier to taste and smell, and for controlling dissolution. AFM has been used mainly to acquire topography information, where it has the advantage over SEM in that it can provide quantitative surface roughness and surface area measurements and allow for their study in realtime; data on compositional distribution and porosity may also be discerned. Scientists investigated tablet surfaces using different imaging and roughness techniques, including AFM, and concluded that KCl tablets were smoother than NaCl tablets. A summary of surface roughness parameters that can be readily obtained from AFM topography profiles.

The coating materials of tablets are frequently studied as free films so that the effects of the tablet core can be eliminated. AFM imaging was used to examine the surfaces of pharmaceutical tablets that were coated with different aqueous hydroxypropyl methylcellulose films. AFM has been used to assess the quality of montmorillonite/poly(styrene)/poly(butyl acrylate) films prepared using dispersion methods; the influence of composition on the form and arrangement of polymer droplets, and also the uniformity of the polymer film surface on the tablets were investigated. AFM can also be used to investigate areas on surfaces that have different properties, such as crystallinity and chemical composition; these are important parameters concerning the dissolution of a tablet.

The mechanisms and dissolution rates of the cholesterol monohydrate (001) surface, of relevance to the removal of gallstones, were investigated. The dissolution rate was found to be closely related to local variations in topography. the dissolution rates of the (001) and (100) planes of aspirin crystals (0.45 and 2.93 $nm \cdot s^{-1}$, respectively) in 0.05 $mol \cdot L^{-1}$ HCl were measured. The (001) crystal plane dissolved by receding step edges, whilst the (100) surface showed crystal terrace sinking. Such studies are important as in vitro crystal dissolution is proportional to in vivo drug absorption.

9.2.2 Crystal Growth and Polymorphism

Drug crystal growth, particle characterisation and tablet coatings are critical elements in the manufacture of solid dosage forms. Thus, microscopic examination is important for the design and evaluation of a pharmaceutical product after the steps in the drug formulation process have been taken.

AFM was first introduced into crystal growth studies by Durbin and Carlson, who detected the growth of steps on the surfaces of lysozyme crystals. AFM was used for monitoring the crystal surface and they reported the step velocity of egg-white lysozyme crystal planes on a nanometre-scale using a sealed vessel in the AFM and noted the consequences of controlling the supersaturation. Then An in situ AFM was used for investigating the growth and activity of dislocation sources as a function of supersaturation during canavalin crystal growth. They reported that growth occurs on monomolecular steps generated either by simple or complex screw dislocation sources, and also visualised 2D nucleating islands that form onto the surface before spreading laterally as step bunches. AFM was also used to investigate the topography of hydroxyapatite single crystals that had been synthesised from hydrothermal solution, and reported that growth proceeded through a layer-by-layer mechanism. And the crystal growth rates of nifedipine at the surface of amorphous solids with and without polymers using AFM were determined; It has been found the technique to be useful for studying the crystallisation kinetics of amorphous solids by targeting the crystals at the surface. AFM was also utilised to assess the growth on the (001) face of aspirin crystals at two supersaturations, elucidating both the growth mechanisms and kinetics at each supersaturation.

Then the capability of AFM to follow the structural transformations of crystals that can occur in unstable pharmaceutical compounds was assessed. Other scientists used AFM to observe the phase changes at crystal surfaces where the transformation is supplemented by changes in the spacing between layers of molecules, and analysed the thermodynamically stable form of the caffeine-glutaric acid cocrystal continuously in situ using intermittent-contact mode AFM.

Different polymorphs have different physicochemical properties, which could affect the solubility, dissolution and stability, and therefore polymorphic characterisation is an important parameter in pharmaceutical industry. AFM can be applied both in situ and ex situ to study the growth of crystals from solution, and in particular for investigating the crystallisation of proteins, nucleic acids and viruses. Some scientists used AFM to identify the polymorphic forms of insulin and other mapped the distribution of polymorphs on the drug cimetidine. Another group (DL) have investigated the nanocrystalline growth of dibenz[a,c]anthracene using a FastScan AFM (Fig.9.4).

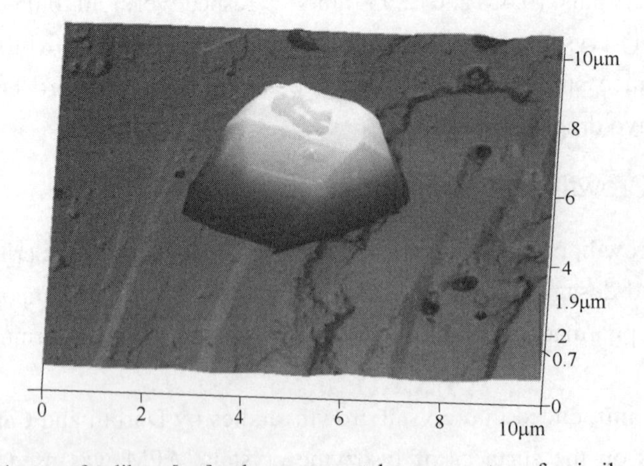

Fig.9.4 AFM image of a dibenz[a,c]anthracene crystal grown on top of a similar structure crystal.

9.2.3 Particles and Fibers

AFM offers particular advantages over TEM and SEM for the characterization of particles and fibers, such as height measurement, minimal sample preparation, the ability to operate under atmospheric pressure and in liquids, and the acquisition of nanomechanical/adhesion data. These advantages facilitate the study of loaded and empty delivery systems.

AFM is an excellent technique for visualizing particles with sizes ranging from 1 nm to 10 μm, allowing quantitative particle size measurements. These are not prone to problems experienced when using SEM, such as conducting coating thickness, astigmatism, penetration depth and absence of height information.

SEM and dynamic light scattering (DLS) may also cause slight deformations of soft particles, such as liposomes (Fig.9.5). Since x, y and z distances may be recorded using AFM, parameters such as diameter, volume and surface area can be calculated. AFM is one of the most important techniques for the characterisation of lipid drug delivery systems, which have been successfully used as drug carriers for the treatment of many cancers.

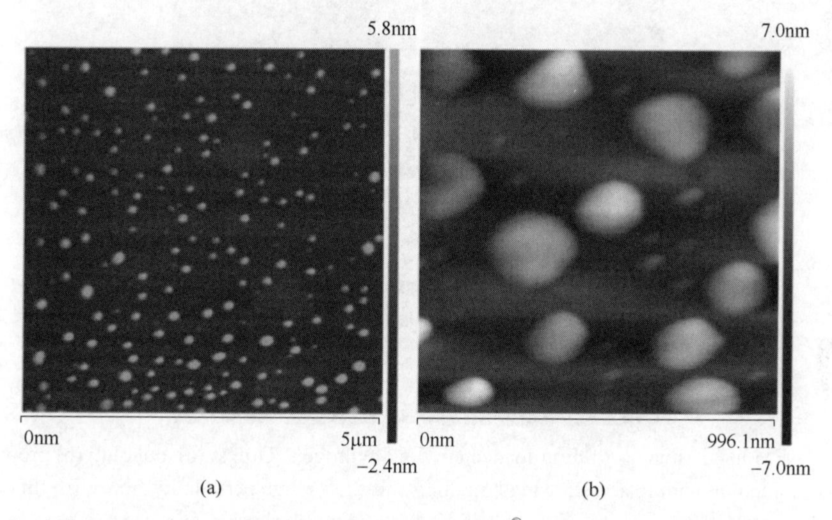

Fig.9.5 AFM topography images (using PeakForce QNM® mode in air) of cholesterol stabilised, dipalmitoylphosphatidyl choline (DPPC) Stealth liposomes. (a) unloaded, (b) sirolimus-loaded.

Recently, AFM has been used to study the scale-up and shelf-stability of curcumin-encapsulated poly(lactic acid-co-glycolic acid) (PLGA) nanoparticles (NPs), which were found to be stable for periods up to 6 months; the particles were spherical and had smooth surfaces. AFM has been also used to characterise Au NPs with sizes 25 nm, 55 nm and 90 nm, for investigating the stability of the naked, PEGylated, and Pt-conjugated NPs as a function of time under various conditions.

AFM colloid probe techniques were developed to mount a 1~3 μm spherical polycrystalline drug particle on a cantilever to measure adhesion to an α-lactose monohydrate layer for developing formulations for dry powder inhalers (DPI). While A similar arrangement was used to investigate adhesive and cohesive force characteristics of DPI systems containing budesonide or salbutamol sulphate to α-lactose monohydrate.

AFM has recently been used for the analysis of amyloid fibrils, an important research area in diseases such as Parkinson's, Alzheimer's, and type Ⅱ diabetes. For example, drug-loaded lysozyme amyloid hydrogels, prepared by misfolding lysozyme in the presence and absence of drugs, such as atenolol, propranolol hydrochloride or timolol tartrat (Fig.9.6), were analysis by AFM. Different amyloid fibre structures were formed depending on the type of drug used. Therefore, AFM can contribute to research of amyloid fibrils by providing important information concerning fibril structure and fibrillation processes, and also to analyse some important properties of amyloid fibrils, such as their strength and Young's modulus. The contour length is also a very useful structural parameter of amyloid fibres that can be determined from AFM imaging; the property can be used to interpret the cellular response to the presence of amyloid fibrils of different sizes after fragmentation. Shorter fibrils have been found to have enhanced cytotoxicity compared to longer fibrils.

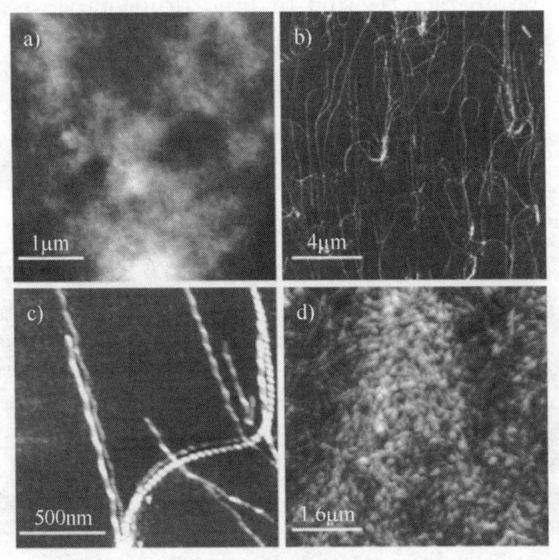

Fig.9.6 AFM height images of drug-loaded amyloid hydrogels. Drugs: (a) atenolol; (b) propranolol; (c) propranolol, with image showing both long and short range periodicity fibres; (d) timolol.

9.2.4 Nanomedicine

Nanomedicine is an interdisciplinary field encompassing the detection, prevention and treatment of diseases at the nanometre scale, which includes the longer terms goals of producing personalised medicines. Materials used in this application area include polymer coatings and nanoscale drug delivery devices, respectively.

AFM phase imaging, in conjunction with TEM and other techniques, has been used to characterise PEGylated lipoplexes for siRNA drug delivery. Post-PEGylation was found to yield improved homogeneity with regards to PEG coverage. Tapping Mode® AFM was used to assess the morphology, surface roughness and cytocompatibility of conducting polymers as nano-coatings for tissue regeneration for cardiovascular implants.

The size and morphology of liposomes for transporting boronated compounds for use in boron neutron capture therapy (BNCT) for targeted cancer treatments have been investigated by Tapping Mode® AFM. Carbon nanotubes show promise as drug delivery devices owing to their dimensions, biocompatibility and ease of chemical functionalisation. Tapping Mode® AFM has been used to show lipids wrapped around single-walled carbon nanotubes to increase their dispersion in aqueous media. Adhesion forces between hematite NPs and *E. coli* immobilised onto a tip-less cantilever have been measured in phosphate buffer solution. This fundamental study investigated the interaction forces and contact mechanics of the system, and a new model to describe the interaction was devised. And the use of PeakForce QNM® was investigated for improving the developments in the field of nanomedicines, by measuring the effect of particles into various tissues (e.g., liver, kidney and small intestines). They also described how this detailed imaging approach may also help scientists address growing concerns in nanotoxicology.

9.2.5 Nanotoxicology

Nanotoxicology is a relatively new field, developed to study the toxicological effects of NPs (natural or engineered)/nanomaterials in the environment, which can differ markedly from their bulk materials due to their small particle size and large surface area. With the rapidly increasing use of nanomaterials, currently over 1000 commercial products, there is an urgency to determine their toxicity and to control exposure. There are a number of in vitro techniques that can be used for testing, such as proliferation assays, reactive oxygen species (ROS) generation analysis, flow cytometry, DNA damaging potential assays, gene expression analysis, genotoxicity and microscopic evaluation, including SEM/EDAX, TEM, fluorescence microscopy, MRI and AFM. For example, AFM was used to investigate the persistence of citrate-capped Ag NPs (20 nm) in natural freshwaters and synthetic aquatic media; single-walled carbon nanohorns and SiO_2 NPs buried in cells were observed by using various AFM oscillation techniques; the morphology and particle size of TiO_2 NPs, which are being increasingly used in catalysis and as a pigment, were characterized; the nanotoxicological effects of graphene on human plasma were studied, which found low molecular weight proteins to have a high affinity for the nanomaterial; parallel investigations found decreased nuclei numbers and increased ROS after prolonged incubation with Hela and Panc-1 cell lines.

9.2.6 Drug-Protein and Protein-Protein Interactions

The study of drug-protein and protein-protein interactions are key topics in pharmaceutical sciences and are critical for targeting drug delivery. Proteins can be used as adapters conjugated to NPs (chitosan, Au, liposomes, silica, self-assembly) for targeting drug delivery. Furthermore, the interaction between protein molecules with drug carriers and cell surfaces is crucial, since cell adhesion to surfaces depends on the availability of specific protein-binding sites. Protein-material interactions also play a significant role in biosensors as a diagnostic tool since ligands can be immobilised on a probe surface and used to analyse the corresponding integrin. In addition, proteins encounter a wide range of surfaces during processing, each of which has the potential to affect their structure if adsorption takes place. If a loss of structure takes place upon surface adsorption, or even a small change in the native fold, subsequent protein-protein interactions may occur, resulting in the formation of aggregates and thus potentially an immunogenic response. Scientists have identified challenges that need to be addressed when AFM adhesion measurements are to be used to study single drug particles interacting with proteins or cells. Some researchers studied the interactions between monoclonal antibodies and albumin to surfaces of varying functionality and hydrophobicity by monitoring the adhesion over time and found a two-step interaction process involving an initial, rapid perturbation of the protein surface on contact with the surface, followed by relaxation and unfolding. The nanomechanical and adhesion properties of 2S albumin and 12S globulin were investigated. Differences in tip-protein interaction strength with regard to the nature of the protein and pH of the aqueous environment in terms of protein unfolding were observed. Protein-protein interactions have been also studied, where B-cell/CD80 was immobilised on an AFM tip and T-cell/CD28 was immobilised to a surface and forces were measured before and after adding cynarine.

9.2.7 Live Cells

The ability to image, probe ligand-receptor interactions and obtain nanomechanical information, all in physiological media, makes AFM particularly suited for studying cells in a therapeutic context. For example, scientists have obtained contact and intermittent contact mode images of the surfaces of red blood cells (RBCs) that had been treated in vitro with various agents (hemin, furosemide, chlorpromazine and zinc ions) to investigate blood intoxication. Images were filtered using a Fourier transform algorithm to detect poorly seen structures on RBC membranes. The study showed that blood intoxication affected the nanostructures present on the RBC membrane surfaces.

Surface roughness has been suggested to provide a diagnostic measure of the health state of cells. Widespread application of this method has been limited by scan-size dependence on surface roughness.

SMFS was used to map CD20 molecules on surfaces of cancer B cells (fluorescently labelled; mapped area 500×500 nm^2) obtained from patients with B-cell non-Hodgkin's lymphoma (NHL; marginal zone lymphoma; Fig.9.7). CD20 can be targeted therapeutically with monoclonal antibodies (mAb), such as rituximab. These antibodies were covalently linked to an AFM tip via silanisation and a PEG linker; the density was such that only one CD20-rituximab complex was formed per force curve. RBCs, which do not express CD20, were used as controls. These studies are useful in understanding mechanisms, and developing new anti-CD20 mAbs especially where rituximab resistance becomes a problem.

Numerous nanoindentation studies of cells have appeared in the literature, many describing differences in elastic behaviour between cancer cells and benign equivalents, where the Young's modulus of the affected cells is often cited as being ca. 70% less stiff than non-cancerous cells. The differences are usually attributed to rearrangements in the cytoskeleton network and have been suggested to be of early diagnostic value.

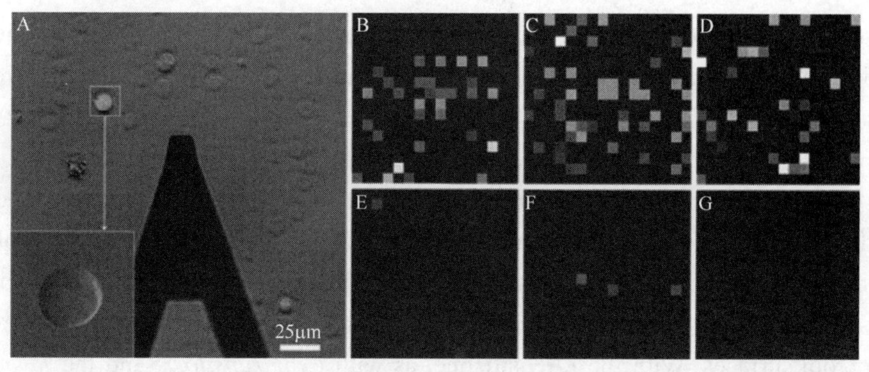

Fig.9.7 Measuring specific CD20-rituximab interactions on cancer cells using a rituximab functionalised tip. (A) Bone marrow cancer (inset) and cells fluorescently labelled for easy identification; (B)~(D) CD20 distribution map (xy: 500×500 nm^2, z: 0~100 pN, 16×6 pixels); (E)~(G) CD20 distribution map after blocking (same scale).

A combination of AFM cell stiffness measurements and cell-substrate adhesion studies, together with calcium imaging and migration studies, were used to investigate prostate cancer metastasis. The Young's modulus was found to be larger for more metastatic cells, in contrast to but also in keeping with other studies, suggesting that mechanical studies cannot be used solely as a biomarker for metastasis. Indeed, a complicated relationship is found with adhesion and calcium dynamics also playing important key roles.

Quantitative Imaging® (JPK Instruments), a rapid tip modulation technique that acquires nanomechanical/adhesion data simultaneously with topography, was used to investigate a number of different cell types, some of which had very little adhesion to the substrate. Force-volume elasticity and adhesion maps could be obtained rapidly and at high resolution for weakly adhered cells.

PeakForce Tapping® (Bruker), similar to Quantitative Imaging® for rapidly obtaining nanomechanical data at high resolution, was used to study the increase in stiffness of HaCaT keratinocytes (as a model for skin cancer) brought about through exposure to the herbicide glyphosate. A concentration dependence was observed and the addition of quercetin, a common flavonoid considered to provide protection against oxidative injury and inflammation, reversed this process.

9.2.8 Bacteria and Bacterial Biofilms

There are numerous reports on the use of AFM for investigating bacteria and bacterial biofilms. For example, Some researchers investigated the adhesion of pathogenic microorganisms, *Candida parapsilosis* and *Pseudomonas aeruginosa* (chosen for their clinical relevance), to biomaterials including the *P. aeruginosa* biofilm to unmodified silicone rubber. The attractive force between *C. parapsilosis*, adhered to a probe tip, and a bare silicon substrate was 4.3×0.25 nN, comparable to the smaller attractive forces between *C. parapsilosis* and the *P. aeruginosa* biofilm (2.0×0.4 nN), although the tip experienced repulsive forces 75 nm away from the biofilm surface (2.0 nN). The magnitude of the attractive forces, both towards silicone rubber and the biofilm, led the authors to suggest that they may allow adhesion and colonisation of these surfaces which, in a clinical setting, would increase the risk of death and disease.

The adhesion of *E. coli* to modified silicones using SEM and AFM was investigated. The aim was to find a bacteria-resistant surface by varying hydrophobicity through modification. Octadecyltrichlorosilane (OTS) and fluoroalkylsilane (FAS) were tested against hydrophilic mica, which acted as a bacteria-adhesive control surface. Adhesion was investigated by adhering *E. coli* cells to the AFM probe tip and force measurements were taken from the approach to and retraction from surfaces. With the FAS silicone, as with *C. parapsilosis* and the *P. aeruginosa* biofilm, a repulsive force was observed at close proximity and the force required to remove the tip was low. In contrast, however, when the tip was coated with heparin, the attractive forces to FAS were high on both approach and retraction. This led the authors to conclude that hydrophobicity of a material alone is not enough to predict bacterial adhesion.

'Microbead Force Spectroscopy (MBFS)' was used to investigate the viscoelasticity of biofilms. A glass bead coated with biofilm was attached to a tip-less cantilever and used as a probe against a flat glass surface. The properties of a *P. aeruginosa* wild-type biofilm were compared with a lipopolysaccharide (LPS)- deficient mutant strain, wapR. Biofilms at different levels of maturity were also compared. When immature, the wapR strain adhered to the glass slide with far more force than the immature wild-type (ranges of 2~13 nN and 0~3 nN, respectively). As they matured, adhesion forces decreased in both strains. AFM has also been used for investigating real-time visualisation of the antibiotic azithromycin with various lipid domains in solution.

9.2.9 Viruses

Viruses consist of an oligomeric protein head, called a capsid, which contains the viral geonome; in more complex virus structures, the capsid may also contain other macromolecules, such as proteins and molecular motors. AFM was first used to image viruses in 1992, almost immediately after the instruments became commercially available, although these studies were mostly confined to exploiting their welldefined geometry to measure AFM tip radii. An interest in the biological structure of viruses, however, soon followed, with contact mode imaging of head and tail components of bacteriophage T4. Tapping Mode® in liquid rapidly became the mode of preference, due to reduced lateral forces and a more relevant imaging environment although contact mode is still widely used. Topography imaging has been used to distinguish capsomers, determine the triangulation number (T) of capsids, revealing exposed nucleic acids subsequent to capsid stripping and to observe viral budding from living fibroblast cells. Treatment of pinostroin, an antiviral, was shown to cause severe disruption to HSV-1 virus morphology, as evidenced from topography and phase imaging.

In addition to imaging, force *vs.* distance curves, both tip approach (indentation) and tip retract (adhesion) cycles, have been acquired from immobilised viruses. Head, collar and tail regions of $\phi 29$ phage virions were found to exhibit different elasticity values. A decrease in

Young's modulus (stiffness) was reported for *E. coli* after infection with filamentous phage M13. Concerning adhesion events, spatially-resolved force mapping was used to examine single influenza virus particles (glycoprotein hemagglutinin, HA, surface) using an anti-HA derivatised tip.

Recent developments in high-speed AFM imaging at high resolution offer even greater opportunities for studying dynamic virus-cell interactions.

9.3 AFM Combined With Optical or Spectroscopic Techniques

In the last couple of years, instruments that combine AFM with spectroscopic techniques, such as IR and Raman, and/or improved light microscopy for studying cells conveniently in physiological medium have been made commercially available. These instruments offer considerable potential for in vitro studies, where high-resolution imaging can be combined with biochemical measurements. A few examples are provided here.

It has been showed there to be excellent agreement between conventional IR spectroscopy with nanoscale combined AFM-IR data from polymer samples. By combining AFM imaging with mid-IR spectroscopy, some scientists have examined the micro- and nanostructure and chemical phase composition of felodipine/poly(acrylic acid) blends. Oil inclusions in streptomyces, without the need for staining, were mapped using a combined AFM-IR instrument. A high-speed AFM for nano/mesoscale analysis of living cell surfaces (HeLa and 3T3 fibroblasts) has been combined with fluorescence microscopy. High-resolution fluorescence imaging (super resolution stimulated emission depletion, STED) has been combined with nanomechanical (stiffness) mapping to investigate COS-7 cells with immunolabelled microtubuli. Cell biological aspects such as dynamics of mitochondrial movement and drug uptake have also been investigated.

9.4 Summary

AFM is a high-resolution imaging technique that can be used to study a variety of samples under physiologically relevant conditions. The ability to chemically or biologically functionalise AFM probes combined with various modes that permit the acquisition of spatially-resolved force data allows for the study of systems at the single-molecule level. AFM technology, now frequently using high-speed acquisition, interfaced with spectroscopic modes and light microscopy methods for live cell imaging offer huge potential for the pharmaceutical sciences.

Appendix: Obtaining an AFM Contact Mode Image in Air

As a practical demonstration, this section outlines a typical sequence of the steps necessary for the acquisition of the simplest of AFM operations: a contact mode image to be obtained in air. Most of the details that make the sequence particularly relevant for a specific instrument have been excluded deliberately. Bacteria on a mica surface has been chosen as an example. Mica is an ideal substrate for many AFM studies since it is atomically flat (glass coverslips can appear quite rough for many high-resolution studies); fresh, uncontaminated surfaces can be also prepared, without the need for cleaning, by simply attaching adhesive tape and peeling away the top layer from this layered material. Mica is negatively charged and so improved adhesion to often negatively charged biological specimens, such as DNA, can be achieved by derivatising the mica surface with a suitable polycation, e.g., poly-L-lysine.

1. Turn on the AFM instrument and computer, and open the software.

2. Place a piece of mica (1 cm^2, cut with scissors) on a nickel stub (1.2 cm^2) using double-sided adhesive tape. Press it on firmly.

3. Cleave the mica with adhesive tape. Derivatise the mica, if required.

4. Add an aliquot (10 μL) of the solution containing bacteria to the mica surface. Leave the drop of solution in place for 2 min.

5. Carefully rinse the treated mica plate with distilled water to remove buffer salts, which might mask any biological sample features.

6. Allow to air-dry or carefully use a jet of nitrogen gas.

7. Place the sample on top of the AFM scanner; the magnet will hold the nickel disc of the sample in place.

8. Select a contact mode probe (of low spring constant k, ca. 0.06 $N \cdot m^{-1}$) and fix into the AFM head above the sample.

9. Line up the laser (according to manufacturer's instructions).

10. Move the sample and/or probe to select imaging region of interest.

11. Select a required scan range (say, 20 μm) and set the scan rate to 1 Hz. Use an image resolution size of at least 512×512 pixels. Select the integral, proportional and derivative (PID, external scanner feedback) settings outlined by the manufacturer (these will depend mostly on the scanner being used and whether air or liquid is the medium).

12. Lower the probe to just above the sample surface and use the automated approach.

13. Slowly increase the PID settings to maximise image contrast to just below the level that produces noise (piezo ringing). It should also be possible to reduce the applied load (reduce deflection) on the cantilever to improve image quality.

14. Once image settings are optimised, obtain a complete image and save (capture) it.

15. The next typical options will either be to zoom in, move to a different area or change the sample.

References

[1] R. S. Abendan, J. A. Swift. (2005) Dissolution on cholesterol monohydrate single-crystal surfaces monitored by in situ atomic force microscopy. Cryst Growth Des 5:2146~2153.

[2] J. Adamcik, R. Mezzenga. (2012) Study of amyloid fibrils via atomic force microscopy. Curr Opin Colloid Interface Sci 17:369~379.

[3] T. Ando, T. Uchihashi, N. Kodera, et al. (2008)High-speed AFM and nano-visualization of biomolecular processes. Pflugers Arch 456:211~225.

[4] P. D. Antonio, M. Lasalvia, G. Perna, V. Capozzi. (2012) Scale-independent roughness value of cell membranes studied by means of AFM technique. Biochim Biophys Acta 1818:3141~3148.

[5] S. Arora, J. M. Rajwade, K. M. Paknikar. (2012) Nanotoxicology and in vitro studies: the need of the hour. Toxicol Appl Pharmacol 258:151~167.

[6] M. Baclayon, G. J. L. Wuite, W. H. Roos. (2010) Imaging and manipulation of single viruses by atomic force microscopy. Soft Matter 6:5273~5285.

[7] L. Bastatas, D. Martinez-Martin, J. Matthews, et al. (2012) AFM nano-mechanics and calcium dynamics of prostate cancer cells with distinct metastatic potential. Biochim Biophys Acta 1820:1111~1120.

[8] P. Begat, D. A. V. Morton, J. N. Staniforth, R. Price. (2004) The cohesive-adhesive balances in dry powder inhaler formulations I: direct quantification by atomic force microscopy. Pharm Res 21:1591~1597.

[9] D. Belletti, M. Tonelli, F. Forni, et al. (2013) AFM and TEM characterization of siRNAs lipoplexes: a combinatory tools to predict the efficacy of complexation. Colloid Surface Physicochem Eng Aspect 436:459~466.

[10] A. Berquand, M. P. Mingeot-Leclercq, Y. F. Dufrene. (2004) Real-time imaging of drug-membrane interactions by atomic force microscopy. BBA-Biomembranes 1664:198~205.

[11] A. Bianco, K. Kostarelos, M. Prato. (2005) Applications of carbon nanotubes in drug delivery. Curr Opin Chem Biol 9:674~679.

[12] G. Binnig, H. Rohrer. (1982) Scanning tunnelling microscopy. Helv Phys Acta 55:726~735.

[13] G. Binnig, C. F. Quate, C. Gerber. (1986) Atomic force microscope. Phys Rev Lett 56:930~933.

[14] G. R. Bushell, G. S. Watson, S. A. Holt, S. Myhra. (1995) Imaging and nano-dissection of tobacco virus by atomic force microscopy. J Microsc 180:174~192.

Chapter 10

Rheology in Pharmaceutically Used Polymers

Rheology is the science of flow and deformation of matter. Particularly gels and non-Newtonian fluids, which exhibit complex flow behavior, are frequently encountered in pharmaceutical engineering and manufacturing, or when dealing with various in vivo fluids. Therefore understanding rheology is important, and the ability to use rheological characterization tools is of great importance for any pharmaceutical scientist involved in the field. Flow can be generated by shear or extensional deformations, or a combination of both. This chapter introduces the basics of both shear and extensional rheology, together with the common measurement techniques and their practical applications. Examples of the use of rheological techniques in the pharmaceutical field, as well as other closely related fields such as food and polymer science, are also given.

10.1 Introduction

Traditionally variables such as pH, temperature, surface tension, ionic strength, osmolarity, and composition have been used to characterize liquid or semi-solid formulations. However, complex liquids are dynamic systems with related variables that are hard to distinguish. Rheology is the science of flow and deformation of matter and is a well-known method in the food industry where it is applied on various food products to analyze e.g. mouth feel, chew ability and the ease to swallow. Most pharmaceutical systems contain one or more macromolecular excipients and their physicochemical properties are important for the performance of the end product, and thus rheological behavior becomes a critical quality attribute. Additionally, as the human body contains on average 60% water and thus is mostly composed of various complex fluid systems, physicochemical properties like the rheology of the complex in vivo fluids can play a crucial role in the outcome of any pharmacotherapy. Hence, the knowledge on the rheological behavior of many pharmaceutical systems is crucial in order to understand the stability, bioavailability and absorption of the active pharmaceutical ingredient (API) in the human body.

This chapter will give a short introduction to the theory of rheology and give practical examples of its utility in the field of pharmacy and how both shear and extensional rheology can help in characterizing various pharmaceutically relevant systems.

10.2 Theoretical Background and Fundamental Concepts

10.2.1 Shear Rheology

Rheology is the study of flow and deformation of matter. When external forces are applied to a system, it can either change form or change volume or a combination of both effects. Simple elongation (simple extension) in general involves both effects [Fig.10.1(c)]. Shear deformations are characterized by a change of shape but no volume change. Simple shear and simple elongation give the same information about the material properties as long as the volume does not change during deformations. When forces are applied to a system the deformation response depends on area and height of the system. In order to obtain information about material characteristics, relative forces and deformations are therefore needed. Shear stress is defined as the shear force divided by area of system, $\sigma = F/A$, and the relative deformation is given by the strain defined by deformation divided by height, $\gamma = \Delta x/h$ (Fig.10.1).

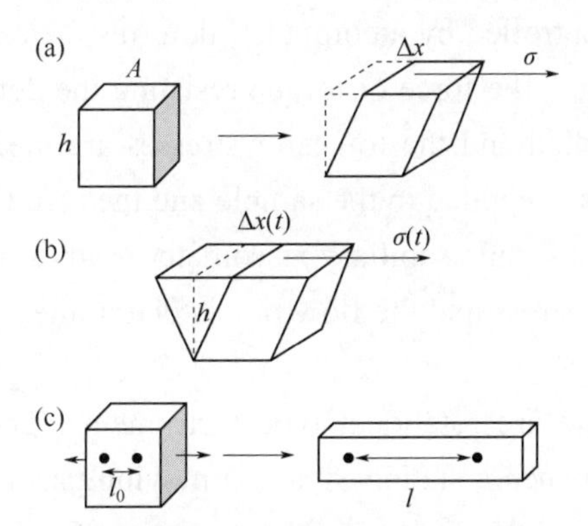

Fig.10.1 Deformation (Δx) in (a) simple shear and (b) simple oscillatory shear of a system, with height h, surface area A, and shear stress σ and (c) extensional deformation (extension ratio l/l_0) in simple uniaxial stretching.

A simple shear geometry (cone-and-plate geometry) is illustrated in Fig.10.2(a). This type of deformation is used in most modern rotational rheometers. In this geometry all parts of the sample experience the same strain and the deformation is homogenous. The shear rate, $\dot{\gamma}$, is defined as the time-derivative of the shear strain. In the flow of a liquid between two plates, the shear rate is simply given by the rate of the moving plate divided by the distance between the plates shown in Fig.10.2(b). Flow through capillaries and pipes are also used to measure shear deformations, but deformations are inhomogeneous with zero shear rate in the center of the pipe and maximal rates at the pipe surface. Cone-and-plate and concentric cylinder tools [Fig.10.2(c)] ensure homogenous deformation in rotational rheometers, whereas the plate-and-plate tools result in inhomogeneous shear deformations.

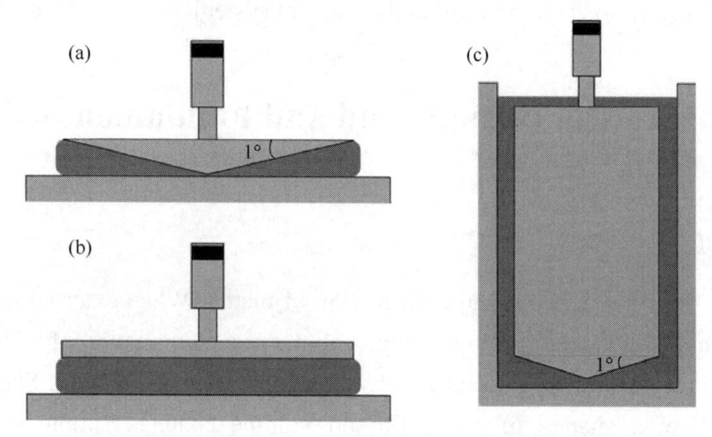

Fig.10.2 Geometries used in shear rheology measurement on rotational rheometers (a) cone-and-plate, (b) plate-plate, and (c) concentric cylinders (Couette).

A rheometer is any instrument that enables determination of rheological properties. Very simple rheometers such as capillaries have been used for centuries to determine flow of liquids. Modern rheometers allow a more precise quantification of material properties in well-defined geometries, and they can be divided into two main types. In a controlled rate type (CR) instrument, a motor, often controlled by a computer, deforms the sample in a controlled manner and a force transducer monitors the force or torque resisting the deformation of the sample. The strain or strain rate is controlled and the forces or stresses are measured. In a controlled stress (CS) type instrument, a stress is applied to the sample and the resulting deformation is monitored. The flow of a liquid through a simple capillary viscometer is an example of a CS type instrument where gravity determines the stress and the flow rate is determined from the flow time of a fixed volume of the flowing liquid.

Most modern rheometers are rotational type instruments. The sample to be measured is confined in a narrow gap between a stationary and a moving part of a measuring cell. Different measuring cells are used depending on the sample properties. The most common measuring cells are cone-and-plate, parallel plate (plate-plate), and concentric cylinders (Couette).

Two signals from a rotational rheometer, the angular position of the moving part and the torque, are used to compute strains and stresses. A position sensor registers the angular position of the moving cell part. The angle of rotation, ϕ, is proportional to the strain and the proportionality constant depends on gap size and geometry. The stress is proportional to the torque (force times lever arm length) and the proportionality constant depends on the cell geometry and especially on the surface area of the measuring tool. Tools with a large area are used for soft materials and low viscosity liquids, whereas smaller tools are used for harder materials. A summary of proportionality constants for various geometries can be found in the book "Viscoelastic properties of polymers".

10.2.1.1 Ideal Model Systems

Many of the materials encountered in pharmacy are complex systems displaying both viscous and elastic properties. In order to understand such systems knowledge of the characteristic properties of simple model systems is important. These extreme models are the ideal elastic material and the ideal Newtonian liquid. A spring is an example of an ideal Hookean elastic material where the resulting extension of the spring is proportional to the force applied. Ideal isotropic elastic materials in rheology follow Hooke's law with proportionality between stress and strain

$$\sigma = G\gamma \tag{10-1}$$

The proportionality constant G is the (elastic) shear modulus. Typical values are 10^3 Pa for many gels, 10^6 Pa for rubber networks, and about 10^{10} Pa for wood, glass and metals. For an ideal elastic material G is independent of strain, time, and shear rate, but it will often depend on temperature and to a lesser extent on pressure. A Hookean solid in a deformed state will remain deformed at constant strain as long as the applied stress persists and recover completely when the stress is removed. The energy needed to deform the elastic solid is stored in the sample and can be recovered. Knowledge of the G value enables prediction of the strain or stress in any type of deformation for an ideal elastic system.

In contrast, the other extreme model system is an ideal Newtonian liquid where a constant flow will persist as long as a stress is applied. For an ideal Newtonian liquid the flow rate and hence the shear rate will be proportional to the stress applied

$$\sigma = \eta\dot{\gamma} \tag{10-2}$$

where the proportionality constant, η, is the shear viscosity, which is independent of strain, strain rate, and time for an ideal Newtonian liquid. It does, however, depend on temperature and to a lesser extent on pressure. The SI unit of viscosity is seen from Eq. (10-2) to be Pa·s, and in older literature Poise (P) with 10 P = 1 Pa·s.

Typical values at room temperature are 10^{-3} Pa•s for water and 1 Pa•s for glycerol. When a constant stress is applied to such a liquid it will flow at a constant rate determined by the viscosity of the liquid and the applied stress. When the stress is removed flow will cease and there will be no recovery in strain. The energy needed to maintain the flow will be dissipated as thermal energy in the liquid. Eq. (10-3) shows that the energy needed per volume and time, $\dot{E}$, to maintain a shear flow is given by

$$\dot{E} = \eta\dot{\gamma}^2 \tag{10-3}$$

Energy dissipation, and hence a resulting temperature increase, should be considered for liquids with high viscosities and at high shear rates. An effective temperature control is therefore especially needed under such conditions. Knowledge of the η value enables prediction of the rate and strain in any type of deformation for an ideal viscous system.

10.2.1.2 Non-Newtonian Liquids

Real materials are, except for very simple systems, in general neither ideal solids nor ideal liquids. The deformation response depends on how fast the material is deformed. If deformations are fast, molecular rearrangements in materials cannot take place and materials will be deformed elastically. However, if the same material is deformed slowly, the molecules can reorganize and relax and often flow like a liquid. Materials, which exhibit both elastic and liquid-like properties, are called viscoelastic. The practical use of rheological experiments is to quantify the viscoelasticity of materials over as wide a range of time, rate and deformation scales as possible, and often to relate these viscoelastic properties to the molecular structure of the material and different components in formulations.

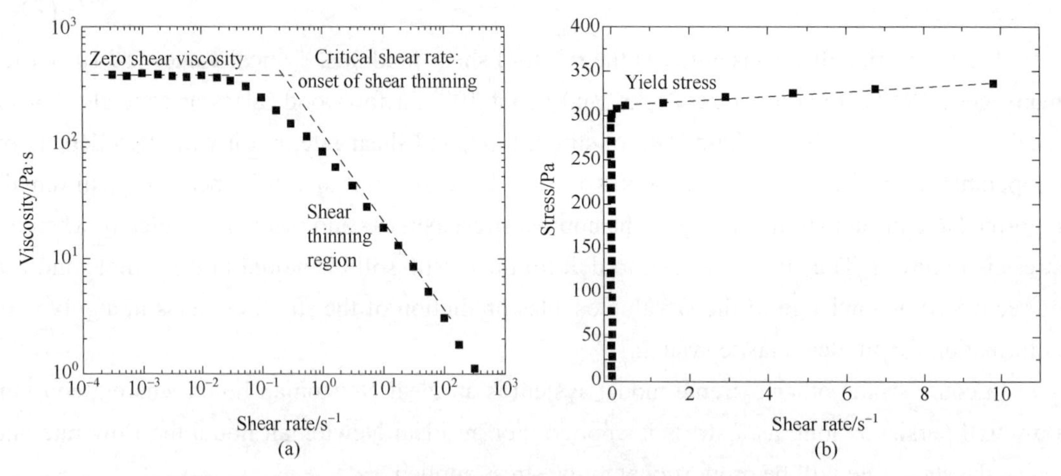

Fig.10.3 (a) A flow curve of a high molecular mass polymer in semi-dilute solution. (b) A stress ramp is applied to a micellar suspension and shear rates are measured. A vanishing shear rate is observed at low stresses. Above a critical stress the suspension flows and the slope is the Bingham viscosity. The intercept at vanishing shear rates is a measure of the yield stress.

Non-Newtonian liquids are liquids that cannot be characterized by a constant viscosity. Many pharmaceutical systems, including suspensions, emulsions and polymer solutions, show a constant viscosity only at low shear rates (zero-shear viscosity, first Newtonian plateau) but a decreasing viscosity (shear thinning) above a characteristic shear rate. Flow curves are plots of viscosity against shear rate, most often on log-log scales, and allow determinations of the zero shear rate viscosity, the critical shear rate (onset of shear thinning), and how rapidly the viscosity decreases at high rates. A linear decrease is often observed at high rates on a log-log plot, which shows that the viscosity follows a power-law. An example of a high molecular mass polymer in a semi-dilute solution where the polymers overlap and are entangled is shown in Fig.10.3(a). The exponent should always be between 0 (Newtonian liquid) and -1 (limit of indeterminate flow).

The critical shear rate is related to the inverse of the longest relaxation time of the system. At low shear rates the friction in this solution is high because of this entangling of chains. At higher shear rates the polymer chains disentangle and orient in the flow with less friction and hence a lower viscosity.

Flow curves can also be used to determine yield stresses of elastic materials. If the shear rate is measured with increasing steady stresses (stress ramp), the elastic properties will dominate at small stresses, with a vanishing shear (flow) rate [Fig.10.3(b)]. At stresses above the yield stress, structures will be broken and the system will flow. The example in Fig.10.3(b) is typical for many systems and is referred to as a "Bingham liquid", with a Bingham viscosity, η_B, which is the slop of the curve at higher stresses and a yield stress corresponding to the intercept of this curve at vanishing shear rates. Empirical information about the amount of stress that a sample can withstand before it is damaged and flows, often called a yield stress, is generally useful for studies of samples such as suspensions and gels.

10.2.1.3 Oscillatory Measurements

When a constant stress is applied to an ideal elastic material, a constant strain is obtained [Eq. (10-1)] and corresponds to zero shear rate (or zero flow). Ideal elastic materials are therefore characterized by an infinite viscosity [Eq. (10-2)]. As a result, viscosity measurements of elastic and many viscoelastic systems under steady shear are of limited use. Furthermore, in many cases deformation caused by a steady shear rate will destroy fragile structures in the sample and provide little information about the unperturbed properties of the system.

Viscoelastic materials are therefore frequently studied by applying small ampli┐ tude oscillatory stresses or strains to the sample rather than steady flows (Fig.10.4).

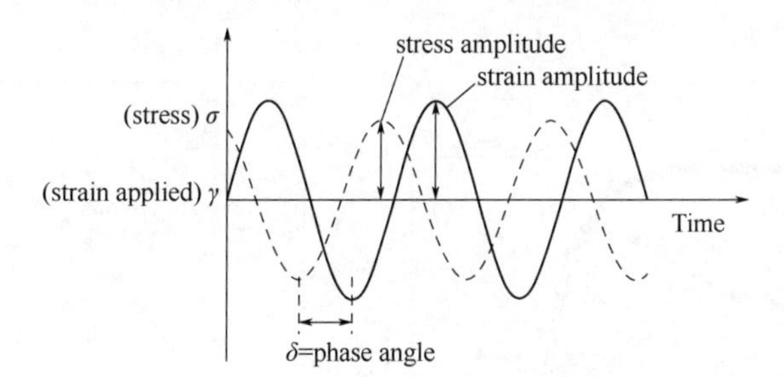

Fig.10.4 Schematic illustration of the applied force and strain in an oscillatory shear measurement.

If an oscillatory strain deformation with an amplitude γ_0 and an angular frequency ω is applied, the stress will also oscillate in time t with the same frequency but it will be phase shifted by δ, the phase angle, relative to the strain

$$\gamma(t) = \gamma_0 \sin\omega t \tag{10-4a}$$

$$\sigma(t) = \sigma_0 \sin(\omega t+\delta) \tag{10-4b}$$

where σ_0 is the stress amplitude and the angular frequency ω (in rad•s^{-1}) equals $2\pi f$, where f is the frequency in Hz. The phase shift δ is always between 0 and 90 degrees. For an ideal elastic system the phase shift is 0 as seen from Eqs. (10-1) and (10-4a), and for an ideal Newtonian liquid it is always 90 degrees as seen from Eqs. (10-2) and (10-4a) (differentiation of the strain with respect to time in Eq. (10-4a) will result in a shear rate which depends on cos ωt, which is phase shifted by 90 degrees relative to sin ωt) (Fig.10.4). Materials with phase shifts between 0 and 90 degrees are viscoelastic and the stress in Eq. (10-4b) can be written as a sum of elastic and loss contributions

$$\sigma(t) = \gamma_0(G'\sin\omega t + G''\cos\omega t) \tag{10-5}$$

In this equation, G' is referred to as the elastic storage shear modulus because it is associated with energy storage. Likewise, G'' is the loss shear modulus because it is related to the viscous properties and associated with energy loss in the sample during an oscillatory deformation. For an ideal Newtonian liquid $G'' = \omega\eta$ and $G' = 0$, which shows that no energy is stored in the sample. For an ideal elastic system $G' = G$ and $G'' = 0$, which shows that no energy is lost during deformation.

For viscoelastic materials G' and G'' will in general depend on angular frequency and the elastic properties will often dominate at high frequencies. Small amplitude oscillatory measurements are especially useful for monitoring structure development as a function of time or temperature at a fixed frequency.

According to Eqs. (10-1) and (10-4a), stress is proportional to strain or strain amplitude. This is valid for all materials at small strains or amplitudes, and is called the linear elastic or viscoelastic range. At larger strains or strain amplitudes, stress and strain will not be proportional and the material will be in the non-linear range.

An example of a stress amplitude sweep is depicted in Fig.10.5(a) where the linear elastic or viscoelastic range is seen at stress amplitudes from 10 Pa to 60 Pa, followed by the non-linear range above 60 Pa.

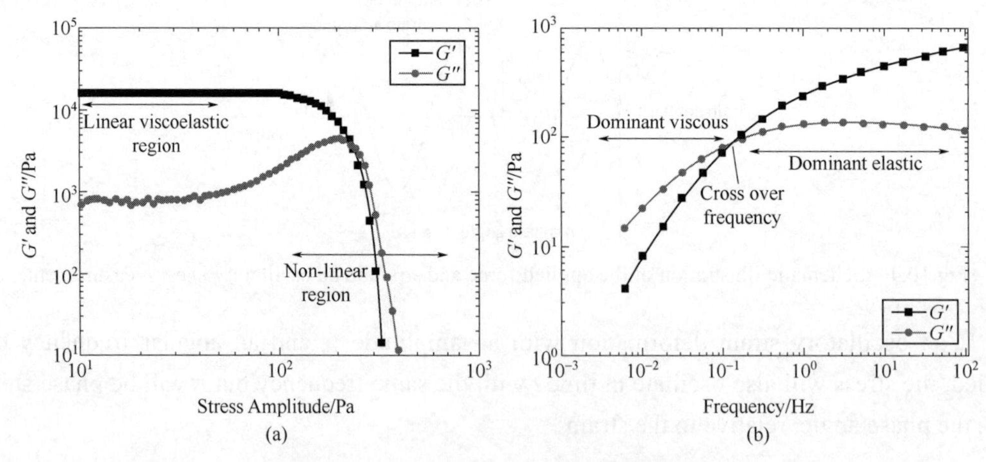

Fig.10.5 (a) A stress amplitude sweep of a viscoelastic micellar gel and (b) a frequency sweep of a viscoelastic polymer solution.

Oscillatory measurements at low strain or stress amplitudes allow a determination of G' and G'' as a function of frequency. Such measurements are important because they can give information about both structure and dynamics. For viscoelastic liquids G'' dominates at low frequencies. At higher frequencies, relaxation of structures cannot take place within the oscillation cycle, resulting in an increase in G'. An example of a frequency sweep is given in Fig.10.5(b), which shows the frequency dependencies of a high molar mass hyaluronic acid solution. The figure illustrates that G'' dominates at low frequencies, where G' is related to the zero shear rate viscosity and angular frequency through $G'' = \omega\eta$. At higher frequencies G′ dominates and a plateau is nearly detected, as expected for an ideal elastic material. The crossover angular frequency is approximately the inverse of the longest relaxation time, which for long polymer chains is strongly dependent on molecular mass and concentration.

10.2.1.4 Creep and Relaxation Measurements

Steady shear and oscillatory measurements are probably the two most important types of rheological measurements. However, other types of measurements can give more relevant information for some pharmaceutical systems. Suspensions experience a sustained force due to gravity, and it is therefore also useful to monitor how such systems deform under a constant load or stress. This type of measurement is called a creep experiment, and in such an experiment the strain $\gamma(t)$ is monitored as a function of time for a fixed applied stress σ_0. The compliance J is then obtained from the time dependence of the strain as

$$J(t) = \gamma(t)/\sigma_0 \tag{10-6}$$

After a certain time period under a constant stress, the stress can be removed and the system's ability to recover towards the original un-deformed state can be investigated in a creep-recovery experiment. Creep experiments are of interest primarily for viscoelastic materials. The elastic properties are seen in the rapid deformation and in the recovery when the stress is applied or removed, respectively. The slow subsequent deformation is characteristic of the viscoelastic time-dependent processes in materials. If this deformation (strain) increases linearly with time, one can obtain the viscosity at a low shear rate based on the inverse relationship between viscosity and the slope. The viscous or plastic deformation can also be calculated from the non-recoverable compliance at long times. Creep tests are of special interest for viscoelastic systems with long relaxation times, because they can be used to measure flow at very low shear rates.

Creep experiments are generally performed on a controlled stress (CS) type instrument, which applies a defined stress and measures the resulting changes in strain with time. A closely related type of experiment, stress relaxation, can be performed on controlled rate (CR) type instruments. In a stress relaxation experiment, the sample is rapidly deformed to a fixed strain, γ_0, and the stress is monitored as a function of time. The shear stress relaxation modulus, $G(t)$, is then calculated as

$$G(t) = \sigma(t)/\gamma_0 \tag{10-7}$$

The relaxation modulus contains information about how rapidly structures can reorganize and relax to relieve the stress in the system. For an ideal elastic system $G(t) = G$ at all times, and for an ideal Newtonian liquid $G(t) = 0$ since the shear rate and hence the stress is zero at constant strain. For a viscoelastic material, $G(t)$ will decrease with time and the decrease will occur on a time scale which is determined by the distribution of relaxation times of the sample. At long (infinite) times the relaxation modulus will either relax to zero (characteristic of all liquids) or reach a constant value which will be the equilibrium elastic modulus of the system characteristic of a viscoelastic solid.

10.2.1.5 Capillary Flow: Capillary Viscometers

Capillary viscometers have been used for centuries to measure viscosities of especially low viscosity liquids, where even modern rheometers lack their sensitivity and precision. Flow through capillaries, dip-cups, extruders and pumping through pipes are important in delivery systems and in production. Measurements at low concentrations furthermore enable determination of intrinsic viscosities, which are sensitive to solvent-solute interactions and molar mass of solutes.

In the general case a pressure-drop ΔP is applied over a capillary or pipe with radius R and length l. The flow rate (volume per time) Q of a Newtonian liquid passing through the pipe is given by

$$Q = \frac{\pi R^4 \Delta P}{8\eta l} \tag{10-8}$$

where η is the shear viscosity. A major problem with capillary and pipe flow is that the flow rate and shear rate are not constant in the liquid. There is maximal flow rate but zero shear rate of the liquid in the center of the pipe, whereas the shear rate is maximal and the flow rate zero at the wall of the pipe. The maximal shear rate at the wall is given by

$$\dot{\gamma}_{\max} = \frac{4Q}{\pi R^3} \tag{10-9}$$

The use of capillary viscometers for shear thinning solutions is therefore problematic, since the shear rate is not defined, and plug flow can be expected. Well known examples of plug flow are squeezing pastes and cremes from a tube. In a classical capillary viscometer gravity forces a liquid through the capillary. The time it takes a fixed volume to pass through the capillary is measured. This time is proportional to viscosity divided by density, the so-called kinematic viscosity with units of Stokes. The kinematic viscosity of water at room temperature is 1 cS or 1 $mm^2 \cdot s^{-1}$.

10.2.1.6 Capillary Flow: Capillary Rheometers

For the pressure-driven flow, capillary rheometers with circular or rectangular slit die geometry are commonly used (Fig.10.6). The sample, usually in granules, pellets or powder form, is fed into the rheometer barrel that is pre-heated to the desired temperature, and the molten material is then extruded through the die at a defined piston speed. When a circular capillary die is used, the melt pressure is recorded in the barrel above the die entrance, and this requires taking into account the extra pressure drop caused by the contraction of the flow from the barrel into the capillary. In a slit die, the flat wall geometry enables measurement of pressure directly in the slit where the flow is fully developed, thus making exact determination of the pressure profile possible. Slit dies are, however, more difficult to assemble and to disassemble and the cleaning of the slit edges requires more effort.

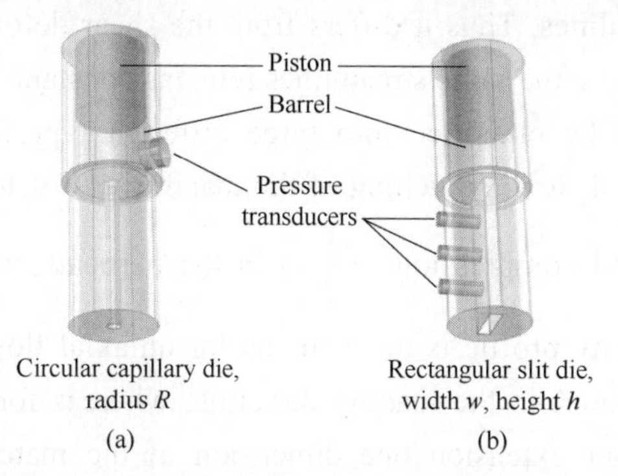

Fig.10.6 Principle of capillary rheometer with (a) circular capillary die, where the pressure is measured in the barrel before the entrance into the capillary and (b) rectangular slit die, where the slit geometry allows placing the pressure transducer(s) directly on the slit wall.

The volume flow rate in the barrel can be calculated from the barrel/piston radius R_p and piston speed V_p as $Q = \pi R_p^2 V_p$. The apparent shear rate, that is, the shear rate for a Newtonian fluid, at the capillary wall can be determined in a circular capillary die by Eq. (10-9). However, for the rectangular slit die the apparent shear rate at the wall is

$$\dot{\gamma}_{wa} = \frac{6Q}{h^2 w} \tag{10-10}$$

where h and w are the slit height and width. The term "apparent" is used because of the plug-like flow profile typical for shear thinning fluids, which means that the shear rate is higher than predicted for Newtonian fluids with a parabolic velocity profile. The true shear rate is calculated by multiplying the apparent values in Eqs. (10-9) and (10-10) by a correction factor achieved using the Rabinowitch correction procedure. From the pressure measured at the die entrance, Δp, and the additional pressure drop caused by the contraction flow at the capillary entrance, Δp_e, determined using a so called Bagley correction method, the true shear stress at the circular capillary die wall is calculated as

$$\tau_w = \frac{\Delta p - \Delta p_e}{2(L/R)} \tag{10-11}$$

where L/R is the length-to-radius ratio of the capillary die. Bagley correction requires measurements with at least two dies that have the same diameter but different length. For a slit die the true wall shear stress is calculated directly from the pressure profile $\Delta p/l$, the pressure difference between two pressure readings on the die wall at distance l from each other.

$$\tau_w = \frac{h}{2}\frac{\Delta p}{l} \tag{10-12}$$

Conventional, industrial scale capillary rheometer tests usually require a large sample amount (barrel volume ~ 25 cm^3) particularly for the measurement at high shear rates.

10.2.2 Extensional Rheology

In extensional flow (also called elongation or elongational flow) the material undergoes stretching along the streamlines. Thus it differs from the shear deformation where the distance between two fluid particles on the same streamlines remains constant.

Extensional flow can be classified into three different types. Uniaxial extension is the simplest form of extensional flow: stretching of the material at a velocity, v_1, imposes the strain rate, $\dot{\varepsilon}$, in direction x_1 and compression $(-\frac{1}{2}\dot{\varepsilon})$ in the perpendicular directions x_2 and x_3. In biaxial extension the velocity profile is the same as for uniaxial flow, but the extension rate is always negative (compression) in the loading direction, whereas for uniaxial flow it is always positive (tension). In planar extension one dimension of the material is extended while the second one is maintained constant and the third one compressed (Fig.10.7).

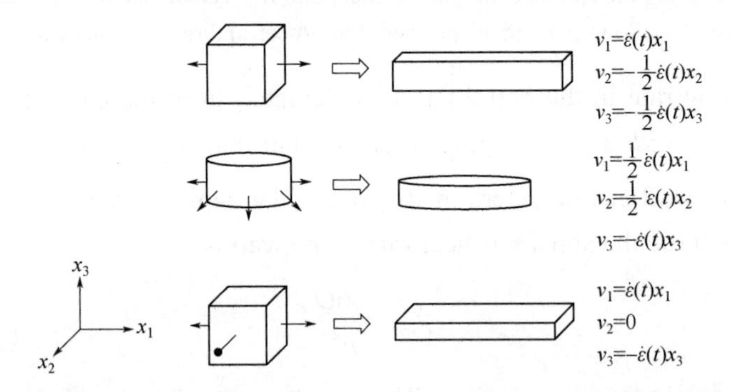

Fig.10.7 Uniaxial, biaxial, and planar extension and their velocity distributions

In uniaxial extension the particles accelerate, so that the distance between the particles on the same flow stream lines [depicted in Fig.10.1(c)] increases exponentially with time. Thus, typically extension produces a rapid deformation, compared to shear flow:

$$\frac{l}{l_0} = e^{\dot{\varepsilon}t} \tag{10-13}$$

where l_0 is the initial sample length and l the length at time t. Stretching at a constant strain rate $\dot{\varepsilon}$ results in a final logarithmic strain $\varepsilon = \dot{\varepsilon}t = \ln(l / l_0)$; note that "strain" in this context is a natural logarithm of the extension ratio, $\ln(l/l_0)$, whereas in the study of solid materials, the simple extension ratio l/l_0 is called "strain". Usually, in order to distinguish between these, ε and $\dot{\varepsilon}$ are called Hencky strain and Hencky strain rate, after the German engineer Heinrich Hencky (1885–1951) who committed remarkable pioneering work in the field. The extensional viscosity η_e can be determined analogously to shear viscosity as

$$\eta_e = \frac{\sigma}{\dot{\varepsilon}} \tag{10-14}$$

where σ is the extensional stress. The relationship between viscosity of Newtonian fluid shear viscosity and steady state uniaxial extensional viscosity is called Trouton's ratio, and it is determined as

$$Tr = \eta_{e(steady)} = 3\eta_0 \tag{10-15}$$

This relation is valid also for non-Newtonian fluids, as long as the deformation is sufficiently small, i.e. within the linear viscoelastic region. Thus at small Henck strain rate ($\dot{\varepsilon} \to 0$), the uniaxial extensional viscosity equals three times the Newtonian shear viscosity. Extensional properties are usually measured in transient start-up flow, and expressed in curves of transient viscosity, η_e^+, *vs*. Time [Fig.10.8(a)]. In these plots Trouton viscosity, $3\eta_0$, marks the linear viscoelastic envelope (LVE), obtained from start-up shear experiments at a shear rate within the Newtonian flow region, and is denoted as $3\eta_s^+$.

Uniaxial extension is the simplest extensional flow type to generate and measure under laboratory conditions, and allows most efficiently the detection of non-linear viscoelastic behavior: Strain hardening or strain softening of the material can be seen as deviations from the LVE; in the first case the transient viscosity curve will rise above, and in the latter case below the LVE as described in Fig.10.8(a). After the transient phase the flow should level off to the steady-state, where the viscosity becomes independent of time. However, the steady-state flow may be difficult to observe experimentally, as instability of the sample and limitations of the test device often impair these experiments.

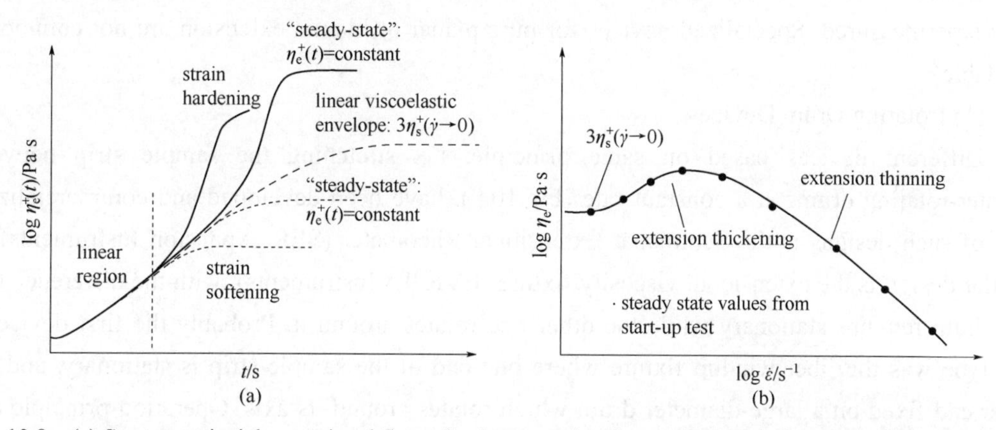

Fig.10.8 (a) Start-up uniaxial extensional flow of polymer melts. Curves for a typical strain hardening polymer are represented by solid lines and for strain softening polymer by dotted lines, and the dashed line represents the LVE, and (b) Typical curve of extensional viscosity *vs*. Hencky strain rate for an extension-thickening polymer.

Strain hardening polymers have a characteristic "hump" or "ridge" in the plot of steady-state extensional viscosity values (from transient start-up flow) against Hencky strain rate; first the flow typically exhibits a constant $\eta_e(\dot{\varepsilon})$, then extension thickening at increasing $\dot{\varepsilon}$, followed by an extension thinning region in a similar manner to shear flow [Fig.10.8(b)]. However, achieving the true steady-state in extensional viscosity measurements can be a challenge and is currently under debate.

Extensional rheology can tell more about intrinsic properties and structure of macro-molecular materials than mere shear viscosity measurements; a common example is from the polymer industry: two polymers can have overlaying shear viscosity curves but yet behave differently when they are processed. If extensional viscosity measurements are performed it can be observed that one of them shows strain hardening due to its more branched chain architecture and therefore its elastic properties are more pronounced than for the more linear polymer.

Many industrial processes involve both shear and extensional deformation types. Extensional deformation appears in free-surface flow and in closed pipe or channel flow whenever there is a change, either constriction or dilation, of the flow channel cross-section. Shear flow, however, is easier to produce in laboratory conditions and is thus the most commonly used way of characterizing flow behavior.

10.2.2.1 Extensional Rheometry: Uniaxial Flow

Being a relatively new experimental area, routines for generating and measuring extensional flow are not as well established as for measurements in shear flow. Thus many of the experiments described in the scientific literature have been performed with custom made unique rheometers which are only accessible to a few research groups. Uniaxial extensional flow, also called simple extension or simple elongation, is a standard rheological flow like simple shear, and it is also the easiest type of extensional flow to generate. Therefore it is the most common flow type measured. Specialized devices for pure planar or biaxial extension are not commonly available.

(1) Rotating Drum Devices

Different devices based on same principle, i.e. stretching the sample strip between counter-rotating drums at a constant rate (Fig.10.9), have been developed and commercialized. One of such designs is the Sentmanat Extensional Rheometer (SER, Xpansion Instruments). A similar design is the extensional viscosity fixture (EVF, TA Instruments), with the difference that one drum remains stationary while the other one rotates around it. Probably the first device of this type was the fiber windup fixture where one end of the sample strip is stationary, and the other end fixed on a large-diameter drum which rotates around its axis. Operation principle and calculations related to the SER device are presented here. Usually the experiments are made in un-steady start-up extension which results in transient viscosity curves presented in the Fig.10.8(a). The Hencky strain rate at constant drum rotating speed ω is

$$\dot{\varepsilon} = \frac{2\omega R}{l_0} \tag{10-16}$$

The transient extensional viscosity can be calculated as

$$\eta_e^+(t) = \frac{F(t)}{\dot{\varepsilon} A(t)} \tag{10-17}$$

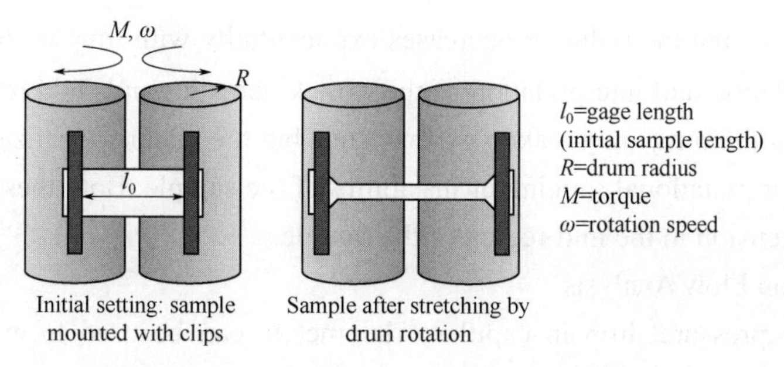

Fig.10.9 Operation principle of a counter-rotating drum device.

The stretching force F is calculated from the measured torque, $M = 2FR$, and the change of the sample cross-sectional area upon stretching is expressed by $A(t) = A_0 \exp[-\dot{\varepsilon}t]$, where A_0 is the initial sample cross-sectional area. Reaching the true steady-state flow in transient extension can be challenging. With the SER device the limiting factor in practice is often sample necking and rupture. The maximum drum rotation angle can also be a limiting factor for achieving steady state flow: If the drums are allowed to rotate over one full revolution, 360 degrees, the sample strip starts to wind up on the clamps, which causes an erroneous peak in the transient viscosity curve. Due to these limitations, the maximum Hencky strain achieved with the device is $\varepsilon_{max} < 4$.

(2) Filament Thinning Experiments

In the filament thinning experiments (Fig.10.10), performed for example with the capillary breakup extensional rheometer (CaBER), the sample is stretched vertically between circular plates by applying a rapid step-deformation with an exponentially increasing separation profile $l(t) = l_0 \exp(\dot{\varepsilon}t)$, after which the plates are held at an axial final separation and the thinning of the mid-filament region is observed during the necking and breakup.

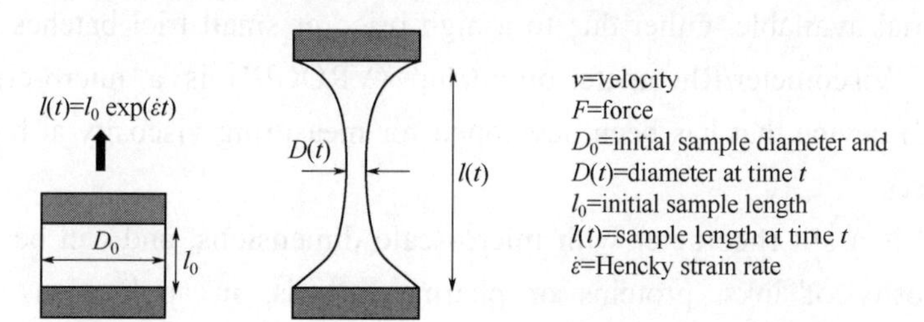

Fig.10.10 Principle of filament thinning experimental setup.

The Hencky strain in the filament thinning experiment is $\varepsilon = 2\ln(D_0/D(t))$, and the maximum strain achieved by capillary break-up rheometers $\varepsilon_{max} \approx 10$, but can be even above that. The CaBER is designed for capillary break-up experiments of low to mid-range viscosity fluids, where the thinning of the filament is monitored by a laser micrometer. The experiment itself is relatively straight forward, but the analysis may be complicated due to the dynamic nature and time dependence of the fluid flow.

In the filament stretching rheometer (FISER or FSR the experimental setting is basically the same as in CaBER, but the plates are separated at a constant deformation rate

$\dot{\varepsilon} = -(2/R)(\mathrm{d}R/\mathrm{d}t)$ and the radius R decreases exponentially with time in the mid-point of the sample. The calculation and interpretation of the flow kinematics and fluid response are simpler than in the filament thinning and breakup experiments, but the practical realization of the test can be difficult due to gravitational sagging or instability of the sample. Both these methods produce a pure uniaxial extension in the mid-region of the sample.

(3) Contraction Flow Analysis

The entrance pressure drop in capillary rheometers can be used to evaluate extensional viscosity. The contraction flow analysis method is based on the assumption that the pressure drops due to shear and extensional deformation can be calculated separately and that their sum is the total pressure drop. Viscoelasticity of the material causes circulating corner vortices in abrupt contraction (Fig.10.11) and their size is proportional to the extensional deformation. Additionally the general assumptions of capillary rheometry, are applied in the contraction flow analysis as well. Cogswell and Binding analyses are the most popular techniques for estimating the extensional viscosity from capillary rheometry data.

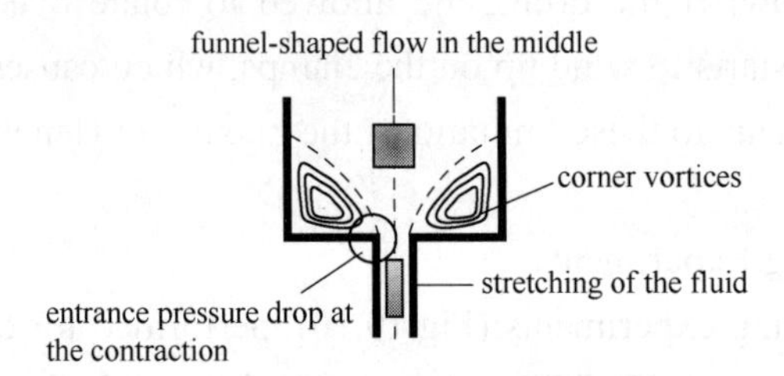

Fig.10.11 Schematic picture of the contraction flow analysis.

(4) Testing in Miniature Scale

In pre-screening of pharmaceutical formulations, the biggest limitation is often the limited amount of material available, either due to a high price or small trial batches manufactured in laboratory-scale. Viscometer/Rheometer-on-a-Chip (VROC™) is a micro-electro-mechanical systems (MEMS) device that has been developed for measuring viscosity at high shear rates at very narrow passes.

Basically it is a slit rheometer with micro-scale dimensions, and can be used to measure high-shear viscosity of inks, proteins or pharmaceuticals, in applications where the flow geometry dimensions are very small, such as in microfluidic devices or syringes. This is also an attractive option when only a very small amount of sample is available. Recently, also extensional microfluidic rheometers based on a flow through a hyperbolic die and a cross-slot geometry have been used for characterization of biological samples in extensional flow.

10.2.2.2 Extensional Rheometry: Biaxial and Planar Flow

Robust systems for measuring biaxial and planar extension are not commonly available, although some custom made devices for these exist. At least approximate planar extension can be produced in cross-slot devices or lubricated dies by two opposing fluid streams that meet in the middle creating a stagnation flow point, where purely planar elongation is expected to occur.

Biaxial flow can be produced for example by sheet inflation and lubricated squeezing, which is presented in the following section.

In lubricated squeezing, biaxial extensional flow is created between plates that are lubricated constantly to prevent the effects of friction between them and the sample. The compression of the sample can be performed using for example a universal material tester or a texture analyzer. Thus the method is widely available and the test setting is relatively simple, but if continuous lubrication cannot be supplied during the measurement, the friction between the sample and the plates will affect the results. Moreover, it is not possible to react very high extensional strain ($\varepsilon_{max} \approx 1$) because of the thinning of the lubrication layer. Lubricated squeezing flow has been used in food rheology for example to analyze structural changes in cheese due to composition, process parameters and storage time, or fermented milk products, such as kefirs or yogurts.

10.3 Practical Issues of Rheological Measurements

10.3.1 Shear Rheology in Practice

10.3.1.1 Choice of the Right Geometry

The most common measuring cells are cone-and-plate, parallel plate (also called plate-plate), and concentric cylinder (Couette) geometries. The cone-and-plate and Couette geometry are often chosen for liquid samples as all parts of the sample experience the same strain and the deformation is homogenous. The cone-and-plate geometry has the advantage over the Couette geometry in smaller sample size and easier cleaning. However, bigger challenges with evaporation exist, compared to the Couette geometry. Additionally, the gap between the cone tip and the lower plate is very narrow, which must be taken into account when measuring systems that contain particles, as the particle size must be much smaller than the gap size. In the parallel plate geometries the gap can be freely adjusted, thus they do not pose the same strict limitation to the maximum particle size in the sample. However, their disadvantage is that the samples confined between the plates are not deformed to the same degree throughout, because the strain depends on the distance from the center of rotation. The maximum strain is obtained at the perimeter of the measuring plate, whereas the strain is zero at the rotational axis. The surface properties of the measuring cell are also important, since the sample has to adhere to their surfaces during measurements. Slippage between sample and cell surface e.g. due to syneresis, can be a problem for some pharmaceutical systems. If slippage occurs the sample may not be deformed and the entire deformation or flow may occur in the liquid film layer.

Sample evaporation can be avoided by using solvent traps. Some of these are designed to create saturation of the solvent in a little chamber placed around the sample and will only work after a certain evaporation has taken place. Other solvent traps are designed to hold a low viscosity oil at the sample surface, directly preventing the evaporation. These, however, can increase the adsorption of amphiphilic chemicals to the oil-water interface and lead to a wrong interpretation of the results. Additionally, the solvent trap with oil can only be used with fluids that are totally immiscible with it, so that none of the sample components can diffuse into the oil phase.

10.3.1.2 Steady Shear Measurements

Flow curves should represent steady state values of the viscosity against shear rate. This means that viscosity should be independent of measurement time. It is important at each shear rate to allow sufficient time for the stress to reach a constant value before the stress is measured. Otherwise the calculated viscosity will not only depend on shear rate but also on time and past history of the sample. Systems for which the viscosity depends on history are called thixotropic.

At high shear rates the flow may no longer be laminar and turbulent flow may appear, and the viscosity is no longer given by Eq. (10-2). An apparent higher viscosity will be seen due to the extra energy dissipation caused by turbulence. For the common steady shear power law dependence illustrated in Fig.10.3(a) the negative slope must be between 0 (Newtonian liquid) and −1. This means that the viscosity power law exponent, *n*, in $\eta \times \dot{\gamma}^{-n}$ must be between 0 and 1. Values of *n* greater than 1 correspond to indeterminate flow, in which case an increase in shear rate will demand a smaller stress, as seen from Eq. (10-2), resulting in inhomogeneous flow.

10.3.1.3 Oscillatory Shear Measurements

Modern rotational rheometers typically cover at least a frequency window from 0.001 Hz to 100 Hz. G' and G'' are only defined in the linear range where both strain and stress are simple sinusoidal curves (Fig.10.4). In order to ensure this, material should always be tested as a function of stress or strain amplitude as depicted in Fig.10.5(a). In the linear range G' and G'' are independent of these amplitudes. Oscillatory tests can be performed on both CR and CS instruments. At high frequencies mechanical resonances in the instrument may occur, and it should be ensured that G' never decreases with increasing frequency.

10.3.1.4 Creep Measurements

Creep tests are also of interest for viscoelastic systems with long relaxation times, because they can be used to measure flow at very low shear rates. Care should be taken to ensure that the experiments are performed in the linear range where the compliance is independent of the applied stress magnitude. Creep measurements should be performed on CS type instruments.

10.3.1.5 Capillary Viscometry

Capillary viscometers enable very precise viscosity determinations. Flow times can be measured with an accuracy of typically 0.1s and for typical flow times in the range of 100~200 s, a viscosity accuracy of 0.1% is possible. However, a good temperature control is needed and any larger particles in solution should be removed by filtration or centrifugation. The flow times are very dependent on the radius of the capillary, as seen from Eq. (10-8), and a suitable viscometer with flow times greater than about 100s should be used. Intrinsic viscosities, $[\eta]$, can be determined from a series of measurements at low concentrations. Flow times for the most concentrated solution should be about twice the flow time of the solvent, where $[\eta]_c \approx 1$, and the most dilute solution about 5%~10% of this concentration. The maximal wall shear rate in the capillary is given by Eq. (10-9) with typical value of $\approx 100\ s^{-1}$. For non-Newtonian liquids, as illustrated in Fig.10.3(a), zero shear rate viscosities cannot be determined unless the critical shear rate exceeds the shear rate in the viscometer.

10.3.1.6 Capillary Rheometry

Correctly performed capillary rheometer measurements can be quite arduous: In order to achieve the true shear stress, the entrance pressure drop (in case of a circular capillary) needs to be determined, and this requires several measurements with different dies. Moreover, determining the true wall shear rate requires post-experimental data analysis. If these corrections are ignored, the results are merely "apparent" and can only be used for example for internal comparison and quality control purposes. Sometimes the effect of pressure or viscous heating (increase of the temperature due to friction) is relevant and complicates the interpretation of the results. Moreover, the theory of capillary rheometry assumes that the fluid adheres to the die wall (the so called no-slip condition). However, this is not always true: with some materials wall slip can occur when a critical shear stress is exceeded.

10.3.2 Extensional Rheology Measurements in Practice

Despite the variety of available experimental settings for uniaxial extension, reliable measurement is still not a simple task and each technique has certain limitations. Results between devices based on different principles and even between different laboratories, due to their sample preparation techniques, can differ from each other. Moreover, many of the extensional rheometers are custom built and therefore not available for everyone. Sample uniformity can have a tremendous effect on the results achieved by uniaxial extension tests: The samples for rotating drum devices, for instance, should be cut out from a homogenous film that has a uniform thickness. Depending on the material and the sample preparation technique, a certain degree of molecular and/or particular orientation may be introduced into the sample, and the results will depend on whether the measurement is done in perpendicular or parallel direction to the orientation. Moreover, any inhomogeneity, such as air bubbles or impurities, can cause a premature sample rupture below the maximum achievable strain. Therefore it must be kept it mind that the maximum transient extensional viscosity reached in the measurements, as shown in Fig.10.8(a), can actually be far from a true stead state flow. In fact, it seems that a true steady-state extensional flow cannot be established with many of the extensional devices that are limited by their maximum extensional strain. In the rheological field a lot of debate has been going on about the steady-state extensional viscosity.

Gravitational sagging of the sample during the heating phase before the test starts can inflict inconsistency in the uniaxial extension measurements by rotating drum devices. Keeping the sample under a pre-defined small tension in the heating phase can be carried out to avoid the sagging. In this case, the effect of the pre-tension on the initial sample cross-sectional area must be taken into account in the calculation. Moreover, thermal expansion of the sample when heated to the test temperature causes a change in the sample density, and addressing this correctly will further improve the measurement accuracy. If the device design allows, immersion of the testing system into a buoyant liquid can be used to prevent sagging. In this case proper fixing of the sample is essential to avoid slipping or loosening from the test drums.

In lubricated squeezing flow, the greatest problem is the lack of lubrication, and different techniques have been introduced to improve the lubricant supply between the sample and squeezing plate surfaces, such as porous plates through which lubricant can constantly be added during the test.

10.4 Application of the Technique

10.4.1 Shear Rheology of Dosage Forms

For many years, rheology has been used for characterizing polymeric systems, emulsions and suspensions in different industries, e.g. plastic production, food industry, cosmetics, cement industry, etc. Rheology has not been utilized to its fullest extent in the field of pharmaceutics, however, the examples given below can easily be translated to the use within pharmaceutics. Rheology has been used in the formulation development and quality assurance for e.g. gels, creams, lotions, ointments, suspensions and pastes. Additionally, most dosage forms contain one or more macromolecular excipients and their physicochemical properties are important for the performance of the end product, and thus rheological behavior becomes a critical quality attribute.

Understanding the rheological behavior of emulsions has been of great interest for its strong relationship to many properties of emulsions that are vital for various industrial applications. Some groups studied the roles of droplet deformability, internal fluid circulation and surface mobility using steady shear rheological measurements (cone-and-plate geometry), and revealed that droplet deformability plays an important role in controlling the shear thinning behavior. A soft emulsion was less shear thinning than a hard emulsion at low volume fractions, due to high level of structural flexibility. At high volume fractions, however, the soft emulsion exhibited increased shear thinning behavior, presumably due to lateral distortion of droplet structure.

Recent studies have also reported that controlled heteroaggregation of oppositely charged lipid droplets can be used to manipulate the characteristics of emulsion-based products. By mixing two emulsions, one containing positively charge droplets and another one with negatively charged droplets, the oppositely charged droplets interact with each other leading to the formation of micro-clusters. A three-dimensional network of aggregated droplets can be formed at sufficiently high particle concentrations leading to elastic-like behavior. These materials may be useful for commercial applications, such as pharmaceuticals, food, and cosmetic products. A steady shear rheological method (Couette geometry) was used to measure the influence of particle size, ratio of positively charged to negatively charged particles, and free protein content in such materials and found that the rheological behavior could be tuned by varying these parameters.

The specific semi-solid nature and unsteady behavior of edible gelled system have resulted in a unique texture experience beyond that of other food textures. This product performance could be transferred to pharmaceutics, for the use within the design of functional foods, pediatric medicines etc. Fundamental aspects of rheological properties of many polymer-based systems are well described in the literature. However, multifunctional gelled products require precise choice of the ingredients such as gelling agents, sweeteners, colorants, flavors, and some functional and/or technical additives, which can highly influence the properties of the final product. Both steady shear and oscillatory shear rheological measurements (cone-and-plate geometry) were used to show that pectin-based gel systems have suitable rheological behavior to be considered as a functional, low-calorie gelled dessert. However, sensory and psychorheological evaluations should be carried out to correlate the results to the overall acceptability of this new product.

Furthermore, the conformational parameters of polymer excipients, such as polymer coil radius (R_{coil}), overlap concentration (c^*) and Martin constant (K_m), can be determined from intrinsic viscosity measurements of systems using capillary viscometry, which are dependent on the solvent used to dissolve the polymer. This is crucial in the process of particle design.

Recently, the focus has also been on the formulation of in situ gelling systems where the rheological behavior is of interest. An ideal in situ gelling system should be a free flowing, low viscosity liquid at room temperature, that allows the administration either by syringe or as droplets (eye drops) and undergoes gel transition at physiological conditions. The control over the gelation can thus be achieved by activation of the formulation by an internal stimulus (e.g. pH, salts, temperature etc.) or by an external stimulus (e.g. light, ultrasound, electricity, magnetism etc.). Thermal gelling properties of different formulations have been studied both by steady shear and oscillatory shear measurements.

10.4.2 Shear Rheology of Biological Samples

Physicochemical properties like the rheology of the complex in vivo fluids can play a crucial role in the therapeutic outcome of any pharmacotherapy. The physico-chemical characteristics of in vivo fluids such as saliva, gastric juice, and mucus (intestinal, cervical and pulmonary) have been studied and an increased interest has been on the rheology of the body fluids, since their properties can have an impact on the performance of administered drugs.

Diffusion and dissolution characteristics of small molecules, i.e. drug molecules have been described by the Stokes-Einstein [Eq. (10-18)] and Noyes-Whitney equation [Eq. (10-19)]:

$$D = \frac{k_B \cdot T}{6 \cdot \pi \cdot r \cdot \eta} \tag{10-18}$$

where D is the diffusion coefficient, k_B is Boltzmann constant, T is absolute temperature, η is viscosity, and r is radius of a sphere.

$$\frac{dM}{dt} = -D \cdot A \cdot \frac{C_s - C}{h} \tag{10-19}$$

where dM/dt is the rate of dissolution, D is the diffusion coefficient, A is the surface area, C_s is the saturation solubility, C is the apparent concentration of drug, and h is the thickness of the boundary layer.

The Stokes-Einstein [Eq. (10-18)] and Noyes-Whitney equations [Eq. (10-19)] predict that an increased viscosity decreases the diffusivity and dissolution rate of drugs. Increased media viscosity has been shown to significantly delay tablet disintegration. The viscosity of the luminal content can have an impact on wettability of the drug and thereby the dissolution of the drug in the luminal fluids, through its effect on diffusivity, mixing, and flow patterns in the gut.

Mucus is a complex aqueous mixture of mucins, lipids, salts and cellular debris, covering many of the epithelial surfaces in the human body. Mucus lines the epithelium of the gastro-intestinal tract (GI-tract) serving as a lubricant facilitating the passage of food and chyme, and forming a protective barrier against contact between the epithelium and e.g. pathogens and pepsin. Consequently, mucus of the GI-tract constitutes an important barrier to oral delivery of therapeutics, and it is one of the many obstacles for successful oral delivery of biomacromolecules.

Mucins are present in human gastro-intestinal fluids, such as saliva, gastric fluid, and intestinal mucus. Mucin oligomers consist of polymerized glycosylated proteins that exist as an entangled network. The mucins are to a large extent responsible for the rheological profile of the GI-fluids. Mucin is a viscoelastic, non-Newtonian gel with shear thinning properties. The mucins form physical fiber entanglements, and the resistance to flow is exerted by individual fiber segments, and the non-covalent intermolecular interactions are responsible for the rheological behavior of mucins.

When analyzing the rheological properties of biological samples one will observe large heterogeneity within and between individual samples because of the complex nature of the mucus structure and the wide variety of the other substances in the samples.

The steady shear rheology of human whole saliva has been studied and was found to exhibit shear thinning behavior. Although it contains more than 99% water, saliva is composed of a variety of electrolytes, immunoglobulins, enzymes, mucins and other proteins as well as nitrogenous products such as urea and ammonia.

Human gastric fluids, sampled from patients in the fasted state, have also been studied; both by oscillatory and steady shear measurements, where they were found to be highly elastic with a strong shear thinning behavior. The mucus rheology can be affected by the composition of mucins and their glycosylation, which varies with age, diet, and the presence of specific antigens.

However, the inter-patient variation was not ascribed to any of those parameters.

Porcine intestinal mucus was studied by oscillatory and stead shear measurements and were able to design a biocompatible in vivo like mucus, with similar rheological properties to mucus, exhibiting shear thinning and dominant elastic behavior. Thus an in vivo representative in vitro model was established, where the effect of the mucus layer on drug uptake can be studied by combining biocompatible in vivo like mucus the with a cell monolayer of mucus deficient epithelial cell lines.

Mucus is also present in pulmonary, cervical and vaginal fluids. The rheological properties of these can be used in the diagnostics of various diseases; e.g. the clearance of mucus is impaired when mucus secretions are too thick, as in patients with cystic fibrosis or chronic obstructive pulmonary disease, while protection against infection is compromised when secretions are too thin, as in women with bacterial vaginosis.

In a study looking at the diagnostics and therapeutic effects of patients treated for cystic fibrosis the oscillatory shear rheology of sputum was found to give good indications on the patients' response to the treatment. The elastic modulus, G', was found to be dominating for all the patients, with the exception of two cases at high frequencies, and successful treatment seemed to decrease the elastic modulus of sputum.

It has been showed that the shear rheology, both oscillatory and steady shear, and the microstructure of fresh, ex vivo human cervicovaginal mucus are largely stable across a wide range of physiological pH values. Additionally, the studies suggested that this pH stability was likely due to complex biochemical interactions between physically entangled and cross-linked mucins as well as other mucus components, such as lipids, ions, and proteins. The authors speculate that the microstructure and bulk rheology of other human mucus secretions may also be relatively insensitive to pH. Thus reconstituted gels of purified mucins, where these components have been removed, will not give the characteristic rheological and structural properties of native mucus.

10.4.3 Extensional Rheology of Dosage Forms and Biological Samples

In the pharmaceutical industry, extensional flow can occur in many processes, such as filling of viscoelastic liquid dosage forms, manufacturing and use of dosage forms that employ spraying, stretching, or free surface flow, such as coating of tablets, fiber spinning, film casting or extrusion. However, extensional viscosity is not commonly reported for the reasons discussed earlier in this chapter. Some examples of the use of extensional rheology are presented here, both in the pharmaceutical field and in other fields, from where they can easily be adapted to characterization of dosage forms or biological samples.

Spraying processes involve very high shear rates. The flow properties of different nasal spray formulations were investigated by capillary rheometry: the shear viscosity of the investigated formulations was extremely low at the high shear rate, and implies formation of fine atomized droplets upon spraying. However, the extensional viscosity of the same formulations, estimated using contraction flow analysis, was relatively high. The researchers suggested that high extensional viscosity, potentially together with surface properties, leads to larger droplet size that prolongs the residence time on the mucosa, which is important for the therapeutic effect of nasal formulations.

Hydrogels consist of water-soluble polymers that after cross-linking can swell and absorb large amounts of water. They are used for example in tissue engineering, wound care, and cosmetic applications. To support tissue regeneration the properties of the hydrogels must match those of the surrounding tissues. The study that the extensional rheology of different copolymer hydrogels in uniaxial extension using a rotating drum device (EVF) with modified sample fixing was carried out. The method was suggested as complementary to the lubricated squeezing flow method, more commonly used to characterize hydrogels.

Hydrocolloids are among the most common materials for wound dressings. Rheology of hydrocolloids has, however, been examined particularly for foodrelated applications, where different hydrocolloids are used for example as thickeners, or for increasing the elasticity of bread dough to improve its baking properties. In gluten-free baking, hydrocolloids such as xanthan gum, hydroxypropyl methylcellulose, guar gum, or psyllium husk are used to imitate the gluten networks of the wheat based bread dough, improving the dough elasticity, extensibility, resistance to stretch, mixing tolerance, and gas-holding ability. These properties are largely related to viscoelasticity and extensional rheology. Some research teams have studied the effect of different hydrogel combinations on the wheat bread dough properties in uniaxial extension by texture analyzer and in bi-axial extension using dough sheet inflation, and they found that the significance of hydrocolloids together with enzymes and/or emulsifier for the dough quality could realistically be assessed by rheological parameters, including extensibility, resistance to the extension, strain hardening and stress relaxation properties. A capillary-breakup device was used to characterize the extensional properties of hydrocolloid food thickeners formed using plant-seed based galactomannans from *Gleditsia triacanthos* and *Sophora japonica*, of which the latter is also known for its hemostatic properties and used in some hemorrhagic conditions.

For digestible fluid or gel like formulations, the mouth feel and the swallowing process are largely affected by both shear and extensional properties of the formulation, and correctly tailored extensional viscosity of the substance is important especially for people suffering from eating and drinking disorders (dysphagia) that can be present in various diseases. The extensional properties of dysphagia drink thickeners in a filament break-up device were investigated, and the results proved that although the different thickeners showed very similar behavior in shear, their extensional properties differed from each other significantly.

In electrospinning, very thin fibers are spun from a polymer solution. The process subjects the polymer to a large degree of chain orientation and the extensibility of a polymer is dependent on its rheological properties. The electrospinning was used to produce drug-loaded nanofiber matrices on a polymeric backing film for oralmucosal drug delivery. Extensibility of the solution was estimated from dynamic oscillatory shear tests and the biaxial extensional properties of the ready fiber-laid films characterized by cylindrical probe penetration using a texture analyzer.

Biological fluids, like most other materials, have primarily been characterized using shear flow. However, the function of many biological fluids is closely related to their extensional rheology, such as circulation of blood, ability of the synovial fluid to lubricate the joints, or saliva flow in the mouth. It has been proved the suitability of a cross-slot flow device specifically developed for very small sample amounts, for studying the extensional rheology of human saliva.

Changes in the extensional properties of the fluids can indicate an abnormal health status or changes in an already diagnosed disease: For example, the molecular mass of the hyaluronic acid in synovial fluid decreases in patients with degenerative joint diseases. Since the lubrication of joint surfaces involves extensional flow, the relationship between the extensional properties and the molecular weight of hyaluronic acid was investigated by using the aforementioned cross-slot flow device and flow induced birefringence measurements.

In some cases changes of the extensional flow properties of biological fluids can also, directly or indirectly, cause health complications: One reason for a preterm birth can be a bacterial invasion of the uterus during the pregnancy, and this is normally prevented by the cervical mucus that creates a hydrogel plug which helps to maintain the sterile conditions in the intrauterine cavity. Some researchers studied the extensional rheology of cervical mucus in capillary break-up test and shear rheology in dynamic oscillation in plate-plate rheometer, and found a correlation between the high risk of preterm birth and high extensibility and weak gel-formation of the cervical mucus.

10.5 Summary

Rheological measurements can add significant knowledge on the behavior of complex fluids that the pharmaceutical field has to deal with. Many rheological techniques can readily be applied from other related fields, such as polymer or food industry. However, choosing the appropriate measurement and sample preparation techniques are crucial for the correct results and their interpretation. Shear rheology has been used for decades for characterizing polymeric systems, emulsions, and suspensions in different industries, and the field of pharmaceutical sciences is catching on. The ability to translate the use of the technique between different fields is of great advantage. The field of extensional rheological techniques is still evolving because of the several limitations of the existing devices. However, recent development of devices for miniature scale testing for both shear and extensional rheology, provide interesting options for the fluid characterization in the pharmaceutical field.

References

[1] J. Aho, V. H. Rolon-Garrido, S. Syrjala, M. H. Wagner. (2010) Measurement technique and data analysis of extensional viscosity for polymer melts by Sentmanat extensional rheometer (SER). Rheol Acta 49(4):359~370.

[2] A. Allen, W. J. Cunliffe, J. P. Pearson, L. A. Sellers, R. Ward. (1984) Studies on gastrointestinal mucus. Scand J Gastroenterol Suppl 93:101~113.

[3] N. J. Alvarez, J. M. R. Marin, Q. Huang, M. L. Michelsen, O. Hassager. (2013) Creep measurements confirm steady flow after stress maximum in extension of branched polymer melts. Phys Rev Lett 110(16):168~301.

[4] S. L. Anna, G. H. McKinley. (2001) Elasto-capillary thinning and breakup of model elastic liquids. J Rheol 45(1):115~138.

[5] E. Antoniou, P. Alexandridis. (2010) Polymer conformation in mixed aqueous-polar organic solvents. Eur Polym J 46(2): 324~335.

[6] S. Anwar, J. T. Fell, P. A. Dickinson. (2005) An investigation of the disintegration of tablets in biorelevant media. Int J Pharm 290(1~2):121~127.

[7] C. Atuma, V. Strugala, A. Allen, L .Holm. (2001) The adherent gastrointestinal mucus gel layer: thickness and physical state in vivo. Am J Physiol 280(5):G922~G929.

[8] D. Auhl, D. M. Hoyle, D. Hassell, et al. (2011) Crosslot extensional rheometry and the steady-state extensional response of long chain branched polymer melts. J Rheol 55(4):875~900.

[9] A. Bach, H. K. Rasmussen, O. Hassager. (2003) Extensional viscosity for polymer melts measured in the filament stretching rheometer. J Rheol 47(2):429~441.

[10] E. B. Bagley. (1957) End corrections in the capillary flow of polyethylene. J Appl Phys 28 (5): 624~627.

[11] M. K. Bain, D. Maity, B. Bhowmick, et al. (2013) Effect of PEG–salt mixture on the gelation temperature and morphology of MC gel for sustained delivery of drug. Carbohydr Polym 91(2):529~536.